THE FORGOTTEN RISING SONS

The true story behind the worst training accident in the history of the Australian Army, and how, for fifty years, an entire nation forgot it ever happened.

Andrew Johnston

First published in 2014 by Barrallier Books Pty Ltd,
trading as Echo Books

Registered Office: 35-37 Gordon Avenue, West Geelong, Victoria 3220, Australia.

www.echobooks.com.au

Copyright ©Andrew Johnston

National Library of Australia Cataloguing-in-Publication entry.

Author: Johnston, Andrew, 1967- author.

Title: The forgotten rising sons : the true story behind the 1945 Kapooka tragedy / Andrew Johnston.

ISBN: 9780992530181 (paperback)

Notes: Includes index.

Subjects: Military engineers--Accidents--New South Wales--Kapooka. Military education--Accidents--New South Wales--Kapooka. Explosives, Military--Accidents--New South Wales--Kapooka. Military training camps--New South Wales--Kapooka. Kapooka (N.S.W.)--History, Military.

Dewey Number: 358.2250994

Book and cover design by Peter Gamble, Ink Pot Graphic Design, Canberra.
Set in Garamond Premier Pro Display, 12/17 and Minerva, Small Caps.

www.echobooks.com.au

Contents

21 May 1945	vi
Note to readers:	ix
Preface	xi
Introduction	xix

Part One—Kapooka

The 'Rising Sun'	1
Chapter One—Just Jack	3
Chapter Two—Daily Routine	17
Chapter Three—Fixing & Firing	33

Part Two—Monday 21 May 1945: Tragedy

Chapter Four—Long Day Ahead	51
Chapter Five—Reveille	61
Chapter Six—Squads	73
Chapter Seven—Morning Delivery	89
Chapter Eight—Discretion	97
Chapter Nine—Jack's Growing Squad	101
Chapter Ten—A Handful of Detonators	109
Chapter Eleven—'… The Dugout's gone!'	119
Chapter Twelve—Faint Hope!	127
Chapter Thirteen—A Makeshift Morgue	141

Contents (cont)

Part Two—Monday 21 May 1945: Tragedy (cont)
    Chapter Fourteen—A Case of Identity 147
    Chapter Fifteen—Front Page News 171
    Chapter Sixteen—Court of Inquiry 183
    Chapter Seventeen—A Terrible Procession 209
    Chapter Eighteen—Just Theories 221
    Chapter Nineteen—Something Strange Happened 237

Part Three—The Story Of Their Lives
    The Last Post—Note to readers 243
    Chapter Twenty—Our 'Jack' 245
    Chapter Twenty-One—Jack's Squad 253
        Victorians 257
        West Australians 265
        South Australians 283
        Queenslanders 297
        New South Welshmen 313
    Chapter Twenty-two—A Fond Farewell 387

Conclusion 393
Author's Post Script 403
Acknowledgements 419
Index 429

21 May 1945

21 May 1945 Jack Pomeroy rose early 'twould be a long drive
Kapooka it sat on the far side of town
He was off to train with the first battalion

And the Pomingalarna it rang to the sounds
Of 8000 feet there marching around
And the camp it was the pride of the town
Its part in the second Great War

Down to the bunker they strode in a line
It was seven feet high and was nineteen feet wide
Awaiting the explosives with which they would train
Filled with confidence youth and bravado

And a flash and a rumble the trees a quiver
And their lives tumble out like the leaves on the river
And twenty six men they would never come home
To loved ones, or friends or to family

And the lesson was ready, the stores were in place
The fuses and monobel were prepped and were safe
And Kendall left the dugout to review the wire
And it saved his life that morning

And the ground it shook and it cast him down
And a cloud reached up and the earth fell down
And the parts of men well they fell all around
It was one hell of an explosion

Wagga remembers that terrible procession
They lined Edward Street in a sombre reflection
And four flag-draped trucks such a terrible dimension
So many caskets to mourn for

And the paper opined that they died for their nation
They died for their King and for God's own creation
And far from the war they had come to train for
In the farmlands called Riverina

The blame was apportioned the story consigned
To time and the memory of those it defined
That terrible morning in 45
When 26 men met their maker

Words by Group Captain Bob Rogers

Dedicated to 'all' servicemen and women who served, or are still serving the country as members of Australia's Military Forces, especially those who've paid the ultimate sacrifice during training, also, to their families who've dealt with the lonely legacy of loved ones existing in painful memories
as their
'forgotten heroes'

Note to readers:

The Forgotten Rising Sons is an incredible true story based largely on the unknown tragic events which took place at the Royal Australian Engineer Training Centre (RAETC)—'Kapooka Camp', located nine miles from Wagga Wagga, New South Wales, Australia on Monday the 21st of May 1945. On that fateful day, without warning, a single blinding flash of highly explosive gelignite violently detonated in an underground training bunker. In less than a second, 26 of Australia's promising young soldiers were killed. Many mutilated beyond recognition. Miraculously, one scared 18-year-old trainee survived the supersonic blast to become the unofficial face of the tragedy. Just two days later, following an unprecedented outpouring of public emotion and grief, including a Royal message of condolence to affected families, the largest military funeral and coordinated interment of troops in Australia's history took place. Then, inexplicably, for the next fifty years, Australia chose to forget the tragedy ever happened.

Describing the hazardous and realistic demolitions' training at the camp, the coincidences contributing to the fatal explosion, a ceremonial procession of flag-draped coffins reducing an entire town to tears, and the solemn and embarrassing aftermath, the story of the Kapooka tragedy is largely based on eyewitness accounts. It comes from testimony provided to the Australian

Military Forces Court of Inquiry by the very men and women who were there when this tragic event unfolded and therefore provides a chilling insight into exactly what went wrong. However, for the purpose of dialogue and story, it does contain passages of fictional adaptation of events interpreted by the author from research and personal interviews with surviving WWII veterans and their families.

Preface

After two decades of proud and dedicated military service in the Australian Army, a chronic injury, one that had plagued my twilight years of soldiering, finally brought an unwelcome and slightly undignified end to my career. But when I think back upon those twenty years, I do so with pride. My entire military career, including the amazing experiences, unimagined opportunities and working alongside the most inspirational workforce in the country, my career as a soldier remains the most professionally satisfying period of my 46 years. Mind you though, closing the door on a profession as a Senior Non-Commissioned Officer (SNCO), and of all things a Military Police Investigator, I did welcome the change from a regimented life to one that was somewhat less organised. And, dare I say it, with a little less discipline. After so many years as a soldier, I desperately yearned for a life that many would consider normal. Nevertheless, upon my discharge I wasn't actually in a rush to find appropriate meaningful or orthodox work in my new post-service life. Instead, I became withdrawn; more interested in just shutting out the rest of the world for a little while. I wanted time to do my own thing without being told when, why and how to get it done.

My withdrawal from a regimented life did however initiate my first ever self-imposed work hiatus. Regardless of the fact that it felt strange not to be employed, it was an opportunity to recharge the batteries; after all they had been running flat-out for all of those years. My post-service 'break' did mean one thing; spending a hell of a lot of time at my desk. It was there I spent many long days staring at the small 'apple' symbol at the bottom of the monitor trying to imagine what the future looked like for me. I knew my next two decades weren't exactly planned with the same military precision as the previous twenty years; this time I was a slightly broken individual in need of a different focus and purpose.

However, as a soldier, I was generally self-motivated, but boredom and a lack of focus soon got the better of me. To break the cycle of despair, I made a life-changing decision. I chose to combine my appetite for research and a long held passion for creative writing. So, in the spirit of adventure, I commenced an almost rudderless search for an interesting topic. It was to be a topic so engrossing that any anxiety I may have about the future would be 'consigned' to the simple things in life—like the night's dinner menu. I was kind of anticipating, more hoping and praying, that this researching and writing focus would become my therapy. Therefore, eager to find something within my interest for the non-fictional, I craved that at least one undiscovered story would magically appear. Not just any story, one that would not only get me excited in the research, but optimistically give me a spiritual 'compass bearing' of re-discovery directing me towards a new purpose in my new life out of uniform. It had to have an underpinning theme of human interest. It had to have heartbreak, a moving sub-plot, characters I could relate to and the story needed to be anonymous.

Needless to say, with those requirements, I was soon bogged-down. Scrolling through endless websites, that one inspirational story I longed for to kick-start my inner drive was eluding me. The speed at which it eluded me almost had my search going in reverse. At that juncture, after almost three months of featureless, uninspiring searching, I finally arrived at my 'A-hah' moment.

Regardless of the fact that much has been written over the years about Australia's involvement in WWII, including countless stories of courageous battles, heroes and heroines, I discovered one particular story that today still remains relatively untold. Hidden amongst statistical details of Australian disasters by death toll, unremarkably tucked away in a nondescript website, I'd found my research focus and literary Rubicon. Registered between the 1840 shipwreck of the *Maria* off Coorong in South Australia and the 1954 Gold Coast and northern New South Wales cyclone, both of which claimed 26 lives, one emotive entry simply read:

> 'Kapooka Army base soldiers received instruction of demolitions work in a bunker which was three metres below ground when an explosion took place which resulted in Australia's largest military funeral'. 26 killed; the date was 21 May 1945.

Needless to say, the details moved me in a most unexpected way. Appreciating the scale of loss, in an instant my heart became heavy. 26 dead. It was more than an entire Australian Rules football team wiped out in less than a second. Being an ex-soldier, having experienced first-hand recruit training at the Kapooka Army base in 1991, as soon as I read the words 'Kapooka' and 'explosion' I knew I'd found my motivation. But how true was it? As a soldier for a very long time, and a military investigator for fifteen long years, I'd witnessed and investigated some of the most heinous crimes and incidents the Australian Defence Force had experienced. I'd heard many more stories of tragedy and death within Australia's military system over the years, but this Kapooka story wasn't one of them. Slightly embarrassed; why didn't I know about it?

Annoyed, I started burrowing deeper. I delved into the depths of the story in an effort to not only validate the source of my discovery and newfound knowledge, but to understand why I didn't know about it. I was already familiar with some of Australia premier military history research sources, so I knew to proceed immediately to the National Archives of Australia. There I was able to locate military file MT885/1 titled: *Explosion-1 RAE Training Centre, Kapooka,*

21 May 1945. Reviewing the archived documents, I was both captivated by its historic context, but more concerned by its simplistic content. After all, it was the largest loss of life in a single training accident in the history of the Army. I thought the file would be bigger and the investigation more thorough. I felt a little deflated. That was until I started reading the details. The official government correspondence littered a 252-page dossier and amongst the statements, sketches and reports were a series of black and white photos. Although slightly out of focus, and looking every bit like photos that had been tucked away in a drawer for half a century, they recorded a ghastly outdoor accident scene of twisted wire, timber, and dirt. Importantly, the bundle of photos was attached to the file supporting the military's report into the incident which had caused such large-scale carnage. The report titled: *1945 Military Court of Inquiry*, recorded names of victims, details of the devastating event and the outcomes of the entire 1945 investigation. It quickly became the crucial document I was looking for—I had my starting point.

As I pondered my next move, for further details I turned to Australia's premier military history source, the Australian War Memorial. Scanning official records, I soon discovered that the authorised history of World War II, as chronicled by the Australian War Memorial in its extensive five-part series *Second World War Official Histories: Army; Navy; Air; Civil; and Medical*, contains no mention of a tragedy at Kapooka on the 21st of May 1945. Consulting historians of the Royal Australian Engineers (RAE), including the RAE Museum at the School of Military Engineering (SME) in Sydney, I found, alarmingly, the greatest single loss of life to befall this proud and historic Corps, was also overlooked in its honored and well-respected history. But this isn't just any tragedy. This was a tragedy so great, its scale and magnitude were unheard of, not only during WWII, but in the entire proud history of the Australian Army. By today's comparison, what happened at Kapooka is so unique and horrific; the loss of life in training is unparalleled in all of Australia's history.

Understandably I was perplexed. The scale of the tragedy and its horrible aftermath became even more unbelievable when I realised mainstream Australia is not even aware it occurred! I was equally concerned by the fact that the very custodians entrusted to promote and protect our military heritage, those working in museums and memorials, had also somehow forgotten. When a small rural New South Wales town was shutdown to bury 26 of their military dead, how did a country, which reveres its military personnel with such respect, so easily forget these men?

Initially embarking on this journey, what I didn't expect was just how much this story would touch me. After many months of research into the unsettling tragedy, including a systematic unlocking of the early lives of all 26 men killed, painful stories of the past stirred deep emotions within. I knew then it was time to bring the tragic events of that fateful day out of the dim recesses of Australia's forgotten military past. Based on my findings, I became convinced that this was an extraordinary military story that would touch all Australians. In the beginning, the event at Kapooka was the type of story I *wanted* to tell. It met all my pre-determined criteria. However, after speaking with families of victims, and the many ageing witnesses who've lived with the painful memory of loved ones and mates killed in a manner not fitting a proud soldier, son, father, brother, uncle and friend, Kapooka became the story I *needed* to tell.

In view of my responsibilities to tell this story, I've researched detail from various and diverse sources; many of which included open public information, archived military records, family interviews, digitised and hard copy printed publications, journals and from my own military experiences. In compounding such a plethora of material, I have neither footnoted nor referenced my sources—as is normal for historical writing. Cluttering sentences and pages with footnotes and references would not only change the aesthetic appearance of this poignant story, but more importantly, readers would be distracted from following the real heroes of this tragedy.

However, those readers interested in tracking down my sources can do so of their own volition, or through me. I'm more than happy to share the toils of what has been a labour of love in researching the lives of the victims, their patriotic endeavours eventually leading to their untimely mortality. For all the long hours of research, quintessentially I've arrived at a conclusive postulation; Australia should never forget the sacrifice of any of our military personnel—regardless of how their sacrifice was obtained.

As I reach the end of my preamble for what is my very first attempt at a non-fictional novel, I find myself reflecting over what I've so far managed to achieve. Many months of toil, tears, and late nights are contained in the pages of the product you hold in your hands today. It's a product I'm proud of. Not only because it's the first and only book written on the catastrophic events at Kapooka, but also for its contribution to Australia's military history. I'm especially proud of the fact that the book recognises previously forgotten soldiers who paid the ultimate price. In remembering their sacrifice, I've hopefully contributed to preserving a small piece of military history, thereby adding to Australia's national heritage and identity. The tragedy at Kapooka should stand as a reference point in Australia's military history—but sadly it doesn't raise an eyebrow on the military calendar. Had these men died in combat, they would've been remembered forever, but they didn't, and they weren't. Their deaths were consigned solely to the memories of distraught families. But like combat fatalities, they were also Australian men killed on duty. They were still the very same men raised in the shadows of the Great War who embraced the ANZAC spirit after a country at war forced them to become soldiers. I'm therefore humbled to be in a position to present to you their life in stories as young Australians, and importantly, as iconic WWII Diggers.

It's therefore with enormous honour and pride, both as a father and former soldier, that I'm able to present what has been for me an emotionally-charged journey. It's my attempt to explain the unbelievable events of Kapooka and how, for over fifty years, an entire country forgot the sacrifice of its brave victims. As a mark of respect to all parents of Australia's sailors, soldiers and airmen who've lost loved ones to military training misadventures and accidents over the years, where your loved ones may also have been forgotten, this work is titled in your honour.

Buoyed also by my novice literary endeavors, and inspirationally driven by words of encouragement from family, friends and members of the serving and ex-military community, I proudly present *The Forgotten Rising Sons*—the unbelievable true story of the 1945 Kapooka tragedy, its inglorious aftermath, and the short lives of 26 amazing young Australian soldiers.

Introduction

The Pride of a Town

> '... Fellow Australians, it is my melancholy duty to inform you officially that in consequence of a persistence by Germany in her invasion of Poland, Great Britain has declared war upon her and that, as a result—Australia is also at war.'
>
> Robert Menzies, Prime Minister of Australia,
> 3rd of September 1939

This story occurs in a time when, for the first and only time in the history of a young and developing Australia, the very fabric of her existence and identity was being challenged by an unexpected invading force. The courage displayed by Australian soldiers during WWII, including versatile and resilient combat engineers—iconic 'Sappers'—not only defended and saved a country, but their achievements forged a grateful ongoing national immeasurable reverence towards its entire military forces.

Resting peacefully in the Adelaide River Cemetery, a 113 kilometre drive south of Darwin on the Stuart Highway in the Northern Territory of Australia, the remains of the first two casualties of Australia's involvement in World War II lie in Commonwealth war graves. Disillusioned by a conflicted legacy surrounding the enormous loss of life in the Great War (WWI), seven million Australians tuned into the wireless on 3 September 1939 to hear Prime Minister Menzies declare that the country, once again, was at war. Quickly realising it was on the brink of yet another senseless battle, parents who swore that their children should never know and bear the weight of a gun like they did, faced another frightening and deeply distressing uncertainty. Really Britain was at war, but Australia felt compelled. Guided by its first wartime Prime Minister in more than two decades, Australia carefully went about raising a volunteer force to support Great Britain in its fight against the German invasion of Poland. It was common knowledge at the time that this new war would provide the nation with the supreme test of its own destiny. But they ended up fighting in two wars—one was against Germany and Italy, as part of the British Commonwealth in the European theatre and North Africa campaign; the other was against Japan as part of an alliance with America and Britain in the South West Pacific theatre.

Two days after the declaration of war, 23-year-old Flying Officer Arnold Dolphin from New South Wales and 28-year-old Corporal Harold Johnson from Victoria were members of the Royal Australian Air Force ferrying a Wirraway aircraft to Darwin for coastal patrol missions. Sadly, the aircraft stalled and crashed killing both men. The rest—as they say about Australia's involvement in WWII—is history. Or is it?

Australia's well-chronicled record of WWII, describing how and why a country has such an ongoing national reverence towards its military force, contains volumes of stories recollecting the bravery and courage of Australia's Second Imperial Force—the 2nd AIF. As a result, details containing the how and why of recruitment, the process of enlistment and discussions on conscription, don't necessarily need greater attention. Suffice to say though, nearly one million Australian men, just like Arnold and Harold, were rushed to fight in a war, and many paid the ultimate price. A total of 39,649 Australian souls were lost during Australia's 1939-1945 WWII experiences.

And, understandably, not every one of those deaths is recorded in finite detail. Countless stories of Australia's involvement in the war, and the thousands of senseless casualties, still remain anonymous—they're the unknown soldiers and for many known only to God. But there is also a growing quantum of soldiers who've added to Australia's casualty rates. But they're not treated as one of the unknown; they are regarded as the insignificant. Their deaths are equally painful, but they've been branded insignificant simply because they didn't even make it to the battle. Instead, their lives came to an end by errant misfortune whilst on home soil training for the honour.

On the 15th of February 1942, Australia and her allies lost Malayan soil to bike-riding Japanese soldiers, eventually leading to the fall of Singapore (Great Britain's bastion against invasion). When Singapore fell, an invasion of Australian soil by Japanese forces seemed very, very real. Four days later Japanese planes bombed Darwin and simultaneously commenced landing waves of ground force troops on New Guinea; the war was on Australia's doorstep. As the Pacific campaigns raged on, Australia and her allies soon realised that to successfully repel the Japanese advances towards Australia through New Guinea, they needed a never-ending supply of soldiers. General reinforcement troops of the 2nd AIF continued to volunteer. Quickly processed to support Australia's growing casualty rate, before they knew it they were issued with equipment and sent to Recruit Training Battalions like the Army's largest Recruit Training Centre at Cowra in NSW and after further specialist training, shuffled off to the front.

Amidst the challenges of encouraging a growing number of volunteers, the war-fighting landscape had changed. Fighting moved from the desert heat of the Middle East to the oppressive heat and humidity of the New Guinean jungles. To prevail in the changing conditions Australia needed a new type of soldier; one capable of fighting this new type of war in lush jungle terrain, and equally capable of engaging with the enemy in hand to hand combat all the while deploying demolitions and booby-traps. In WWII, the Army created a new type of jungle-ready soldier. One part infantry and two parts engineers, they would be known as combat engineers—so-called 'Sappers'. Generally men who became Sappers were allocated to Engineers because they showed a greater potential than the average soldier. Sent to complete advanced and highly dangerous engineer training, they were the cream of the volunteer crop. From an extraordinary display of commitment by brave volunteers and conscripts joining multivariate Corps and trades like infantry and artillery during WWII, more than 35,000 volunteers and conscripts were identified and rose to meet the challenges of soldiering as Royal Australian Engineers (RAE). But the Army had a problem. In order to meet this changing demand for warfare, and train almost 35,000 men in new engineering techniques, they needed a new camp.

A shortfall in the government's strategic military engineering plan eventually led to the Army establishing the country's largest military camp in rural New South Wales. It was a visionary camp whose purpose was to centralise and consolidate the country's dispersed military engineering training locations from around Australia into the one more manageable location. The centralisation of engineer assets meant greater command and control over training. Greater command and control meant an improved level of troop preparedness in the unique environments they now confronted.

In what seemed like an overnight transformation, a dynamic and vivacious camp was established at Wagga Wagga in the Riverina district of regional New South Wales. Wagga Wagga was a bustling rural township established on the banks of the Murrumbidgee River. The town had gained some notoriety around the world for its unique double-barrelled name gleaned from the local indigenous Wiradjuri people, meaning ' a place of crows'. The new camp was the vision of Commander-in-Chief of the Australian Military Forces—General Thomas Blamey, a local from

Wagga Wagga, and his military engineering supremo, Major-General Clive Steele. It was to be known as the Royal Australian Engineer Training Centre (RAETC). The two experienced officers and their advisers purposefully established the unorthodox engineering camp in the hills of the Pomingalarna Range, approximately nine miles from Wagga Wagga. Using the natural and man-made resources found in the district, such as the nearby Kapooka railway loop and existing roadways, the Army capitalised on the generous commerce supplied by the regional business community. Not only that, the camp was perfectly positioned equidistance between the Army's key recruiting cities—Melbourne and Sydney. The townsfolk referred to their new military camp as sitting on 'the far side of town'. The Pomingalarna Range was ideally located in the vicinity of the Sturt Highway. It linked Sydney to Adelaide. This made it possible for recruits to be transported in from the north and east, and the Hume Highway could run troops in from the south. The area also boasted close proximity to the Forest Hill, Uranquinty and Coolamon grazing airfields. The airfields became strategic military assets later becoming the foundation for Royal Australian Air Force bases in the region. For General Steele the Riverina district was an ideal location for his robust engineer-training regime. The large and growing inland city of Wagga Wagga provided the ideal lifeline both fiscally and, with the presence of a number of pubs in town, made the location socially ideal for the never-ending thirst of the camp's occupants. Thanks to Major-General Steele, Kapooka Camp, as it became known, commenced its long military life as the blood and bones of Australia's rich military engineering identity. It was, and still is today, the pride of this outback town.

Controversially, the outgoing Director-General of Engineering, Brigadier Murdoch, commented in late 1941, ' ... the state of Australia's current military camps made life too luxurious'. Compared to the conditions experienced by soldiers on active service overseas, who were living out in the open and surviving in primitive conditions, there was a clear disparity in preparedness back in Australia's comfortable camps. Declaring that military camps should be simple to allow soldiers to acclimatise to these so-called comparable 'primitive conditions', Brigadier Murdoch suggested that conditions in camp should be harsh, but not so harsh that it would affect a soldier's physical condition. Essentially he wanted fresh soldiers to be hardened before they arrived in theatre. To achieve this he believed military camps needed to replicate the uncertainty of combat and be devoid of any comforts of home. That meant being placed in tents, not cushy barrack blocks with mattresses and pillows.

Kapooka's rudimentary and character-testing conditions appeared to initially appease Brigadier Murdoch's wishes. At Kapooka, a camp of this magnitude, servicing 8,000 men, meant living conditions were at best uncomfortable and downright inhospitable. Echoing Brigadier Murdoch's sentiments, the underdeveloped nature of Kapooka meant that most of the troops destined for the camp were expected to sleep in six-man tents, not purposely-built accommodation huts with the comforts of home. Not only would they be tired and hungry, they endured living in cramped tents sleeping on straw mattresses. At times, six-man tents had up to twelve men living in the overcrowded unhygienic surrounds. In the summer months the heat inside the tents was unbearable; at least the men could roll up the sides to let fresh air in. However, rolling up the sides meant privacy was a high price paid for little comfort. Conversely, during bone-chilling winter months, tents would rock in the winter night winds and would creak and leak during wet days and nights. Stepping out of bed to tighten guy-ropes and batten down the hatches was routine drills when you lived in a tent for long periods. Kapooka was no exception.

By late 1942, at the height of Japan's attempted infiltration of Australian soil, all WWII military engineering training was being taught at the newly established Kapooka Camp. Under proudly worn (if not slightly weathered) slouch hats adorned with the distinctive 'Rising Sun' badge resting proudly on its upturned side, either fighting or training to fight, almost 33,000 of the 47,000 or so engineers trained during WWII were trained at Kapooka. Representing the Corps of Engineers they all performed with the same ANZAC spirit displayed by their engineering forefathers in the first Australian Imperial Force. Although the figure of 47,000 men of the RAE may have been small in comparison to the larger Corps and overall total number of soldiers serving in the 2nd AIF, their well-respected enthusiasm, expertise, commitment and professionalism were never undervalued. They earned and maintained their status and reputation as the hard men of the Army through hard work. The type of life and work they experienced at Kapooka merely cemented their iconic status amongst other Corps. They were men who made a living doing backbreaking work. They thrived on challenges, and with an almost bullish endeavor, they met every challenge big or small front on.

One of those accepting the challenge was father of four Herbert John 'Jack' Pomeroy. Fittingly, his individual journey, from civilian to soldier, provides the indispensable inspirational background to this story. The 31-year-old Victorian volunteered to join the Army in order to provide for his young family. As the son of an English immigrant, making a life in a comparatively unsettled Australia, the man they called Jack perhaps had a life which symbolically represented the hundreds and thousands of other soldiers who made the difficult decision between serving their country and caring for a family. Men like Jack, who courageously made the decision to leave family behind, were filled with a necessary mixture of confidence, youth and bravado. Displaying unparalleled zeal, promise, passion, purpose, above all else, they were fighting for a future for the very families they'd left behind. Taking their oath, in overcrowded enlistment centres and

Town Halls, to diligently serve King and Commonwealth in the 2nd AIF, they uttered the forewarning words of commitment—'... so help me God'.

Becoming members of the 2nd AIF, they paraded in uniform as proud sons of Australia. They would soon be fighting with the same spirit their fathers fought with in the Great War. Acknowledging that the Rising Sun hat badge insignia adorning the uniform of Australia's Imperial Force was their 'badge of honour'—we colloquially refer to these sacrificial patriotic military volunteers as Australia's very own- 'rising sons'. But when the war arrived these aptly titled 'rising sons' of Australia, like Jack Pomeroy, didn't just become 'Diggers', they forged reputations, performed heroic deeds, and when they did it well—they became legends. Coming from a mixture of ordinary blue-collar backgrounds, even from proud military stock, they were sons of former soldiers who gave birth to a well-known cultural reverence Australian's refer to as the ANZAC spirit. Although strangers, once in uniform they became one. Side-by-side they trained together and fought together. And sadly, on battlefields of foreign soil, side-by-side they died together. Finally, when white wooden crosses were laid out side by side on foreign lands, these rising sons of Australia were never far apart. Mates to the end, even in death.

Meanwhile, when places like Kapooka were instituted to train the next wave of legends for WWII, troops weren't necessarily walking into ideal training conditions. For instance, Kapooka only consisted of approximately thirteen rudimentary roughly-constructed buildings during its early stages in 1942. They certainly weren't environments and facilities that we'd contemporarily refer to as 'state of the art' training facilities. Kapooka was never destined to be a permanent camp, and thus it was certainly devoid of creature comforts. The majority of buildings were low wooden hut types with skillion roofs of corrugated iron. Later, larger more dedicated structures such as the headquarters buildings, dining halls and medical facilities, were built with some hint of permanency in mind. This collection of semi-permanent amenities formed the majority of the important dwellings at the camp; these were well thought out and soundly built by the very engineers who occupied them. All along the Pomingalarna landscape countless temporary huts, which were used for training stores and instructor-type shelters, also appeared. In addition to temporary dwellings and difficult living conditions, the uninviting terrain of the Pomingalarna Range meant most of the camp was built on the side of hills. Apart from the small number of well-constructed buildings, the uneven landscape quickly became dominated by the countless rows of pitched tents. Dotting the Range like snow-capped mounds, the tents, with flooring levelled by house bricks to combat the unevenness of the land, were tightly packed together and provided sleeping quarters for tens of thousands of Army engineers who'd eventually transit through the camp. Meanwhile, Officers, Non-Commissioned Officers, and permanent staff were afforded some semblance of luxury—they were quartered in huts—on beds!

Even before the camp had completed final construction it started accepting engineer trainees. Waiting for them were veterans of Syria and Tobruk campaigns who took charge of the formation and initial administration and logistics. Arriving troops were soon allocated into companies, formed into platoons and further allotted down to sections and down as far as squads. Assigned their respective tents, the men were soon introduced to other trainees whom they would soon enough get to know quite well. Men, who started out as strangers, with endearing nicknames reflecting physical attributes of the time like Lofty, Bluey, Porkie, Brickie, Shorty, Curly, Rocky and Snowy would become new best mates courtesy of virtually living on top of each other in cramped tents. A welcome distraction and relief from the uncomfortable tent life was the actual training itself. Within days of their arrival from recruit areas like Cowra and other units, training started. But it wasn't all basic military training like weapons instruction, bayonet drill and grenade throwing, many of the early trainees were seconded to acquire material and building supplies and put to work completing the finishing touches at the camp such as latrines and rifle ranges. In the chaos of construction and training, engineer bridges would appear and disappear on local streams, all the while the camp became a sea of barbed-wire entanglements, troops and most of all—organized confusion.

Engineers processed at Kapooka during WWII came from dissimilar backgrounds from the far reaches of Australia and various occupations. Much like the way modern recruits, regardless of their origins and experience, assemble at Kapooka to undergo recruit training, in WWII they were similarly drawn to Kapooka for the honour of being called a Sapper. With that honour came the opportunity of an overseas tour in either the Middle East or New Guinea where there was a chance to fight the invading Japanese. But training recruits in the early days of the war was difficult. It was especially difficult when you consider the problems faced by the government kitting out soldiers and providing them with the necessary resources so soon after Australia's crippling Depression. In the initial stages, there wasn't even enough money to clothe recruits and trainees let alone contemporarily train them for immediate insertion into combat. For some of Kapooka's earliest trainees half of their initial training was held in their own civilian clothes. Using broomsticks as rifles, it was a far cry from reality. Arguably, it was making a mockery of what the Army was attempting to achieve; enticing more volunteers to join the engineering ranks.

Over the war years the supporting township of Wagga Wagga was invaded by thousands of engineering and support troops of the 2nd AIF. At its peak, in July 1943, with more than 9000 troops, drawn from Royal Australian Engineers, Royal Australian Navy, Royal Australian Air Force, US

Army Corps of Engineers and the US Army Air Corps, all training and living at the camp, Kapooka was the largest military camp in the Commonwealth.

Capable of processing more than 8000 men at any given time, Kapooka camp may have been visionary, strategic and exteremly busy, but it was also highly hazardous. Compounding the bone-aching chill of winter and the oppressive heat of the summer, three months of training at Kapooka was purposely designed to be hard, dangerous and as realistic as possible. It was where fresh-faced recruits and trainees stared down their own fears and personal demons whilst discovering a previously hidden inner strength of patriotism, camaraderie and mateship. At the camp, inexperienced Sappers participated in a combined regime of basic military skills and specialist engineer training. Specialist engineer training which was purposely designed to be realistic—replicating the horrors and dangers they'd perhaps soon face in combat. Therefore, to inoculate young Australian men, the realistic training was highly dangerous and regrettably sometimes with fatal consequences.

Some of the most dangerous training conducted at the camp occurred on the demolitions range. Regular like clockwork, explosions at the camp caused cascades of brown Riverina soil to mushroom high into the air in the training minefields and on the demolitions range. On a dusty barren range, inexperienced trainees, many with less than three months experience as soldiers, found themselves doing small demolition tasks and setting booby traps with volatile and deadly high explosives. As early as the fourth week of their training (commonly known as demolitions week), men who were previously shopkeepers, factory hands, labourers and even the odd hairdresser, who didn't necessarily know one end of a rifle from another, were handling volatile and unpredictable high explosives and mines. By the fourth week of the syllabus men were training with gelignite, TNT and a new type of plasticine explosive called 'quarry monobel', all of which were capable of causing unimaginable damage, especially in the hands of a nervous novice. For most men what took place on the demolitions range in week four was a baptism of fire. With a destructive force of nothing they'd ever seen before, especially when things went wrong, trainee engineers approached the week with a mixture of excitement and sheer terror.

As a consequence of this feverish training regime, Kapooka, and its transient military microcosm, which amazingly still exists today, immeasurably changed the public face of Wagga Wagga. The camp, and its Military presence, now sits at the epicentre of the town's colourful identity. The general façade reflects its historical alliance to the original spirit of military community; a legacy of WWII, and now, overwhelmingly, Wagga Wagga is officially referred to as a 'garrison town'.

But this isn't just another military story about the how and why of Kapooka's establishment, or how successful the camp was in training and developing robust and professional engineers for WWII. Nor is it a story about Wagga Wagga and its historical relationship to the camp; a relationship which developed one of the most important contributions of a regional town, not only for WWII, but the ongoing contribution to the Australian Army today as the 'The Home of the Soldier'. This is in fact a sad story. It's about one of the greatest military tragedies in Australia's history that occurred at Kapooka in the long drawn-out shadows of Australia's involvement in WWII. It's a tragedy so great and of such military significance that when it happened, the reaction displayed by the entire Riverina district towards its itinerant military workforce became the unintended envy of the entire country.

In 1945, although the war in Europe was drawing to a close, the Japanese were still wreaking havoc in the Pacific. Kapooka was still essential. It was needed to continue steeling young engineer trainees with newfound strength, courage and determination using combat realism delivered by men like Jack Pomeroy. You could say many young men, guided by Jack and his experienced colleagues, discovered their manhood at Kapooka. Their maturity was fuelled by their patriotic obligation forged in the romanticism of becoming Australian 'Diggers'. Whether it was on the legendary obstacle course dodging flying debris, or down on the camp's dry, dusty and dangerous demolitions range blowing up Riverina soil under the tutelage of men like Jack, young wide-eyed trainees entered the camp as toothless tigers but departed sixteen weeks later as hardened combat-ready warriors. For many of these young men their thirst for adventure was unquenchable. Sadly, with unenviable regularity that pursuit of military immortality and the chance of battlefield honour came at a very expensive price. Kapooka was an unforgiving tutor.

When Jack Pomeroy and his young family settled into life in Wagga Wagga in late 1944, it was a happy time for the family. But Jack's idyllic life journey was heading towards a tragically brutal unexpected conclusion. A vicarious product of the military fervor to train as many soldiers as possible, Jack's Kapooka experience and eventual mortality became complicated by coincidences, irony, and his good nature. All of which intertwined with disastrous consequences on 21 May 1945. On this day, the day of his 31st birthday, he crossed paths with 27 strangers. But this path wasn't so innocent. As an Army instructor that was his job, but this day was different. In one single preventable incident, the experienced veteran of the Middle East and New Guinea Campaigns and 25 of those 27 young adventurous soldiers in his care, all busy experiencing everything that was Kapooka, were killed instantly. When his life and those of the young soldiers were tragically cut short, in a manner not befitting a soldier's death, the unimaginable incident stirred the nation. For a brief moment the tragedy became front-page national news; briefly the whole country knew Jack's name. Later they would also know the names of every one of the victims. But it wasn't just distraught families who wanted to know the how and why; an entire population of seven million war-weary Australians needed to know how so many men could be killed on home soil in rural New South Wales thousands of miles away from the actual fighting? And three short days later, when the caskets carrying their remains paraded through a rural NSW town in a moving ceremony marking the greatest military funeral cortege ever seen in the country's history, it was expected that the previously unknown 26 ordinary young Australian volunteer soldiers would join the ranks of Arnold Dolphin and Harold Johnson as part of Australia's tragic military story. Regrettably it didn't happen that way. Apart from a period of brief mourning, no formal recognition of their contribution by the Government was forthcoming. The incident should've afforded Jack Pomeroy a greater recognition and status than that of a simple English immigrant soldier killed in an unfortunate manner doing his duty. Jack Pomeroy's role in the Army's darkest hour should have catapulted him to a more fitting status as a national military identity. Instead, the opposite happened. He became a scapegoat for the army's shortcomings in engineering training. Their deaths were consigned to simply a 'lesson learnt'.

Not long after they were laid to rest, Australia's long WWII campaign came to an end. With every white military cross recording the military details of yet another lost life, the nation counted and reflected on the senseless and pathetic cost of war. In the long shadows of victory and reflecting in the magnitude of servicemen and women lying in Commonwealth war graves around the world, Jack Pomeroy and the story of the Australian Army's greatest loss of life in a single military training accident, somehow, became just another untold military incident. For more than fifty years, that's exactly where the story lingered—all but forgotten. How a tragedy of such national significance could ever be forgotten is as mysterious as the cause. You could say the story was systematically airbrushed from a country's rich military history and somehow forgotten—until now.

Part One
Kapooka

The 'Rising Sun'

Proudly worn in two world wars, the Rising Sun, officially known as the General Service badge of the Australian Imperial Forces (AIF), became one of the most recognised and honoured insignia of the allied armies—an integral part of the Digger tradition. Its distinctive shape, consisting of a series of bayonets radiating out in a semi-circle from a crown and worn on the upturned brim of a slouch hat, is readily identified with the spirit of ANZAC.

Chapter One
Just Jack

Of the 47,000 men and women who volunteered, or were forced to serve during Australia's 1939-1945 war, especially those who experienced the highs, lows and dangers as Royal Australian Engineers at Kapooka, one particular engineer's story is remarkable, inspirational and sadly—unimaginably tragic. The military journey of engineering demolitions instructor Sergeant Herbert John 'Jack' Pomeroy best sums up the type of senseless courage displayed by ordinary men who were called upon to perform extraordinary duties during WWII. With unnerving patriotism, Herbert, and his five brothers, all of military age, were submerged in the depths of confusion and uncertainty of war. Coming from a large English immigrant family of six boys and one girl, Jack and his five male siblings, Bill, Ted, George, Charles and Len, were five ordinary Victorian boys. But when the war caused Australian boys just like them to quickly become men, the Pomeroys became more than just proud Australian sons—they became even prouder 'Diggers'.

Much like Jack and his brothers, Australian soldiers forged reputations by performing heroic deeds, selfless sacrifices and they weren't shy of hard work. When they did all those things well together, they became legends. But Jack's individual journey from Victorian country boy to a Senior Non-Commissioned Officer in the Australian Army at Wagga Wagga in rural

New South Wales is poignant for many reasons, and on many levels—the least of which is his incredible role in this unforgettable military tragedy. He's not a legend, but perhaps should be. When the Pomeroy's adopted country called for volunteers, a beautiful bond of brotherhood and mateship allowed them to face the challenges of life in uniform together. But it was never easy, especially on their widowed mother. The legacy of a country at war had profound impact on the Pomeroy boys, especially Herbert. Unlike his siblings, Herbert's uncertainty mirrored the thousands of other volunteers who, as husbands and fathers, had to decide between sacrifice or savior? Volunteering would mean leaving his wife and children to suffer and struggle in silence and isolation. But not volunteering didn't seem right.

As a non-drinking Catholic who enjoyed going to church, Herbert John Pomeroy was a dedicated and loyal family man. As the husband of a loving wife and the father of three young boys and a beautiful baby girl, he was charismatic, professional, dedicated and extremely well-respected. To his extended loving family, Jack the man, the son, brother, and uncle, was larger than life itself. He was the tall, athletic good-looking Pomeroy boy with a heart of gold and sense of patience to match. Along with his father's dedicated work ethic, he toiled hard to gain his reputation amongst friends and colleagues as a genuine 'top bloke'.

In his early adulthood years, Herbert became commonly known by just one name—'Jack'. Reportedly it was his mother Susan's nickname for her brother, Ernest 'Jack' Barlow, a much-loved brother whom she lost in 1918 during the Great War. Perhaps guided by the memory of her brother, Susan Pomeroy spent Herbert's developing adult years lovingly referring to him as just Jack. As her second eldest son reminded her so much of a brother who never came home, in her lost brother's memory she called Herbert 'Jack'. Although in later years Susan would often refer to Jack as 'Johnny', as he grew up and formed his own life as a soldier, somehow, the endearing name stuck. Throughout his entire military career, he was well-known as just Jack Pomeroy.

Although separated by half a decade, growing up together Jack and his little brother George were inseparable. More than just brothers, they were good mates. Apart from playing football, which was quite a popular sport in those days, both boys became interested in cycling thanks largely to Jack's love of the sport. Joining the local cycling club, Jack rode as an amateur for a number of years competing in local races and other competitions throughout the Geelong, Terang and Warrnambool regions on Saturday afternoons. Jack developed into a good strong bike rider. His prowess on two wheels earned him many trophies and accolades in his short amateur career. In the small community of Terang he was gaining a reputation as a fierce competitor. In 1935, at age 21, he was riding off the scratch marker as an amateur rider on Carnival Day. Riding from the showgrounds through the nearby town of Chocolyn, Jack was giving other riders up to 5-minute head start. Before too much longer he was reeling the other riders in one by one; he was that good. Two of the men he reeled in on one day were his good mates— Cecil Lamb from nearby Tesbury, and Norm Parslow from Colac.

In 1937, aged 23, Jack was living with him mum in Grey Street, Terang still working as a labourer when he met and fell in love with Dorothy Hutchinson. The couple later married and their first-born son Barry arrived in 1938.

The outbreak of the war in 1939 meant that many friends, acquaintances and relatives of the Pomeroy boys, all living in the rural farming areas of Terang and surrounding districts, were quickly signing up to volunteer. Many scrambled to enlist as soon as the Prime Minister made his declaration of war on Father's Day, including Jack's cycling mates Cecil Lamb and Norm Parslow. There were also good family friends like the Woodmason brothers, Ed and Bill. At 26 years-of-age, Bill was among the first of the Pomeroy's mates to sign up. Enlisting in November of 1939, Bill went straight into the Infantry Corps. He enlisted at a time when numerous infantry soldiers were seeing direct action and many were not making it home again. This troubled Jack. And when other mates followed Bill's lead and enlisted, including friends Tom McRae from Camperdown and Lawrie Mahon from Geelong, it was evident to Jack that perhaps his enlistment was inevitable. But Jack often wondered if it would be any different if he didn't have kids. The way he saw it these local lads, his mates, had been in it from the start and he watched as other mates who weren't as enthusiastic continued to join as their time came up. Families in the district were soon saying goodbye to sons at an alarming rate. Off fighting as brave infantry soldiers or artillery gunners, sadly, many didn't return to the rich Victorian farming district. Instead of returning to the backbreaking labouring work of farming, surrounded by family and friends, they were now laying in marked graves somewhere in the Middle East—alone.

Deep down, Jack's mother Susan wasn't so enthusiastic about her five strapping six-foot sons joining the Army, and she had a good reason. Her only brother was killed in action serving with the 1st Battalion of the Royal Berkshire Regiment in France in August 1918. Understandably then, Susan was anxious for the welfare of her children. Her husband William had died of pneumonia in 1923, and having raised her five boys and one daughter on her own, her biggest fear was losing them all. However, inspired by the heroics of their good mates, all of her eligible male children were soon delivering her the unwanted news. Leaving in quick succession, it wasn't long before Susan was left at home with only her daughter. Over the war years, the six Pomeroy soldiers never really served together. Tragically, one would never make it home to a worried mum.

By 1940 Jack and Dorothy had welcomed son number two, Frank. By then, aged 26, Jack was working in a munitions factory as an electrical linesman. Married with two children, he was content with his life and by now his cycling career had taken a back-seat to the pleasures of raising his young family.

However, the declaration of war in late 1939 had a different effect on Jack. After all he was a family man with two children and by 1940 was now torn between his obligation to serve his country alongside mates and raising his young family. He certainly couldn't match the enthusiasm of his siblings or mates.

During the pre-war years Bill Pomeroy, at over six foot and weighing 171 pounds, had already enlisted into the 21st/23rd Battalion in September of 1936. However, when the war arrived and with his engineering machinist background, Bill was somehow seconded out of the Army to do more important civilian duties. By 1938 he was forced to manufacture gas-bottled equipment for the Army. Bill never served in the military again. As far as Susan was concerned, at least one of her boys was safe.

Meanwhile Jack's younger brother and cycling mate George, by now a butcher and member of the volunteer Army, decided to momentarily ditch his trade and enlisted in the 2nd AIF in April of 1941. Serving with the 2/24th Battalion in the Middle East, George would later go on to have a mixed military career. Much like his musically-talented mother, George was a born entertainer. Leading up to his enlistment, as a solo vocalist he'd entered talent quests and loved being on stage performing. Trying to be a soldier, strangely his music wouldn't and couldn't wait. A chance meeting with the 9th Division Touring Concert Party at Alamein presented George with a perfect opportunity. With his mother's pipes and talent for a song and dance routine, it wasn't long before George was a member of the 7th Division Concert Party. As a Corporal he would eventually go on entertaining troops overseas. He discharged from the Army in June 1946 after a successful career 'treading the boards'.

Charles, the second youngest Pomeroy son was twenty years of age in 1941. Conscription was fast becoming a certainty. Named for service in the National Call up, he was assigned to artillery with the 2nd Field Regiment.

With so many of the Pomeroy boys away, Susan was left at home with only her daughter Anne for company. Little did she know that Anne was also preparing to leave. Dressed in an ensemble of grey and black with shoulder spray of white roses with matching hat and accessories, Susan Pomeroy, in her husband's absence, had the honour at St Phillips Church of England in Collingwood on 29 May 1942 of giving her only daughter away in marriage. Caught up in a wartime romance, Anne married an Englishman and soldier eleven years her senior. Sadly for Anne, but thankfully for Susan, the marriage didn't last. Anne was soon back home with her lonely mum. But Susan and Anne wanted at least one of the boys at home.

Without a male at home Susan took advantage of a military rule allowing families to claim sons back from the military. Determined to have at least one male presence in the family home, Susan claimed back Charles from the 2nd Field Regiment. Within a year of his conscription, Charles, at just 21, had been in and out of the Army and back in Terang looking after his mother and sister. Susan was relieved. With Charles and Anne at home it provided some relief from the anxiety of her remaining sons now engaged in war fighting somewhere in the world. As a regular churchgoer Susan prayed for the safe return of her remaining serving military sons.

Deep down Jack's own position on the war troubled him. It was especially troubling after seeing his mates in uniform; needless to say he was somewhat envious. Jack had important decisions to make; decisions that constantly played on his mind. Should he evade enlistment and conscription to care for his family in uneventful Terang, or should he join his mates who were no doubt off exploring the world? Initially, his reluctance prevented him from serving alongside his eligible brothers who were signing up. Instead he did the family thing—he chose to work. But finding work was never going to be easy. Such was the economic status of the country at the time, the government was focusing all its resources and labour towards the war effort. Struggling for work Jack pursued the local PMG (Postmaster-General) with an ulterior motive. Supposedly knowing that postmaster work was deemed a 'reserved occupation', Jack knew if he could secure this type of work he'd be exempt from war service. Unfortunately, as the PMG were actually putting people off at the time, Jack, being one of the last to go on the employment list, was suddenly out of work. He knew his engineering skills and experience would soon need to be put to work to support his family. Being unemployed in rural Victoria during the war was difficult. Jack needed more opportunities to provide for his family

Jack did what he knew he had to do. Accompanied by Dorothy—who originally hailed from Melbourne—he packed up the two kids, and his miserly belongings, and returned to the city to live with relatives in Port Melbourne. Later they moved to their own place in Albert Street, but Jack still struggled to find meaningful well-paid work. His future was fast becoming inevitable. Before he knew it he was standing in the volunteer line amongst other 'down on their luck' Victorians. The life of a soldier beckoned; at least for Jack it was reasonably well paid. On 17 June 1941, with a degree of hesitation and trepidation, the man they called just 'Jack' offered himself up for service. Signing his *Oath of Enlistment* into the 2nd AIF to serve the 'King and Commonwealth' at the Melbourne Town Hall, Royal Park, the son of a British immigrant was now an Australian soldier. Listing his occupation as a 'munitions worker' (electrical linesman), with his engineering experience he was quickly seconded as a specialist and released for service into the Royal Australian Engineers. He was aware that his mates, including Lawrie Mahon, were being killed in battle long before he enlisted and now, with news filtering through that other Terang district boys were succumbing in battle at an alarming rate, the reality of war on Jack and Dorothy was frightening. However, it was Lawrie's death which hit Jack and Dorothy hardest. His good mate died doing what he liked best— riding bikes. He collided with a local lorry truck in Gaza in July 1940 and was killed accidently. Much to Jack's distress, Lawrie never made it back to Terang; he was buried in Gaza.

Within a week of his enlistment, Jack had been shipped off to the recruit and engineer training base just outside of Melbourne at Puckapunyal.

At least it wasn't too far from Dorothy and the kids. Later he transferred much further north to Bonegilla for engineer bridging training at what was known then as the Engineer Training Battalion. It was a bit further away from his family, but for now he was safe and not overseas. By January 1942, Jack had completed his entire training, including all required specialist training to qualify him as an engineer. He was finally an engineering Sapper. In February of 1942, back in Puckapunyal, he marched into his new unit, 2/8th Army Field Engineer Company (2/8th Fd Coy) of 6th Division. Marching into a fighting unit he was a little anxious about his operational obligation to travel overseas and fight. Like every soldier, regardless of Corps or rank, overseas service and the thought of death played on his mind. It didn't help his cause marching into a unit which had suffered significant casualties in the Middle East. It was an especially anxious time for Jack as his third son, Les, had also been born. He didn't want to leave his family full stop. Needless to say, duty called. Not long after baby Les' arrival Jack left Dorothy alone, with little support, to deal with the three little boys.

Demand for combat engineers in the Middle East gradually intensified. It soon became expected that the reluctant husband and father would deploy on overseas operations. By March of 1942, just nine short months after signing on the dotted line, Jack left Dorothy and their three children in Melbourne and embarked on the 14-day trip from Sydney to Ceylon (Sri Lanka) as part of the 6th Division Force. 2/8th Fd Coy was assigned to the 19th Brigade, where they fought at Tobruk chasing the Italians right across North Africa from Bardia to Benghazi. From there they went to Greece. Within a couple of weeks of his arrival in the Middle East Jack came across a familiar face which ever so briefly lifted his spirits. He managed to run into his cycling mate Cecil Lamb. Having a good old-fashioned yarn with Cecil about the old days, Cecil broke the news to Jack that their mate, Bill Woodmason, one of the two brothers from Terang, had also been killed in action and buried at Gaza alongside Lawrie. The news hit Jack hard.

Meanwhile, the younger of the Pomeroy boys were still putting their hands up. Jack's brother Ted, previously a member of the Militia with the 23/21st Battalion in both 1934 and again in 1936, decided to join the fulltime AIF. In March of 1943 he enlisted and was placed in charge of bulk stores at Port Melbourne. Issuing supplies and foodstuffs to troops around the world, Ted Pomeroy stayed in this position for a number of years. Len, the youngest of the Pomeroy boys, eventually joined the AIF on 21 March 1943 and was sent to the Armoured Corps in Western Australia. But Jack, as the elder brother with his couple of years of Army experience, took it upon himself to look out for his little brother. At the time there was an Army policy whereby an older brother could claim a younger brother to serve alongside. As a result, Jack claimed Len to join him with the engineer group as a Sapper in the 2/8th. The two Pomeroy men were excited to work alongside each other, not only because they were also good mates, but also Jack could at least look out for his younger brother.

Displaying his combat engineering skills, Jack, the mature 28-year-old Sapper non-drinking Catholic family man, was promoted to Lance Corporal! Fortunately, and with God's blessing, by August of 1942 he was back on home soil and into the loving arms of Dorothy and the children. Jack's Army career, however, fluctuated with a series of promotions and demotions. It was understandable. Engineers worked hard and partied hard and Jack was perhaps no exception. Still, he did have a deep concern for his serving brothers and how they were getting on. He was actually fined two days' loss of pay and once again reverted back to Sapper after going AWOL for more than 24 hours in September just to be with his family. It appeared as if the Army didn't like their soldiers to have families. Seeing the sense of it all, Jack's demotion didn't last long. By October 1942 he was back wearing a Lance Corporal chevron and formally promoted in January of 1943. But an 'undisclosed' incident, which reportedly occurred in December of 1942 in which he allegedly prejudiced the good order and discipline of the unit, resulted in him being severely reprimanded. And, by March of 1943, the highly promotable, yet equally un-promotable soldier, had once again been reverted back to Sapper.

In July 1943, Jack contracted Malaria. Admitted to the 2/2 Australian Field Ambulance, he was kept on the sick and injured list for nearly three weeks. Discharged as fighting fit in September 1943, he was soon on his way to Milne Bay, Papua New Guinea, supporting the 17th Brigade. Initially his unit was tasked as close support to the infantry, which wasn't necessarily appealing to the engineers or utilizing their skills, but that quickly changed. Realising the importance of engineers, before too much longer, they were heavily engaged in their primary role of building and maintaining roads, bridges, constructing airstrips, water holes, building storage huts, kitchens and clearing rivers and cutting channels in the Wewak area. The work of the engineers in New Guinea was phenomenal. In one year they built 118 bridges and over 100 assault bridges, which were later removed to make way for permanent bridges. Contributing to engineering success, Jack had experienced a high tempo of workload in his three months in PNG and was desperately looking forward to getting back home to Dorothy and his boys.

Having survived the oppressive conditions of the New Guinea jungles and the invading Japanese, Jack was on his way back to Australia by January of 1944. During an overnight stay in Cairns, en-route back to Victoria, letting off a bit of steam, the weary engineer and father went AWOL. Slapped with a fine of one day's pay by his OC, Jack accepted the punishment. It was an indiscretion that changed his attitude to service. Jack knew once and for all that he needed to settle down into professional soldier mode. Showing his immense application to the job at hand, by April 1944 his more recent indiscretion didn't prevent him from being promoted to Lance Corporal—again!

Finally arriving home to his family, it was a warm and moving reunion. And it didn't take Jack long to reconnect with his family, especially his young sons. He was a loving husband and father deeply adored by his kids. Jack cherished the time he had with them after his return from New Guinea, but as a soldier he knew another absence wasn't far away. He'd hardly had time to enjoy his young family when duty called yet again. This time though he wasn't going overseas, merely going away to do a course. His hierarchy believed that he displayed the qualities of a good leader. During the period 16th of May to the 9th of June 1944, as a Lance Corporal, Jack spent his 30th Birthday surrounded by strangers of the 17 Australian Infantry Brigade Number 5 Junior Leader's Course. Performing well, he was promoted to Acting Sergeant in November that same year. Following the subordinate leaders course, he was sent on a further course. This time the 107 Second Australian Army Junior Leader's Advanced Course. On this occasion, his Commanding Officer offered him some incentive to do well on the course. Pass the course and he would get a posting to a position in Australia, meaning temporary respite from serving overseas. In true competitive spirit Jack easily passed the course in December of 1944, and, as promised by the OC, was posted to the newly formed RAETC at Kapooka Camp in Wagga Wagga, NSW.

At Kapooka he was expected to take up a position as an engineer instructor. His experiences in the Middle East and PNG, and also as a former miner and munitions factory worker, made him an excellent candidate to become a responsible demolitions instructor. Regardless of the fact he was described by senior officers following his three years and three months in 2/8th Fd Coy as 'thoroughly reliable', Jack knew his new position wasn't going to be easy. But many commanders began to rely on Jack because he epitomised the three principles of demolitions work they were advocating; one, he treated explosives with respect and remained scared through his lack of confidence; two, he certainly wasn't a careless man through over-confidence; and finally, having witnessed the destructive power of his profession in realistic combat conditions, Jack was most careful after seeing an accident or narrow escape from harm himself on numerous occasions. They knew Jack would be the perfect SNCO to teach engineer recruits the art, and, more importantly, to respect demolitions work. It was Jack's military experiences, his maturity, his methodically cautious work ethos, his non-drinking steady hands and adherence to the three principles of demolitions work that perhaps made him a great mentor and educator to up-and-coming Sappers. He wasn't what they called 'bomb happy' like some of his former Field Company mates—mates who'd perhaps played with and misused their fair share of explosives in the past and therefore were not really suited to instruct trainees in the deadly art.

Just a few short months before the family were due to depart for Wagga Wagga, Jack and Dorothy welcomed the arrival of their only little princess, a baby girl named Maureen. The baby girl melted Jack's heart. His life was steadily becoming complete. A beautiful wife, three little men, a baby princess and now a safe posting in Australia, as far away from the war as possible. By the end of 1944 Jack, Dorothy and the four children, including baby Maureen who was less than six months old, found themselves on a train bound for Wagga Wagga. For Jack, he was pleased to be working as an instructor and training engineer trainees at the extremely busy Kapooka camp. It would be a welcome change and a personal challenge that he accepted unconditionally.

Arriving in Wagga Wagga, Jack, Dorothy and the four children moved into a temporary residence with Harry Hickson, his wife Mary, and their three children. Harry had a profound influence on Jack. He was a former cheese maker from England and a decorated veteran of the Great War. Discharging from the Australian Imperial Force in 1919 at aged 24, by 1937, 42-year-old Gunner Harry Hickson was now Police Constable Hickson, living in Tarcutta St, Wagga Wagga. Quietly positioned on the banks of the Murrumbidgee, Harry's house and surrounds were a far cry from French trenches. But not long after settling back into civilian life Harry and his family moved the mile or so to Beckwith Street. Before he knew it he was opening his doors to fellow soldiers who were posted to Kapooka, soldiers like Jack Pomeroy and his family.

The Hickson's new residence was a small and modest home directly opposite the Wagga Showgrounds. When Jack, Dorothy and the kids moved in, four adults and seven children now occupied the humble home. Eventually spilling out of the tiny house, many of the occupants were forced to sleep on the open verandah. I have no doubt that Jack volunteered and felt more comfortable sleeping out on the open verandah. Having endured his fair share of discomfort, the verandah may have been cold and uninviting, as opposed to a warm bed, but at least he knew his family were together. Due to the lack of available housing for new families to Wagga Wagga the temporary housing situation with the Hicksons was fast becoming a more permanent arrangement—at least for the time being. Over the next few weeks the Pomeroys found the Hicksons and their home most welcome and accommodating.

With Harry's influence, the small policing community got to know the Pomeroy family well. The young family settled in and were well-accepted, not only by the local police, but by the whole garrison town. It was in stark contrast to the welcome Susan and William Pomeroy experienced as English immigrants arriving in country Victoria nearly twenty years earlier. This time, Mr Pomeroy was in a 2nd AIF uniform serving his country with pride. Meanwhile, a real sense of military community spirit hovered and kept the Wagga Wagga population indebted to 'their' nearby camps. Jack, Dorothy and the children were a military family and as such were now well and truly part of the town.

Chapter Two
Daily Routine

'*... they taught us how to use explosives it was just not terribly extensive.*
It was enough for what we had to know!'

VX70606 Sapper Edward Thomas Payne,
2/13 Field Company

By late 1944 Jack, Dorothy their three little boys, Frank, Les, Barry and baby daughter Maureen, settled in as best they could into the modest house with Harry Hickson and his family. Over the Christmas period the entire Pomeroy family enjoyed taking some well-earned rest. Jack squeezed in a little bike riding in between playing with the boys and changing Maureen's nappies. Jack and Harry soon formed a good relationship; they spent many warm Wagga Wagga summer evenings on the verandah swapping war stories. It wouldn't be long though before Jack had to return to work and conversations often turned to Jack's upcoming position as an instructor.

When he marched in to the Orderly Room at the Headquarters of the RAETC in mid-February 1945, Jack was allotted to the 1st Bn. With an important role teaching during the very dangerous demolitions week, Jack was adamant he wouldn't disappoint. Making his way through the front gates past the newly-erected guardroom at Kapooka, Jack reported for duty at the RAETC Headquarters to take up his new post. What surprised him most as he surveyed his new surroundings was just how well established everything was. He was overawed at the sheer scale of the RAETC operation. The RAETC Headquarters, which at times appeared working at fever pitch, was fully manned with staff and oversaw the entire operations of camp including four large Training Battalions—the 1st, 2nd, 3rd and 4th RAE Training Battalions (RAE Trg Bn). Each training battalion had a total of about 960 men running around and, combined with the other minor engineering and support units, it was easy to see how the camp gained the title as the biggest in the Commonwealth. There were men everywhere. Although he observed trainees and staff appeared to be getting on seamlessly with daily routine, the sight of 8000 or so men crisscrossing the camp in organised chaos did make him question whether he'd made the right decision to be posted here. How would he fit in with this scene of confusion?

Unlike modern-day Australian Army recruit training, where all-Corps instructors mentor and instruct recruits from day one through to eventual march-out, in early 1945, the RAETC training program was a little unorthodox. Regardless of the fact that Kapooka developed and changed training regimes regularly over the years, engineer training finally settled on an agreed syllabus by 1945. The syllabus was as intense as it was diverse. Broken down for simplicity, it was divided into two phases. Phase One training included infantry-type activities, which included weapon training, anti-gas defences, map reading, and physical training. Phase Two was designed for engineer specific training. It was in the second phase that engineers learnt vital specialist trades such as demolitions, mines, booby-traps, bomb disposal, mining, field craft and field defences. By the time Jack arrived, it was compulsory for trainees to pass progressively through this specialist training in Phase Two on a weekly basis. Trainees had to competently pass the previous week of specialist instruction in order to gain entry into the next week of training. And, with every new week of training, there was of course a new Sergeant or Warrant Officer delivering specialist instruction. This is where Jack would fit in. He would be delivering specialist demolitions instruction in week four of the syllabus.

Regardless of the training phases however, a typical day at Kapooka for trainees and instructional staff ordinarily followed a regular daily routine:

>Reveille 0600h
>Breakfast 0730h
>Parade 0825h
>Training Period 0840h—1210h
>Midday Meal 1230h
>Parade 1330h
>Training Parade 1340h—1630h
>Evening Meal 1715h
>Parade 1855h
>Tattoo 2200h
>Lights Out 2215h

Commander of the 1st RAE Training Battalion in the early months of 1945 was 44-year-old Brigadier Warren McDonald. He was Jack's highest superior officer in his chain of command when he marched into the Centre in February. And, much like his predecessor, the camp's inaugural Commandant, Brigadier 'Porkie' Veale, Brigadier McDonald was a passionate horseman. He resorted to making his way around camp checking on proceedings atop his trusty steed. But unlike the piggery established by the initial Commandant, Brigadier McDonald used the camp as his own private stabling and training facility for *his* racehorses. Gaining unfavourable mention amongst some of the men, it was reported that the Brigadier used the smaller of soldiers on guard duty as his stable of jockeys. Considering the number of men making their way through the camp, the Commandant had an extensive pool of potential jockeys to choose from. Often he handpicked trainees for the guard who was just the right height and weight. Needless to say therefore, during Brigadier McDonald's tenure in charge, Kapooka experienced far too many lightly framed men on guard duty.

Brigadier McDonald's Chief Instructor (CI) at the 1st Bn in early 1945 was 33-year-old Queenslander from Cairns, Major George Macdonnell. Acting as the Chief Instructor since March of that year, George was responsible for the entire engineer-training programme in the early months of '45. His responsibilities included the creation of the training syllabus and the general conduct of all training areas. To achieve all that, he was ably assisted by 36-year-old Major Maurice Berg. Maurice himself was an experienced Field Company engineer from Sydney and took pride in his appointment as the assistant CI to George. As his assistant, if ever George was absent from the camp, Maurice assumed all of George's duties and responsibilities.

Under George and Maurice was a General Service Officer (GSO) of the rank of Captain. They too needed a commissioned officer to order around to ensure the work got done. In early 1945 they had one of the Army's finest GSOs in the 1st Bn; 40-year-old Victorian-born, New South Welshman Captain Edward Merry. For the man they referred to as 'Ed', as the GSO, that also meant he was the Officer Commanding (OC) 'G' Company. His appointment predestined him as the Captain Instructor responsible for weeks four and five of training at G Coy. Before he took his Commission to become an Officer, Ed Merry was a soldier. Much like the men parading before him on a daily basis at G Coy, Ed had come up through the ranks and became a success based on hard work and keeping his nose clean. That fact alone made him a well-respected OC. When a new batch of trainees posted into the Coy, Ed Merry saw himself in many of them. He treated them in a way he would've liked to be treated when he was an inexperienced soldier. He wasn't an old and cranky soldier; he was approachable, fair and understanding. He seemed to take each and every soldier under his wing. He certainly dedicated his time to make certain of their welfare. It wasn't just the trainees of G Coy who looked up to him, the entire RAETC knew of Ed Merry. By the end of his career Ed would've been responsible for training nearly 10,000 men. The men he trained at Kapooka become known not as the men from G Coy, but 'Ed Merry's men'.

Within each RAE Training Battalion there was also a cast of experts in their chosen fields. These specialists were known as Key Instructors (KI) for their craft. Their duties included the complete supervision of training during the daily routine ensuring the busy syllabus was fully carried out and that all safety precautions were adhered to in every dangerous lesson. Working for Ed Merry in May of 1945 was Warrant Officer Class Two Doug McFarlane. Doug was a 30-year-old from Sydney who was appointed as the Key Instructor for the 1st Bn during their delivery of the fourth and fifth weeks of the trainees' syllabus. If however Doug McFarlane was absent, Warrant Officer Class Two Edward Dodds, was his 2IC and would step in to assist. Ed Dodds was an experienced 30-year-old engineer from Sydney who was conscripted and transferred to the 2nd AIF whilst in Port Moresby in 1943. At Kapooka he was quite comfortable barking orders at the trainees when his time arrived.

Upon his arrival, Jack took up his position as an engineer instructor. In the 1st Bn he would work directly for Ed Merry and Doug McFarlane delivering demolitions training during what the Coy called 'Demolitions Week'. Involving some real danger, demolitions work, the military tactic of destroying natural or man-made objects with explosives for the purpose of military advantage, seemed an exciting prospect for young enthusiastic and adventurous men. The danger attracted many Army recruits out of places like Cowra to select their Corps allocation towards military engineering. It soon became more popular than the traditional roles of infantry, or even artillery. It was perhaps the thrill found in the danger of probable harm interspersed with the excitement and pumping adrenalin of destruction using highly explosive monobel, gelignite and guncotton. Either way it made for a white-knuckle military career.

A certain sense of nervousness, the thought of something going horribly wrong, demanded trainees remained grounded in their approach to engineer training—especially the dangers of demolitions work. Of course, the most dangerous aspect of the engineer course at Kapooka were the periods of instruction where trainees were handling, fixing, and firing deadly explosives themselves during demolitions week.

Trainees knew all too well that at least one week, 'fourth week', was dedicated to just demolitions. It was exciting, but for most people who had to handle the volatile arsenal for the first time it was often overwhelmingly frightening. For former miners like Jack he knew an air of caution should always be present when dealing with munitions, and complacency—well, that was an unforgiving enemy. It was a similar view held by the officers at Fort Belvoir, the Engineer Replacement Training Centre in Virginia, USA during World War II. Their rule was simple, '... engineers can't make mistakes ... at least no more than once.' It sent a message that rang true for Australian combat engineers. When treated with disrespect or complacency, the impact of explosive products on human flesh is nearly always catastrophic. Jack knew that when dealing with materials that pack the wallop of a young earthquake, a sure touch comes from good listening, and good listening means hearing instructions right—the first time. It was a lesson he constantly reinforced with his trainee engineers.

Accompanied by the bespectacled and moustachioed thirty-year-old Sergeant Roy Tafe who, with his studious appearance, looked more at home in front of a classroom rather than a demolitions lesson, and 22 year-old James Conwell, both New Guinea Campaign veterans, Jack and the two experienced men provided the most significant contributions as the responsible engineer trainee squad instructors for demolitions training in 'G' Company. However, they weren't going to do it alone. The Companies at Kapooka had a host of less-experienced instructors who acted as assistant instructors during the hectic training weeks. It included men like Lance

Sergeant Colin Kendall and Corporals George Holdsworth, Cooper and a charismatic 25-year-old son of an ANZAC veteran from country New South Wales by the name of William 'Bill' Cousins.

The enthusiastic and laid back Corporal Bill Cousins had recently returned from fighting in New Guinea. After being posted to the RAETC at Wagga Wagga, he was now closer to his parents at Wellington in rural New South Wales but he was also fortunate enough to have the opportunity of escaping the rigours of the camp on a regular basis. Some old family friends, the Titus family, were living in Wagga Wagga and Bill was put in contact with them. Alice Titus was from the Jackson family, a prominent and well-known family in the farming community of Ponto—Bill's hometown. Jack and Alice Titus would become Bill's stand-in parents whilst he was instructing at the Engineer Training Centre. They modestly provided him that touch of family and home; especially after spending two-years fighting in New Guinea, he definitely needed some family around him. He was able to leave the camp and enjoy a beautiful home-cooked roast every Sunday evening provided by Jack and Alice who were living in Morgan Street. Bill relished the opportunity to spend time away from the strictness and often boredom of camp life. No doubt though, Monday mornings would be busy for Bill. He'd be trying to catch a lift back into the camp with some of his fellow instructors, like Jack Pomeroy, who were living in relative comfort with their families in Wagga Wagga, albeit in shared houses, rather than sleep in overcrowded tents. Conveniently for Bill, Jack and his family were still living with the Hicksons just around the corner from the Titus house in nearby Beckwith Street.

As a result, Bill and Jack got to know each other well. So well in fact that Jack became Bill's mentor at the RAETC. Together the two men would later conduct demolitions lessons together—Jack, as the instructor and Bill as his assistant.

Meanwhile, back at Kapooka, due to the inherent dangers of the type of demolitions training Jack, Bill, Roy, and all the other instructors were engaged in, there was an extremely high risk of fatal injuries. To monitor the lessons two to three times a day Ed Merry, in his role as the OC, would visit the demolitions area to check on his staff. The instructors welcomed his visits. He ensured the various safety precautions were being followed and that trainees were complying with the necessary safety precautions during all theory lectures and practical instruction. Ed's visits gave the instructors the required confidence that everything was going okay, and therefore, everyone was safe. Occasionally, at irregular intervals, Major Macdonnell and at inconspicuous times, Brigadier McDonald himself, would visit the demolitions area. Their intention was to confirm the standard of efficiency of the training area and observe whether the training was carried out in accordance with the strict policy and procedures. During the early months of 1945 visits to the demolitions range were becoming more frequent. Their frequency was the result of slight changes to the training program where the instructors were now being issued upwards of more than 100 pounds of deadly high explosives and detonators.

Jack and all the RAETC staff knew the cataclysmic nature of the volatile chemicals and charges they were working with. They certainly knew the dangers of unexpected detonations. They were all skilled engineers who had significant knowledge of the detonation process of explosives' sudden change from a solid form, i.e. from a slab or plug of monobel, to a gaseous state. They knew it took place almost instantaneously throughout the whole bulk of the explosive, and therefore, they needed to ensure safety precautions were always addressed. They knew all too well how the detonation of the explosive creates what is commonly known as a 'shock wave'.

In addition to the pressure effect of high temperature gases caused by the detonation, they'd all experienced in one way or another the power of a shock wave delivering a supersonic over-pressurisation wave.

They'd no doubt also witnessed how, in a millisecond, the blast wave progressed- from the source of the explosion as a sphere of compressed and rapidly expanding gases, robbing the atmosphere of air at a very high velocity. It's no wonder that military safety precautions at the time for a single charge of either gelignite or monobel, up to five pounds, required all personnel to be no less than fifty yards away on detonation. All the staff at RAETC and the highly-experienced demolitions instructors either witnessed firsthand, or were aware, that victims of explosions sustain unique patterns of injury. Depending on the amount of explosive and the circumstances in which they were detonated, more often than not the injuries are permanently debilitating, like amputations, or more likely to be fatal.

When high explosive materials detonate, the victim's injury patterns are synonymous with the amount and composition of the explosive, the surrounding environment, such as protective barriers, and of course the distance between the victim and the blast itself. Needless to say, explosions in confined spaces create greater morbidity and mortality. For example, a single pound of high explosive detonated against a steel rail produces a blow strong enough to cut the rail in half. Consider if you can the effects of 100 pounds. The shock wave alone would be ruinous let alone the mutilating effect of the detonation on any person in its path. Nevertheless, if we combine those devastating effects in an underground environment, such as the underground dugouts found on Kapooka's demolition range, any underground explosion would include shock waves reflecting against walls thereby amplifying the outcomes of a blast up to a ferocious 'nine' times. Any individuals caught between the blast and the confined space generally suffer an intensification of two to three times the degree of injury. Bodily mutilation would therefore be inevitable.

For that reason blast-type injuries are divided into four classes: primary, secondary, tertiary and quaternary injuries. Primary blast injuries in victims occur from the supersonic over-pressurized shock wave. Secondary injuries are injuries caused from flying debris, whilst tertiary injuries are those injuries that were sustained by a victim after being thrown against solid objects or other victims by the displacement of powerful air. Finally, quaternary injuries are those other miscellaneous injuries not included in the first three classes and include burns, crush injuries and respiratory injuries from over-pressurization.

The effects of over-pressurization on the human body are catastrophic. Gas-filled structures in the body, like lungs, the gastrointestinal tract and the inner ear, all haemorrhage and perforate under the pressure of the supersonic wave. Lungs and eyes explode, and the abdominal cavity cannot contain the build-up of gas and explodes through the skin. Meanwhile, air embolisms shock the heart and the central nervous system. Traumatic amputations of the limbs occur with multiple fractures, crush injuries, burns, cuts, severe lacerations and air embolism-induced injuries. For survivors of serious blast, injury trauma includes what is known as 'blast lung'. This is a debilitating over-pressurization of the lungs, often with fatal consequences. The blast causes severe pulmonary contusions, bleeding and swelling of the lungs. It's the most common cause of death among people who initially survive an explosion.

Due to the effects of over-pressurization on the human body, victims of a genuine blast may only sustain fatal primary injuries and therefore display a lack of external evidence of injury. For individuals exposed to frequent blasts, such as military engineers, a loss of memory before and after an explosion can easily manifest itself. If fortunate enough to survive a blast without obvious injuries, concussion after the explosion is expected with recovery due once the brain has recovered; there will more than likely be bleeding from the nose or ears, or injury to the ear drums and the victim would more than likely be coughing up blood and mucous.

Essentially, whilst the victim's internal organs have perforated and ruptured under the enormous pressure, the cause of death is multiple internal organ failure as opposed to more visible and understandable secondary and tertiary injuries. For experienced Sappers, instructors and medical staff at the Engineer Training Centre, they not only knew full well the severity of blast injuries, but also how to triage and treat the unique pattern of injuries.

Understandably, the increase in explosive stores being issued to instructors kept everyone on their toes. Up to 100 pounds of high explosives was being placed in close proximity to trainees, making the potential for catastrophe multiplied by 100. It's no wonder men of the RAETC welcomed visits from the chain of command to ensure training was safe and appropriate. What they couldn't control or treat was complacency, or a lack of safety. During World War II, for Jack and the other instructors at the demolitions wing, their everyday instruction in demolitions week included the use of a new 'unpredictable' product known as 'quarry monobel'. Of course many old favourite explosive compounds were still being taught at Kapooka such as gelignite (also colloquially known as 'jelly') and gun-cotton, but it was this 'monobel' that caught everyone's attention. First made in 1903, monobel was a kind of improved explosive which contained only a small amount of nitro-glycerine. Previously used in England and America, by 1944 it was being manufactured in Australia and supplied to the Army for demolitions tasks. For the RAETC demolitions staff, they had to learn all about its capability in order to instruct trainees safely. What they learnt was that monobel, like many other high explosive products, had a high velocity of detonation (the rate at which the detonation wave travels through the explosive)—for monobel, the detonation wave travels somewhere between 2,500—3000 metres per second.

It was highly potent explosive, however, some concern for Jack and Roy was lessened in the knowledge that it was less susceptible than the gelatines to what was known as sympathetic detonation. Much like a chain reaction, sympathetic detonation of high explosives occurs when the detonation of an explosive charge is caused by the shock wave from explosion of a separate charge in its general vicinity. A good analogy to describe sympathetic detonation could be what we know as the 'domino effect'. But unlike dominos, which touch each other and fall, in explosive terms it would mean a domino falling in the general vicinity of another domino caused it to fall. For Jack and Roy, this meant that monobel was a bit more predictable than first thought. For the instructors receiving greater amounts of volatile munitions to conduct their lessons, it was a matter of safely storing it out of harms way. The larger stores of explosives could be better managed, as long as they kept detonators away from it.

But at Kapooka it wasn't just the Sappers detonating and blowing things up who were cautious and scared, even the truck drivers carrying the deadly loads were frightened of the stuff. One story, from a truck driver recalling his days at the Engineer Training Centre delivering gelignite in his truck and trailer to the demolitions area at the bottom of the hill, he conveyed a good example about how the men feared their dangerous cargo. Driving down to the demolitions area, his truck loaded with gelignite, there was too much weight for the truck's brakes. Stamping and stomping on the pedal, to his horror, the truck wasn't pulling up. Drastically changing gears, whilst physically standing firm on the brakes with his arse out of his seat, he was using every effort to halt the truck before it cannoned into trainees, buildings or even more drastically driving into a dugout. Arriving in the demolitions area bathed in a lather of sweat and a slight stain in his strides, luckily the truck came to a safe halt. Not before he got the fright of his life though.

The uniqueness of the innovative explosive compound, monobel, quickly became a fascinating curiosity for trainees and demolitions staff. It was explosive technology that had never been used before. Primary curiosity surrounded the fact that this new explosive material was like a shapeable plasticine. This meant that you could shape and wrap the explosive compound around almost anything, and once detonated, it made a clean cut. But this wasn't plastic or plasticine; it was highly deadly, extremely dangerous, potent, powerful, slightly unpredictable and fickle stuff. Jack was of course always cautious. Some of his instructional duties were to teach trainee Sappers how to detonate monobel safely. This meant treating it with respect; after all, the detonation of high explosives was the most sensitive part of the entire explosive chain and the hardest part of the demolition operation.

To teach trainees how to detonate monobel safely, they received instruction on the safety precautions and use of the in-service detonators. They were instructed that monobel and gelignite 'shouldn't' sympathetically detonate but 'could' do if care wasn't taken with the detonators. To aid with their instruction Jack and the other instructors used an arsenal of engineer demolitions stores to teach trainees. Their stores included detonators and fuse wire. A detonator is a small copper or aluminium tube, capsule, or case, about two inches long, which contains a small charge of primary and secondary explosive located in its tip. The primary explosive is initiated by a spark or flame which in turn initiates the secondary explosive. The charge is in such a quantity that should one detonator capsule explode it would be sufficient enough to communicate the small explosion to other like capsules or full cases if they were in the direct vicinity— another example of sympathetic detonation. To detonate monobel and gelignite, it requires the purpose-built detonator to be inserted into their plasticine-like compound. Detonators of the time came in three main types: plain or safety fuse detonators; electric; and non-electric. The plain detonator was an aluminium tube closed at one end, manufactured for use with safety fuse and which, upon ignition, explodes with intense local shock.

On small-scale blasting operations, such as those conducted during demolitions week, trainees were instructed how to attach a detonator safely to a piece of safety fuse. Trainees were expected to use a hand crimper tool or a knife to squarely cut off the desired length of safety fuse before inserting it into the hollow end of a detonator. The fuse was then gently pushed into the detonator, not screwed, until the composition in the detonator and fuse came into contact. Then, the detonator is gently, but securely, crimped to the fuse. Much like a pair of pliers, the tool safely crimps the fuse onto the detonator. The use of a hand-crimping tool is the safest way to attach the safety fuse to the detonator safely. Then, using a wooden pencil like tool, a hole was poked in the monobel or gelignite and the detonator inserted. In some instances, to aid with ignition of the safety fuse in adverse or difficult conditions, a match head was attached to the end of the safety fuse by wrapping up in used safety wire. The 'firer' striking the match head with the matchbox striker panel caused it to ignite and burn into the detonator then ignite the fuse. Being dependent on the length of the safety fuse, after a short burn time, it caused the primary charge to explode in turn initiating the detonator and causing the monobel or gelignite to detonate. This was a process that the demolitions instructors repeated every day. But often it didn't work out that way.

Men like Stan Emery, born and raised in the New South Wales bush, was used to blowing-up stumps of old trees. Because of his experience, he treated any munitions with due respect. Enlisting in February 1945, Stan attended the Engineer Training Centre and watched during demolitions training some of the city men who wouldn't know a ' ... hand grenade from an egg.' According to Stan, playing with explosives made some of these men 'bloody dangerous'. Referring to them as ' ... billygoats' he was concerned that many of his trainees treated the deadly material with the contempt of a schoolyard show-off. As Stan watched on with horror at some of the unsafe skylarking going on with his fellow trainees, he was well aware that the most sensitive part of the entire explosive chain was detonation. He knew that detonators had to be treated with the utmost care and respect. But he often witnessed the stupidity of fellow trainees crimping a detonator on to a piece of safety fuse with their teeth, not even bothering to use the crimping tool. For Stan he saw it as tantamount to thumbing your nose at safety.

Stan knew that an incorrectly handled detonator could easily take out the eye or a finger of a less than studious and cautious operator. Certainly the 'billygoats' were taking their lives in the own hands when they crimped detonators with their mouth. Stan knew all too well that doing this would cause unimaginable and horrible disfigurement if detonators exploded whilst a 'show-off' was crimping it with his teeth. It was fortunate for Stan that he never witnessed that occurring. But in his time it became common knowledge at Kapooka that 'cowboys' and 'show-offs' were becoming increasingly dangerous for themselves and everyone else. Thankfully Jack, who was safety conscious and an experienced field engineer, was arriving at Kapooka to take over demolitions lessons

After what he'd witnessed during his time at Kapooka in early 1945, Stan knew it wouldn't be long until something catastrophic occurred during the dangerous demolitions week. He knew whatever it was it would involve trainees and explosives.

In the first few weeks at his new posting to the Engineer Training Centre, Jack Pomeroy met some amazing and inspirational soldiers making their way through the camp as trainee engineers eventually bound for the front lines. But it wasn't just the trainees, he met inspirational men like Ed Dodds, Roy Tafe and all of his fellow instructors. In his circle of new friends and colleagues Jack and his entire family were well loved. Jack, the immigrant son of a cattleman, became an engineering Sergeant that all his engineer trainees imitated. He became well known as the generous and caring Sergeant who followed a well-worn path from English immigrant to an Australian soldier. Whilst his path may have earned him the respect of the 1st Bn, it also earned the admiration from Harry Hickson and the entire Wagga Wagga community. But when he joined the 1st Bn, his family, close friends, serving siblings and colleagues, were totally unaware of the fatal path he was now on.

In the most unimaginable circumstances, in less than three months, they would all discover just how dangerous Kapooka was, especially for Jack and 25 men under his supervision.

Chapter Three
Fixing & Firing

'... we had an assault course (at Kapooka), at least killed one person in a week, or one person in a day, it was a bad assault. ... You put three slabs of TNT on the branch of a tree, light them and go for your bloody life, otherwise you're dead.'

VX118473 Sapper Kevin James Joseph Cassells,
15 Field Company

For Jack, having marched into G Coy about a month earlier, mid-March of 1945 brought problems. His boots had barely touched the ground; it was a case of demolitions lesson after demolitions lesson. Therefore, it didn't take long before the tempo of training and the long days and nights took a toll on the family man. Worn-out mentally, and now worn-down physically, he'd fallen ill. Contracting tonsillitis he required immediate hospitalisation and treatment. By the 22nd of the month, he'd been evacuated to the generously-spaced and fully functional medical facility located within the Kapooka Camp confines known as 54th Camp Hospital—at least there he'd get some treatment and well-earned rest. Subsequently discharged on 03 April, the weary Sergeant was sent home. For more than a week he was able to convalesce with his young family. Being surrounded by Dorothy and the kids, not forgetting of course his 'adopted' family, the Hicksons, Jack was nursed back to health and waited on hand and foot by his loving family. Being surrounded by his young family was a welcome relief from the daily stresses of demolitions instruction. Suddenly life and family were all that mattered. In some quiet moments of solitude Jack's mind wandered; deep down he missed the noise, commotion and confusion of the camp. He was itching to get back.

Returning to full duty by the 12th of April, he was back doing what he enjoyed and had missed terribly during his time off instructing. When he returned he did notice that the Company training tempo was still high. There was still no letting-up, even though the Germans were surrendering in Europe. He appreciated that every Monday morning, without fail, a new group of trainees would be arriving on the troop train from nearby recruit training at Cowra. With military precision, these up-to-the-minute arrivals were processed and sorted into Companies quicker than they were getting there. They had to be. With thousands of men still running around the bustling camp, not to mention the hundreds of scattered tents representing a virtual canvas city, the new arrivals needed to know where they could drop off their kit and more importantly, where their head would be resting at the end of a long day. Jack knew that as long as the trains were still coming, G Coy would be conducting demolitions training.

Training at G Coy was broken down to include eight 'periods' of instruction from Monday through Friday leaving Saturday's set aside for administration and Sundays as a rest day. The entire fourth week of instruction was conducted down on the camp's demolitions area and included lectures, hands-on practical experiences, demonstrations and other activities. The syllabus was designed so as not to cover too much ground, but was considered to be a very sound week's training in the art of demolition for the less than experienced men. A standard schedule ensured the week's training was methodical; however, some flexibility and discretion by the instructors coupled with the often-unpredictable availability of demolition stores, meant that lessons could be rescheduled. But at the very heart of the 1st Bn ethos was professionalism. With men like Ed Merry and Ed Dodds at headquarters, all lessons, regardless of the circumstances, would always satisfy the core competencies required to qualify the soldier safely.

The training syllabus that Jack followed was provided by the RAETC and had been in operation for some time. It was designed generally to cover all necessary elements of demolitions work. A typical week was comprised of the following:

Monday: the first period of instruction was a lecture on the introduction to explosives used for demolitions in the field; second was another lecture on the safety precautions for electric safety fuses, detonating fuses and fuses instantaneous; third period was an introduction on the types of explosives used by British, American and other foreign forces; fifth period was a demonstration and practical application of making-up, fixing and firing single slabs of explosives as per Chapter Three of the précis. The sixth, seventh and eighth periods were dedicated to preparing for the night activity which included the laying of a circuit and the making up of hand charges (taught in lesson five). The final activity of the Monday was the night exercise. It was generally a very long day, but a lot was achieved and the final exercise of firing a barrage of land charges was the highlight.

Tuesday: period one was a lecture on elementary electricity; period two was another lecture on electrical firing and using a circuit (series); three was a demolition test set, testing fusion and continuity; periods five and six provided another practical exercise in testing of circuit and electrical firing, whilst periods seven and eight instructed on another practical lesson of the use of junction boxes and safety and instantaneous fuses.

Wednesday: First and second periods of instruction was a lecture on the calculation and preparation of charges; third was cutting portable charges; the fifth and sixth periods were a lecture on bridge demolitions, and 'recce' of the demolition site plus the construction and use of a main circuit;

periods seven and eight were setting up ring main and series circuits on a bridge, whilst the night activity for Wednesday was a practical exercise in setting-up a bridge demolition.

Thursday: The first and second periods of instruction was a lesson on the theory of hollow bone and beehive charges, considerations and use of charges and demonstration of concussion charges used in field on concrete pillboxes; period three was a lecture on the destruction of ammunition and petrol dumps; period five was a demonstration on making-up, use and firing of a Bangalore torpedo, whilst periods seven and eight included a lecture on the demolition of roads, railways, tanks and guns.

Friday: The final day of instruction in week four started out with a lecture on crater and mine charges; period two and three were construction of cratering by earth auger and compression drill; fifth period after lunch and physical training was a lecture on the use and recognition of Japanese explosives; the remainder of the day was left open to recapture any doubtful training points for the week. The final activity in week four was a Friday night activity, which was a demolitions activity incorporating a compass march.

Whilst Saturday was an administrative day, the morning was set aside to complete any additional testing of trainees who hadn't satisfied the objectives during the week. If however there was no additional testing, Saturday mornings were free, but the afternoons were set aside for dental, medical and any other administration.

Of all the days in the demolitions week training syllabus, the one that worried Jack the most was Mondays. This was not only the first day of the week for new trainee squads, according to the syllabus, it was also the day inexperienced trainees were handling explosives and detonators for the very first time. Never mind the fact that safety precautions were taught during the morning lectures, there was always an air of cautious trepidation during the first period of instruction after lunch on Mondays. It was this period, the fifth period of instruction, which regularly bothered Jack. If something were to go wrong it would probably occur in this period. By this time of the day, the men had conducted morning lessons and were perhaps mentally tired. Lunch would arrive on the range, and following a meal that filled their bellies, a physical training session followed. Shortly after physical training the men went back into the dugouts for more lessons. Perhaps hyped-up after physical training, the men were often a little bit more rambunctious on that first lesson after lunch.

But Jack the instructor was growing in confidence. With every fresh squad of trainees that arrived at G Coy, he was becoming more relaxed. Besides, he hadn't had any safety incidents of note since arriving.

Generally speaking, during all military training, instructors like Jack, Roy Tafe and the other specialists, delivered lessons that were designed and based on a dedicated military doctrine or précis for the activity. They were generally well-written military manuals that guided instructors in how to deliver lessons competently, safely and which covered all the necessary learning outcomes. Based on previous experiences and current methods and practices, the précis or pamphlet became an instructors 'bible'. The guiding précis used by demolitions instructors at Kapooka in 1945 was already three years old: *Engineering Vol IV (Part I), Demolitions 1942*. This particular policy provided demolitions instructors the required technical direction on how to demolish structures like steel, timber and reinforced concrete bridges, tunnels, roads, canals, guns, water tanks, artillery shells, aeroplanes, armoured fighting vehicles and even petrol dumps.

As far as calculations go, the précis also informed instructors that one slab of monobel or gelignite equalled one pound of explosive, and one pound of explosive could cut one inch of metal, ten inches of timber and twenty inches of masonry. Therefore, for an average brick-built house, fifty pounds of explosive was needed to level the building to a pile of rubble.

This was the type of information the instructors knew from experience, but the précis was the document they often referred to for guidance and accountability purposes if something went wrong. During demolitions week, Jack and the other instructors were issued with 110 pounds, consisting of a combination of monobel and gelignite. In the fifth period of instruction the instructors were effectively issued with enough ordnance to level two brick houses. Not only that, in addition to the simple destructive nature of fifty pounds of explosive, if it was buried at a depth of five feet, when detonated it would create a ground crater of no less than fifteen feet in diameter. Using these calculations on Jack's issued 110 pounds, if it was to detonate underground, say for example, in the dugout in which Jack stood giving instruction to trainees, it would create a crater of no less than thirty feet in diameter and no doubt kill everyone in the dugout.

Three general safety-first rules were written on the last page of the précis which every instructor followed religiously; don't smoke near explosives, don't use metal instruments on explosives and the key safety rule— keep detonators separate. It was all the instruction and guidance Jack needed when he started his lessons.

Whilst the latest batch of General Reinforcements (or GRs as they were known) were arriving and becoming familiar with camp administration, routine and logistics, down at the demolitions range Jack was continuing his critical instruction with his next inexperienced and nervous squad. Trainees were always nervously unpredictable when handling high explosives. In turn, their overt signs of nervousness made everyone remember that this was dangerous training and something could go disastrously wrong. Although there were numerous safety checks in place, demolition training was bloody dangerous. The one thing that they couldn't always control was those trainees that Stan Emery referred to as, 'billygoats'. Mind you Jack was becoming comfortable with his Monday routine but he never lapsed when it came to safety. He kept his eye out for the billygoats, after all, he knew at any stage Ed Merry could pay him a visit. Knowing it was his responsibility, Jack maintained his over-cautiousness and remained a little on the edge waiting for Ed's visit.

By Monday, the 7th of May 1945, Jack had delivered approximately ten Demolitions Week Mondays to inexperienced trainees without incident. Apart from the constant flow of new trainee squads, the other aspect of Mondays that changed was Jack's assistant instructors. It felt like every week a new assistant would accompany the new squad. During one of his customary Monday instructions, fellow Victorian Corporal George Holdsworth, was assisting Jack. George had been in the Army for nearly three years and thought of himself as an experienced Field Company engineer.

As far as the squad went on this particular Monday morning, Jack was instructing another normal squad of trainees who'd progressed to the demolitions week. For the benefit of this story the squad was a nameless group of brave men—men who were no less or no more important than the next group or the previous one or even the men who'd just arrived in camp. Needless to say, they all needed the expert instruction from Jack and George. But one man stood apart—Stan Emery.

Stan was about to receive instruction from Jack on day one of demolitions week. As an experienced boy from the bush, blowing tree stumps out of the ground, Stan was keen to see how the Army did it. He kept an inquisitive eye on how safety-conscious Jack was. As a small side issue, Stan didn't know it at the time, but whilst he was on the demolitions range, 30-yearold father of one, Jack Nixon from Cobar in New South Wales, was arriving from Cowra on the Monday morning troop train. Stan and Jack (who was nicknamed 'Porkie' due to his solid appearance) would later become tent mates.

In the morning lessons that Monday, Jack set about teaching his squad in their first 'hands on' practical application with the dangerous and volatile monobel. Regardless of his experience blowing stumps out of the ground with his father in rural New South Wales, Stan listened intently to the experienced instructor who was talking about 'monobel'. Stan was an expert with sticks of jelly but monobel was this new plasticine stuff and he was, as always, cautious. He respected explosives, and he thought little of his fellow trainees who treated detonators like they were billygoats—putting them in their mouths and showing off. He was especially attentive to trainees who thought they knew everything about demolitions— but in actual fact knew nothing. Stan believed these types of 'billygoat' trainees were the real danger, not the explosives themselves.

After lunch that day Jack, George, Stan and the remainder of the squad—including Stan's so-called billygoats—participated in what was the fifth period of instruction. For George, regardless of what level he thought he was, it didn't stop him watching intently as the more experienced Sergeant gave specific and very articulate instruction to trainees during the afternoon. It was Jack's least favourite lesson—how to make-up, fix and fire a single slab of highly explosive monobel.

For the previous nine or so Monday afternoon lessons, instruction always remained the same. Jack would sit the men down on the wooden boxes of the dugout. Ensuring that they could all see him, he started. With the confidence of a ringmaster and the steady hand of his experience, he began,

' ... Righto men listen up, in this period we're going to teach you how to make-up a small plug of explosives, about four ounces, just big enough to sit in the palm of your hand, attach a detonator and fuse to the small plug and then, later in the lesson, we'll go up top and fire the bloody things one by one ... it can be very dangerous so I need you all to pay attention and listen to my directions carefully.'

Now that he had the undivided attention of every single wide-eyed trainee he continued,

'… Everyone will get the opportunity to fix a detonator and fire a small slab, that will be done up top and you will need to pay attention to what I'm doing because you need to do this without any problems in order to be deemed competent.'

'… Men were there any questions?' None were asked.

Jack introduced the trainees to the stores he was going to use and explained each piece of them in detail. Located down and to his immediate right side was a roll of safety fuse, detonators, a small box of gelignite and a box of quarry monobel. To his rear was the dugout blackboard used for additional explanation and his assistant, on this occasion Corporal George Holdsworth was seated close to his Sergeant, ready to jump to his aid if requested. In his pocket, the experienced instructor would be carrying a crimping tool used to cut fuse wire and crimp the detonator by crushing the non-explosive end of the detonator onto the fuse wire. He would either be carrying a box of issued safety matches to light the fuse, or they'd be contained deep within his instructor stores box. He didn't allow trainees this level of luxury.

The lesson continued with the Sergeant picking up the roll of safety fuse. The safety fuse was a fuse that burnt, rather than exploded. It didn't contain its own means of ignition so it was to be lit separately. It contained an explosive in such a small quantity that the burning of the fuse would not ignite other similar fuses if they were to come in contact. They were certainly safe fuses for this type of lesson with inexperienced, fumbling hands. Cutting off and discarding the first six or so inches, so as to get to fresh, safe, undamaged and uncontaminated fuse, he proceeded to cut off approximately one foot of fuse wire. Safety fuse generally burnt at a typical rate of about thirty seconds per foot, therefore using a small controllable plug of monobel, thirty seconds was sufficient time for each trainee to light and move away from the explosion they were about to make.

The Sergeant then carried out the necessary safety precautions on the piece of fuse by testing it. The test included cutting off a small piece of fuse, approximately one foot, lighting it, and timing the burn time on his watch. Either too slow, or too fast, if it wasn't close to thirty seconds, the fuse wire would be discarded and replaced.

The next step in the lesson was to attach a detonator to the safety fuse. The experienced Sergeant grabbed a detonator from his stores and explained to the trainees the correct method of handling the small silver cylinder. Because it contained a small, but still nonetheless dangerous, amount of explosive compound in its solid tip, he showed his intent audience the correct way to handle the detonators safely so as not to activate the explosive compound in the tip. He then took out his detonator nippers (crimping tool) and placed the fuse into the hollow end of the detonator. Using the tool he safely crimped the fuse tight in the detonator. '... Like this men,' He then placed the primed detonator safely in his pocket. '... Never, and I repeat never, crimp the detonators onto the fuse with your teeth ... it will blow your bloody mouth permanently open.' They were stern words from a man who'd seen the damage a detonator could cause. As this was occurring, Corporal Holdsworth was patrolling the dugout ensuring the men were paying attention. Stan Emery at this stage was all eyes and ears. '... Righto men, grab yourself a detonator and fuse and do that.' He added, '... remember the detonator also has a small explosive compound in the tip so don't bloody drop it.'

The Sergeant continued. Holding a small four-ounce plug of monobel in his hand, he continued to explain that the compound was safe on its own and that it needed detonation to be dangerous. Taking the plug in one hand, he demonstrated how a hole was poked into the monobel with a small pencil-like tool in which the detonator sat. 'Like this men'. He then poked a hole in the monobel, explaining the safety importance of the detonator not coming into contact with the explosive until it was ready to be fired. Of course the men didn't have their own plug of monobel, as it was too dangerous; that led to the next part of the lesson, which took place above ground.

The theoretical aspect of the lesson had been completed. To confirm the trainee's knowledge, Jack and his assistant George asked the men a series of questions to ensure they fully understood the process of making and fixing their plug of monobel. The next step of the lesson was for the men to go outside the dugout to make-up, fix and fire their own small plug. It was the dangerous part of the lesson; the part in which Stan Emery knew some of his fellow 'billygoats' couldn't be trusted.

One by one the men prepared their fuse and detonator, just as the experienced Sergeant had showed them earlier. Jack inspected the detonator and affixed a single length of fuse to it. When he was satisfied the fuse had been correctly prepared, as per his previous instruction, he handed the men a small plug of monobel and watched them poke a small hole in the soft compound with the pencil-like tool. Once the trainees had fixed the plug and placed the detonator in it, the explosive compound was live. All it needed now ignition. Jack instructed the trainee to walk approximately forty yards away, light the fuse, and carefully walk—not run—back to his location. Here, both instructor and trainee, urged on by waiting cohorts, watched the small charge detonate in a small puff of Riverina dust.

One by one each man completed the same process until all of the squad completed the task to the Sergeant's satisfaction. Excitingly detonating their own individual charge, it was an introductory step for trainees, enabling them to step up to bigger and more complex use of explosives. After about a dozen or so men, all detonating their own charge with enthusiastic excitement, the period of instruction was complete. As a collective group the men would retire once again to the safety of the dugout. Here Jack and George asked the men if they had any final questions or doubts about the activity; if so, they would be clarified and the period completed. On this occasion, the period of instruction was flawless. There weren't any questions and importantly, no injuries.

After carefully executing their first practical hands-on application of setting an explosive charge, the adrenalin was no doubt pumping through their veins. It was definitely time for a smoke to calm the nerves before the next period. Exiting the dugout, it was more important that the excited men first visited the ablutions for a toilet break and then a follow-up nervous cigarette. Once again clambering for the scant areas of shade on the barren demolitions area, the trainees mustered under the eucalypt to discuss the comparable size of their blasts.

The engineering hierarchy considered that the method of instruction for the fifth period on Monday of demolitions week, this particular lesson conducted by Jack and on subsequent occasions by other qualified instructors, was quite sound. The risk of an accident occurring was negligible. Although the dictionary defines negligible as, 'not significant or important enough to be worth considering', some bad blood amongst instructors and a minority of trainees was creating angst in early 1945 at Kapooka. It was due in part to the allegations by older instructors that over-precautions taken by some instructors, like Jack, in their view, weren't realistic enough for training. Reportedly, many of the older instructors who'd been at Kapooka teaching demolitions for a number of years were told to either move on or at least have a break and let some of the men who'd been overseas, like Jack, Roy Tafe or Colin Kendall, take over demolition instruction. For the older instructors they saw the move as a 'colossal' mistake.

But a different view was shared by many of the men who experienced the demolitions week. According to many trainees and former instructors, safety was fast becoming an issue at Kapooka. Trainees who participated in the notorious demolitions week, and those in critical positions running Kapooka, believed that the instructors who'd been at the camp for years were fast becoming smug—and dangerous. Some of the more experienced men apparently introduced bad instructional techniques that didn't appear in the instructional précis. These older combat engineers who'd seen service and thought they knew everything, apparently brought with them an attitude of complacency and ill-discipline when working with explosives. Allowing trainees to smoke in the vicinity of dangerous explosives, crimping detonators with the teeth and playing jokes on each other during training, was identified as complacent behavior by 'bored' and unchallenged instructors who'd been there too long.

However some of the old and bold instructors complained that the introduction of inexperienced teachers was a backward step in the safety of trainees and the well-trodden processes they'd already established and had been operating with for a number of years. Regardless of what the older instructors thought, many of their trainees thought safety was fast becoming an issue, especially in late 1944 to early 1945—a view highlighted by former engineer trainee Clarence 'Hank' Keenan. Hank was a country boy from Victoria who was working demolitions in a quarry in Horsham when he volunteered for the 2nd AIF. When he attended Kapooka and participated in demolitions week, he had to walk out of the underground dugout. As he sat down to receive instruction with twenty or so other trainees, he observed that the Sergeant instructor not only had his full complement of volatile explosives, including land mines, monobel and gelignite in the dugout, a practice he thought was ludicrous, but he looked around in amazement. The trainees, with a highly potent mix of explosives mere metres away—were flippantly smoking! He thought he was going to die.

'… This is bullshit' he thought to himself and stormed out of the dugout, closely followed by an inquisitive instructor. Explaining to the instructor that he ordinarily worked with explosives and that men smoking in the dugout was bullshit and dangerous, he refused to go back in. Threatened with a charge of insubordination, Hank stood firm. The squad put out their cigarettes, but the volatile pack of high explosive stores was still there. He knew that if it went up no one would survive. In Hank's words, ' ... I couldn't wait to get out of Kapooka—it was bloody unsafe and a bastard of a place.'

It was a similar view held by a former instructor, Corporal Keith Kuhn, from Glen Osmond in South Australia. Many instructors, including Keith, refused to be posted to a training area like Kapooka. They blamed the trainees for the lack of safety. Keith quickly realised that instructing trainees wasn't for him, especially now that safety and welfare were becoming a little too relaxed for his liking. According to Keith he'd end up ' ... choking one or two of the little bastards if they tried that stunt.' Thankfully for the 24-year-old, who'd been in the AIF for a couple of years, he knew how the system worked. He was hard and fast in his convictions about safety and didn't take any bullshit; he wanted out. He found a confidant in the Camp Commandant's secretary. With her help, Keith managed to elude a full-time posting to the Engineer Training Centre, a move that probably saved his life.

Meanwhile the older instructors argued that they were more interested in the job and were performing well because they 'wanted' to be there, unlike men like Jack who were posted there for some respite. It resulted in some friendly banter between the two different castes, but according to the experienced men, the combat engineers like Jack and Roy who'd seen operational service, weren't as committed and just wanted to go back to their units. The older instructors also argued that the new breed didn't imbed themselves into the atmosphere of training. They saw it as a negative situation that many of the men, like Jack and Roy Tafe, chose to live in town with family, rather than intermingle with the troops in camp.

But Jack saw this entire gripe and bullshit differently. He appreciated how safe and meticulous he was. He knew all too well that Roy Tafe, Colin Kendall, and all the other campaign-experienced demolitions experts also thought like him. For them it was always safety—safety—safety. They enjoyed the challenge of instructing their profession to raw squads of men. Specifically for Jack, unsafe practices like crimping detonators with teeth or smoking in the vicinity of explosives, were never going to happen on his watch. Even though he strictly forbade it and mentioned the ill practice during his instruction, he knew it was occurring in other lessons. ' ... Sooner or later,' he thought, ' ... someone's going to blow his face off.

For the fastidious Jack, and many of the other men in G Coy, he took steps to ensure safety was paramount. It was a view amplified by the comments made by former trainee Des Surkitt. Des was a 21-year-old Sapper from Warrnambool in Victoria, when he participated in the same deadly period of instruction during his demolitions training in early 1945. Ironically, Des was raised about thirty kilometers away from Jack Pomeroy's hometown of Terang and, as fate would have it, his fellow Victorian country boy was his instructor that day. Described as a 'top bloke', according to Des, Jack was the quintessential professional soldier. And although he thought Jack appeared very intense, he recalls that he never let up about safety at any stage. For Des, ' ... Sergeant Pomeroy was nothing but safety meticulous during his lessons.'

> Speaking almost seventy years later, Des recalled that Jack made sure every man who entered his dugout was not carrying cigarettes, matches, cigarette lighters or anything else that would cause a spark. Assisted by his Corporal, he made every trainee empty his pockets of sparking material and ensured they were placed in a kerosene tin at the entrance of the dugout. Then, once inside the dugout, the Sergeant was painstakingly conscientious about keeping detonators, explosives and fuses separated by a significantly safe distance.

As expected, Ed Merry visited Jack on the demolitions range. He even went so far as to venture into the dugout to witness first-hand how the lessons were going. Just before he left the dugout in the early afternoon, Ed gave a nod of approval towards Jack and a wry smile of acknowledgement. On that particular Monday in early May of 1945, it was the kind of reinforcement

Jack needed. He wanted Ed to acknowledge that he was doing a good job and making a difference. As expected, the safety-conscious Sergeant completed another successful and safe Monday lesson. Dismissing his trainees that evening, Jack returned to the Headquarters before heading home. It was a bad decision. At the Headquarters, Ed Merry delivered Jack some distressing news from the fighting in New Guinea. Yet another of his mates from back home in Terang had been killed in action with the 27th Infantry Battalion. At only 26 years-of-age, Jack knew Tom McRae was far too young to die. Quickly approaching his 31st birthday, having survived New Guinea and the Middle East, Jack, for one, was thankful to be posted as an instructor at Kapooka as far away from the fighting as possible. Suddenly this angst between original and fresh instructors became trivial bullshit! It certainly meant nothing for Jack! Young men were dying in combat—regardless of 'who' taught them and 'how'. He just knew he had a job to do and Ed Merry's acknowledgement filled him with confidence. He knew that by training them in his style it would be enough to guarantee they could survive for as long as possible in combat.

Returning home after a tough day, both physically and now emotionally, Jack walked through the gate at the Hickson's home with mixed thoughts and emotions. Thankfully the sight of his darling daughter and growing sons allowed his thoughts to wander back to what was immediately important; family. As baby Maureen raced towards her daddy, although only just learning to walk and more than a little unsteady on her feet, she grabbed hold of Jack's huge Army trouser leg. Instantly Jack's mind was back where it belonged. Dressed in her best and most beautiful red dress, baby Maureen was quickly whisked up into the strong powerful arms of her larger than life dad. Staring into the excited eyes of his daughter, Jack wiped the last small residue of tear from his eye, a lasting memory of a good mate. Clumsily hugging her daddy's head in her small powerless arms, Jack found peace in baby Maureen's innocence. The dust of the demolitions range was replaced by the smell of his baby daughter's fresh hair and clothes. Holding her up high, ' ... Now', he says to his pride and joy as she gleefully dangles her feet in his face, ' ... What's mummy cooked for dinner my little princess?'

Part Two
Monday 21 May 1945: Tragedy

Chapter Four
Long Day Ahead

The camp was rarely ever quiet and peaceful. But on Monday the 21st of May 1945 there was greater reason for the camp to be in such high spirits. Just two weeks earlier, the German High Command had signed an unconditional surrender on all fronts. The war in Europe was finally over. At Kapooka, they echoed the same sense of accomplishment being displayed across the entire 2nd AIF. The welcome news put a skip in the step of every Commissioned Officer, Non Commissioned Officer, instructor and trainee alike, those still calling the camp home that day. And although Japan would take a further three months to also surrender its fruitless campaign in the Pacific, engineer training at Kapooka continued unabated with a sense amongst all of a job 'well done'.

Aside from the heightened spirits amongst the entire camp, due to the unconditional German surrender, the Headquarters staff and instructors at G Coy always looked forward to the start of a new training week. For staff like Ed Merry, Ed Dodds, Roy Tafe and Jack Pomeroy, it was the start of a new demolitions week. Fresh faces were always a welcome relief and a change of scenery. Although the staff only had a week to get to know the new trainees, a week was long enough to realise there were some real characters amongst every group. Continuously delivering a mix of young and old, inevitably there was always a larrikin, a leader, a couple of mates, some married men and some dads amongst the groups arriving at the Coy.

In contrast to Jack and Roy's Monday morning excitement witnessing yet another motley crew finally making their way to G Coy, stood the wide-eyed trainees. Coming from a mixture of GRs and ex-Cowra men they were a little more sombre and anxious at the start of a new week. Their nervous faces confirming their anxiety. For them they were heavily entrenched into daily routine and now entering the dangerous and much talked about demolitions week. For men like Jack, Roy, Bill and all the other experienced instructors, the new week always seemed to serve up a batch of nervous but excited demolition trainees. They all looked forward to working with all the trainees and getting to know them individually as well.

After day one of week one, all potential Sappers were welcomed into the camp shortly after stepping off troop trains down at the Kapooka Loop. Marching into the RAETC meant pseudo-membership of the Royal Australian Corps of Engineers. It was a proud moment for any trainee. Although they all knew they had some way to go to march-out of the camp fully qualified, for now simply being referred to as a Sapper was a noble, if not at times novel, change for the better. It was however a title they were prepared to shed more blood sweat and tears for. By demolitions week they were growing in confidence and wearing the title a little more comfortably.

For the trainees starting demolitions week on 21 May, they had only stepped off troop trains in late April and early May, and understandably, with less than four weeks training, they were extremely nervous as they entered the next training phase. Demolitions week was a week that would be one of the most exciting, dangerous and exhilarating of all their time at Kapooka. It was what they were hoping to do all along—blowing things up.

For 25-year old father of one Thomas 'Toddy' Woods from Brighton Le Sands in Sydney, entering demolitions week was going to be easy. He had already experienced first-hand just how hard it was going to be to complete engineer training. For him it really was a case of blood, sweat and tears. Only a couple of weeks earlier, Saturday May the 5th, two weeks into his training schedule with the high-spirited 1st Battalion, he somehow knocked himself out. Perhaps the result of the extreme realism used on the bayonet assault course and hopefully not in a fight over a 'Lady Blamey' in the camp's alcoholic entertainment zone? Either way it was an injury serious enough to have him admitted to the 54th Camp Hospital for treatment. Sustaining concussion and a lacerated face, he'd experienced first hand how real the training was. Perhaps, he thought, it was that one and only injury everyone told him he would get during his time at Kapooka. Thankfully his medical setback wasn't significant enough to remove him from essential training, unlike 18-year old Victorian Kevin Pierce's bout of tinea. Kevin was a skinny-framed former tractor driver turned-soldier who arrived at the camp on the 14th of April 1945 and was posted into the 1st Battalion. However, in less than five days, he found himself in the camp hospital suffering from a mild case of tinea. By early May he'd recovered and was discharged from hospital to rejoin his training squad who were now working towards the exciting demolitions week. Like Kevin, Toddy was admitted to the busy camp hospital; but the hardheaded 25-year-old was released on the same day and directed back to full training.

Week four for the selected squads in G Coy started out much like the previous three weeks of training. But this was no ordinary week. Only the night before, sitting amongst other nervous trainee Sappers listening to stories bantered around the tent lines about demolitions week casualties, the stories were getting bolder, darker, and scarier by the minute. A lost hand here, a disfigured face there, all of sudden it became difficult to get to sleep. But sleep wasn't difficult for those who'd grown up on farms and rural properties blowing stumps out of the ground with TNT, they all slept soundly. But for inexperienced 'city slickers' like 20-year-old Frank Platt, a clerk from Brisbane, or an Allan Flood type, an 18-year-old apprentice engineer brought up on Sydney's streets, their night was a little more restless. They perhaps woke a little pensive—certainly they were a little groggy when reveille was called.

As an experienced demolitions instructor, Jack could've easily approached and considered this week to be yet another 'routine' week at work. After all, he'd been instructing on the hazardous demolitions week ever since he arrived at Kapooka camp in February. Although a bout of tonsillitis temporarily slowed him down, he'd been performing the same instructional techniques for numerous weeks prior to this one. Needless to say he was confident and relaxed. But Jack's respect for all matter of objects that exploded meant that no week was ever 'routine', especially when dealing with explosives. It was even more so when instructing and teaching nervous novices the deadly art—his relaxation was often short-lived. Routine for Jack meant an unchanged approach. He was always thinking safety. He gave clear instruction and accountability for all the trainees in his care, he had respect for his art, and most of all he displayed a distinct lack of complacency.

But the 21st of May, this particular Monday, was no ordinary day—it was his birthday. He tried to get the day off, but Headquarters wanted their best on the job. Turning 31 and celebrating with his young family as a non-drinking man, he perhaps shared a coffee with Harry Hickson in the quiet of the early Wagga Wagga autumn morning. No doubt his regular early start to the working week meant that his birthday cake was shared the Sunday night before between the Pomeroy and Hicksons. Sitting on the verandah of the overcrowded dwelling, the two families laughed and told stories into the mid-evening before retiring for the night. It was a beautiful home-baked cake that didn't last long. Sneakily, Jack knew he'd be on the demolitions range the following day where food was scarce. So unbeknownst to Dorothy and the kids, the last piece of cake was wrapped in brown paper and placed in his pocket.

On this particular morning, it was all about getting up early, trying not to startle the crowded house and then getting to the camp on time. Finishing his tea and gaining some strength and sustenance from a couple of pieces of jam-covered toast and a quick cigarette; he bid Harry a fond farewell and a '...good day ahead'. Before leaving the aromatic kitchen of tea and toast, with a reminiscent smile he picked up his wrist watch from the table, glanced at his neatly-inscribed name on the rear before placing it on his wrist. The watch was a gift from his wife Dorothy, he wore it with pride, but it also had a practical use other than just a reminder of his bride and lovely family. Like every experienced soldier, time was of the essence. For Jack his timepiece was his most important piece of 'kit'. It was a piece of his uniform which was almost as important as the boots he was wearing. He and every other instructor and senior non-commissioned officer in the military couldn't live without a timepiece. The Army ran on timings. From permission to go to the latrine, to the more repetitious often annoying request of '... excuse me Sarge what's the time?' Or the even more popular and doubly annoying, '... Sarge, have I got enough time for another cigarette before the next lesson?' Timepieces were doubly important for the management and control of training programs requiring trainees to move around dangerous demolition areas and rifle ranges. In the end the Army decided to issue them to many officers, instructors and authorised personnel. For Jack and other SNCOs they wore their personal wristwatches with pride. Like a badge of honour, it was a reminder of the life they had outside the Army. For Jack, especially as a demolitions instructor, time was his best friend and also his worst enemy. When it came to explosives and detonation timings—his watch was his lifeline.

Living with the Hicksons remained cramped, but Jack was the quintessential family man. He preferred the noisy atmosphere found in the busy home, surrounded by his loving family and now his second family—the Hicksons. With the constant yelling, running and screaming of young children, he much preferred that type of raucous noise to the banter and tall stories of the men living a dry, dusty, boring life in camp. Described as a 'good living man' Jack no doubt kissed his wife Dorothy and baby daughter Maureen goodbye that morning. Stirring them from their sleep, they quickly woke to give him a special birthday hug. Reluctantly, Jack left the two most special ladies in his life to go back to sleep. It was starting to get a little late and he hadn't yet looked in on the boys.

Moving his bulky, yet sturdy powerful frame through the small dwelling, he looked in on his three sons. They were awake. The creaking of the floor unmistakably warned them that dad was up and getting ready for work. As he opened the door he noticed they were already sitting up. When the huge face of their father glanced around the corner, their drowsiness suddenly became wide-eyed and bushy-tailed. He leant down and gave them all a kiss goodbye. Like well-disciplined sons, they hugged their dad hard and strong. Then individually, they wished him, ' …Happy Birthday Dad', and watched through the bedroom window as their 'larger than life soldier' closed the gate in the front yard of the Hickson home and strode off to work.

With a special birthday dinner planned that night, the kids knew dad would be home early to celebrate his birthday. Importantly also, he'd also be there to help with their prayers before bedtime and later tuck them in. All they knew about his job was that he was in the Army and played with 'bombs'. For the Pomeroy family, although they were a little cramped living with the Hickson family, it was a bonus that dad came home every night. Unlike other kids' fathers, he wasn't a prisoner of the Japanese, or—even worse—killed in action. Instead he came every night to be just that, their dad.

Jack was more than just a loving husband and a doting father to his kids. He was a bloody good competent soldier. He became an inspirational role model for his young charges. As a veteran of the Middle East and New Guinea Campaigns, although he now had a young family to think about, and a baby daughter who melted his heart, he looked upon his regular supply of transient trainee Sappers with the same sense of fatherhood and protection. The majority of men he instructed were just young impressionable men, just like the men he knew his sons would one day become. They often sought comfort and security in Jack's presence and soothing tones. Not only that—those who relied on his professional approach to safely instruct them in such a volatile environment—they got confidence from him as well. What Jack also knew was that many of the trainees were family men as well. During conversations on the range Jack would discover that many had their own children at home who couldn't wait to see them again when all this war business was over. As far as Jack was concerned, he would always look after them all whilst they were in his supervision; especially the ones that needed to get home to be with *their* kids.

Although he cared for his allocated trainees and approached his instruction with implicit professionalism, a certain something was missing in his normal enthusiasm on this particular day. Perhaps with the death of his mate Tom McRae in New Guinea still fresh in his thoughts, and with many of his Field Coy engineer mates still serving in the dangerous theatres of Australia's backyard war, he did momentarily miss the excitement and camaraderie found in the combat units. Earlier in the month, victory had been declared in Europe, so Jack the soldier and theatre veteran was understandably in mixed spirits. He'd earlier received word from Ed Dodds that he might have an opportunity to rejoin his former comrades in June, but today was his birthday and he was looking forward to sharing with his kids when he got home that night.

Nonetheless, regardless of what, or even where, he was soldiering, Jack loved the Army. Brief episodes of boredom in instructing were surpassed by his almost parental obligation to the thousands of men entrusted to him by the engineer hierarchy to keep them safe. His known pride as a virtuous and trustworthy soldier, especially his immense sense of duty, combined to solidify his ongoing professionalism. He'd never let anyone down; let alone young men relying on him just as they would in combat. He knew deep down that he still had a job to do for the next batch of trainees that entered his dugout today. But he'd often go home a worried man. Indicative of his consummate professionalism, he'd discuss his concerns with Dorothy and regularly bend the ear of Harry about some part of training or his instruction that wasn't quite right. At home he'd often appear apprehensive, silently staring into the darkness, worried about the young men he had in his care; were they learning enough to save their lives if exposed to the horrors of battle; would the next squad be complacent and dangerous; what if, during his period of instruction, one of them makes a bad decision and kills them all? For Jack it was a constant source of worry. But it wasn't just him. Every instructor given the responsibility of training young rambunctious men often feared the worst—especially when it came to working with gelignite, monobel and gun-cotton.

When Monday morning arrived, the day he'd turned 31, Jack exited the front gate with an extra-large smile. He could see the busy showground across the road was still active with the regularity of military activity. He knew that one day very soon he'd be back in his unit and the adrenalin-pumping action of busy combat. He'd often ride his bike to work, not only to keep fit, but because he loved the freedom of the open road. It allowed him time to think and ponder. Today he decided to catch the bus out to the camp; probably because he wanted to get home early. After all he had a birthday to celebrate with Dorothy and the kids.

Making his way to the bus stop it wasn't long before the military transport picked him up. Waiting for him on the bus that morning was Bill Cousins. The cheeky yet lovable rogue had just left the Titus' residence after his regular Sunday night roast, bed and home-cooked breakfast, courtesy of Alice. As a picture of professionalism, his boots were polished, uniform starched and his slouch hat adorned his freshly washed hair. But on this occasion Bill left his identity disc behind. They weren't hanging around his neck; they were by the bedside table at the Titus' house. Realising his mistake at being incorrectly attired on the bus next to his superior officer, he couldn't go back to get them. Instead, he thought with ulterior motives in his head, ' ... I'll go back and get them tonight.' With that thought tucked away, his smile, which reached from chinstrap to chinstrap, exuded enthusiasm; he always had the right attitude, albeit this morning he was minus his important identity discs. For Jack sitting alongside the exuberant and confident young Bill, who appeared more than ready to face the day head-on, it was a reassuring bus ride that morning. Jack and Bill would be paired together that morning as instructor and assistant instructor to deliver yet another week of demolitions training to the inexperienced men of G Coy.

The bus delivered the two men at the regular stop outside the RAE Headquarters building. Dodging the flurry of soldiers already engaged in the early morning tempo of the large camp organisation, the two men moved off—Jack towards the 1st Battalion Headquarters, and Bill towards the 2nd Battalion. At the time he was a member of the 2nd Battalion and needed to gather some equipment from his accommodation hut—also to stir up his colleagues about his night at the Titus'. Jack proceeded to 1st Battalion Headquarters to catch up with the other senior non-commissioned officers and report to Captain Merry. As both men departed in their separate directions it was a case of Jack reminding Bill of where he needed to be ' ... see you at headquarters in ten minutes Bill—and by the way, where's your ID discs?', knowing full well he'd left them behind. Both men departed with a wry smile. They knew it was going to be a great day.

Chapter Five
Reveille

0600—0730hours

It was barely daylight at the camp when the Tannoy speakers crackled to life and a bugle call awakened the camp at 0600 hours—'ahh reveille'. For many, it was the best time of the day. Time to enjoy a little peace and quiet; but alas that was always short-lived. Even though many of the speakers dangled precariously from gum trees scattered throughout the camp, they were frighteningly effective. Clear skies on this particular Monday morning bugled a crisp and chilling early morning welcome for the slowly stirring camp. The echoing volume of reveille awakened every creature that might've been sleeping on the Pomingalarna Range. Kapooka, the Army's very own 'creature', had awakened. Ever so slowly its inhabitants extracted themselves from overcrowded and cold tents to greet yet another busy training week. Whilst many profess the 'early morn' is the best time of the day, for the slightly-chilled occupants of the busy camp it became an anxious wait. They sluggishly went about their routine, waiting for the sun to rise and penetrate the landscape through the gum trees bringing much-needed warmth, comfort and motivation.

The Reveille rousing the camp was extremely loud. It needed to be. Not only were tents containing 8000 or so bleary-eyed trainees scattered over the vast camp, the day's activities relied heavily on timings. The bugle of reveille meant the day's schedule had begun. Not only that, to enforce discipline and inculcate a sense of urgency, trainees were required to respond instinctively to the wake-up call and get to the parade ground as quickly as possible. If that didn't occur they'd soon face the fury of an angry Sergeant who'd gotten out of bed early on a freezing morning to make sure no one had gone Ack Willy (Absent Without Leave).

Although his appearance may have resembled a schoolteacher, Roy Tafe was a far cry from a welcoming educator; he was one those angry Sergeants capable of instilling fear with a sideways glace. He was uncompromising, but very fair. Trainees that pulled their weight survived, and for Paddy Cranswick, he stayed on the good side of Roy during his time at G Coy. Once awakened, the men would pick themselves up off the cold floor, dressed in whatever they went to bed in—generally piles of clothes and a greatcoat—throw on a pair of boots or shoes and then make their way to the parade ground for roll call. At the parade ground, some quick impromptu physical training; squats and star jumps got the heart started. Roll call followed and then it was a hasty retreat back to the tents to be prepared for whatever was on the schedule. Preparations included a lukewarm shower and shave in the nearby ablutions block where chilled bones were warmed and then chilled again after stepping from the shower. Men hastily got dressed in the uniform of the day, cleaned boots, arranged their bedspace in a tidy square, and all done before the next parade at 0730 hours—a hearty breakfast.

Reveille not only awakened bleary-eyed and groggy trainees, it prodded cooks in the kitchen to pick up the pace. Their busiest time of the day was fast approaching. Serving rehydrated eggs, looking more like runny custard, was too much for many of the men who were used to eating fresh eggs, which they gathered from the chook pen back home. The sight of a runny yellow substance, something the Army undoubtedly purchased in bulk and called eggs, forced many to steer well clear of the unknown breakfast on offer.

A hearty soldier's breakfast soon became a bit of jam or marmalade on toast, coffee, and to top it all off; a lung-breaking early morning smoke! However, today was a special day in the canteen. The rare sight of fruit could be seen sitting atop a table at the entrance. Accompanied by a small sign detailing the rationing, each man was entitled to one apple and one orange apiece. It was especially important for the men going out at the demolitions area who would be getting their hot lunch delivered. At least fruit could be stashed in their uniform and carried to the range to avoid hunger during the morning sessions.

With time being at the very core of existence for an Army recruit, morning routine needed to run like clockwork. For those who had experienced the distressing recruit regime at Cowra, Kapooka, with a little bit more time to get things done, now seemed a walk in the park. As they woke to greet the bitterly crisp autumn morning there was still little time to dawdle. Breakfast was rushed, tent routine was rushed, and dare I say it, rushing meant forgetting and often men went off to training minus a few important things now and then—like identity discs and rifles!

Life as a trainee engineer, and similarly for many other 2nd AIF recruits, was challenging. For some it was downright uncomfortable, unforgiving and for many inexperienced trainees—personally confronting. Supposedly living in tents suitable for six men, the overcrowded and busy nature of Kapooka camp often meant that up to ten men were living side by side in cramped unhygienic conditions. Space was a luxury. With a small amount of personal area each, there was nowhere for comforting personal possessions. To keep things in order, discipline was handed out to those who'd decided not to fold up their sleeping area correctly, stored their kit neatly, or left dirty laundry hanging in the tent. Kitchen duties weren't necessarily an attractive way to spend their spare time for accommodation rule lapses. It was paramount to keep the tent clean. That was of course only until after inspection. Soldiers quickly learnt that hours spent on immaculately cleaning boots were counterproductive. They knew the first steps they took outside the tent into the dirty dusty surrounds ruined hours of labour. They realised that just doing enough to pass inspection was the requirement and 'brownie points' for clean boots meant for nothing in the long run.

In the G Coy GR tents, after arriving back from a hearty breakfast, ex-Cowra men were busy cleaning and tidying for inspection. In the flurry of the morning, the trainees in one particular tent included Frank Platt an 18-year-old fellow from Queensland and former labourer Stan Morphy, from Charters Towers. Alongside Frank and Stan were many other fresh-faced men who were preparing for demolitions training. Men like 20-year-old milk factory worker Alf Witt from Pinjarra in Western Australia, 18-year-old Denby Grasby, the son of an ANZAC from the Great War, Allan Bartlett who came from a large family whose brother was already serving and a mature—looking man by the name of Ivan Merritt all from South Australia. 26-year-old Ivan was a former tractor driver who was married and expecting his first child. Waking up amongst Denby, Allan, Stan, and Frank, who all looked more like boys than men, Ivan's thoughts often turned to his wife and the life he had left behind. Meanwhile, Kevin Pierce, who'd only recently joined this tent and squad, was a little more active than his colleagues in the mornings. The young Victorian had recovered from a case of tinea, which saw him removed from his initial squad for a fear of spreading the affliction. Passing fit, he was reallocated. Although he'd found himself in a new tent and their specific early morning tent routine and hijinks, Kevin quickly found the rhythm. He was familiar with discipline, cleaning and organizing by virtue of a childhood spent in a Victorian Children's Home as a ward of the state after his mother died when he was young. Now, ably assisted by his new tent mate, 18-year-old Edward 'Teddy' Robson, a dairy hand from Liverpool in New South Wales, together the pair coordinated the task of organizing the tent where the pair took charge during morning routine.

In neighboring overcrowded tents, other G Coy men could hear the after—breakfast hijinks and loud raucous goings-on of their fellow squad members in the adjacent tent through the thin canvas walls. A street-wise Sydney boy, 18-year-old Allan Flood, joined Paddy Cranswick, 18-year-old fruit shop assistant and son of a great War veteran, Geoff Partridge, 20-year-old Les Mather from Nyngan in New South Wales and good mate 18-year-old Joseph Collins, a timber cutter from Nowra, in what was fast becoming a good morning routine system in their tent. Collocated with B Coy men like Allan McPaul, G Coy members also included 18-year-old Kevin 'Brickie' Hurst, a shop assistant from Narrandera in New South Wales. With hair as red as a house brick (hence his nickname) Kev and a couple of other G Coy men such as 20-year-old butter factory worker Colin Hurley from Herons Creek and Terence Moore, a 19-year-old street-wise lad from inner Sydney, the New South Welshmen all contributed to what was normal morning routine for the thousands of diverse occupants in tents littering the Pomingalarna Range.

Although Brickie Hurst was no different to any of the wide-eyed recruits in the tent that morning, one thing set him apart from most—labeling equipment. When it came to labeling and marking his issued equipment; the results were everywhere. In fact, Kev was much fussier than the average soldier. Perhaps it was the sound advice from his father and uncle, both Great War veterans, who reiterated to him that often things 'went missing' at recruit training. Following his father's advice, Kev used a marker to scribe his number and name 'NX205951 HURST' on everything; including a detailed effort on such minor things such as the leather straps of his issued trouser braces. His tent mates couldn't believe it. Naming his equipment appeared to Kevin to be an innocuous act. But little did Kevin know just how important it would be.

In the madness of morning routine, floors needed sweeping—often difficult with a constant stream of dirt and dust entering the tent, lockers needed packing away, paillasses needed re-packing with fresh straw and men needed to use the nearby ablutions for brushing teeth and generally freshening up. In the hundreds of tents sprinkling the landscape, where the majority were fastened to the inclined and unforgiving landscape by a stack of house bricks, similar morning routine scenes were playing out. But when you get a tent full of young men on one of the biggest adventures of their lives, it was always going to promote mischief—especially when there was an older ringleader or a practical joker. In a nearby tent, Colin Boyd, a 21-year-old Cowra-born local was silently trying to slip into his unflattering and unfashionable military issue underpants. Due to their uncomfortable nature, many men refused to wear them. In addition to their prickly and unappealing nature, many men of the time never wore underpants prior to joining the Army; so receiving three pairs of strange-looking garments was somewhat of a novelty. Not for Colin. He'd taken the trouble to label them with his personal details and he thought, '… why not just wear em?'

Of the men who actually knew what underpants were, they chose instead to wear their own. Rather than spend the day in uncomfortable and unfashionable issued undergarments, many men chose a pair they'd not only be more comfortable in, being very well aware that wearing your own clean undies bred an air of confidence. But military underwear was a unique piece of equipment. As an issued piece of useless clothing, the men wouldn't ordinarily wear them. However, when your own fresh supply ran out, as a safety net, you could always rely on the military issue. Affectionately known as 'Romance Busters', they were long, very long in fact, with a drawstring around the waist. If men misjudged their underpants rationing against the training schedule, they could potentially be caught doing rifle and marching drill with these unimaginable underpants. Non-elasticised waist spelt embarrassing danger. If the drawstring decided to loosen and undo during say a drill lesson, it became one hell of a situation.

Owing to the fact that not too many soldiers in the 1940s wore underpants, it was a unique sight for many of the men to actually see someone wearing them. In this particular tent, Colin pulled his out with the intention of wearing them. Nevertheless, living in such close quarters, his somewhat disguised choreographed underwear routine was quickly noticed by his younger and more impressionable audience. Amused at the sight of him trying to find out the front from the back, jibes from his tent mates over his fashion statement came thick and fast. Like a well-fuelled bushfire, one jibe led to another and before too much longer Colin was set upon by the entire cast of younger tent mates. The ringleader in this tent was a young man by the name of Ernie Poschalk. Ernie was eighteen and a former motor mechanic from Townsville. His quick wit and fabulous sense of humour quickly gained him a reputation as the tent comedian. Defiantly deflecting the raucous and smart-arsed comments coming from Ernie, Colin cleverly turned it around on his cohorts. With the theatrics of a well-drilled actor, to the uncontrollable laughter of his new audience, he started parading around the tent in his 'saggy' attire.

Significantly on this particular morning, Colin chose to wear his issued romance busters rather than his own personal supply of underpants. In Colin's case, to the chagrin and ridicule of his tent mates, proudly, with a hint of theatrical overacting, he fastened the drawstring without any hint of embarrassment. His wry smile directed towards his tent mates acknowledged that he accepted their banter in the spirit of fun and jest. Considering his mum was a dressmaker, Colin thought to himself, ' ... geez it'd be nice to get some elastic into this bloody waistline.' Much to the amusement of his earlier hecklers, the arrival of the Sergeant Major, unbeknown to Colin, signaled the show was over even quicker than it had started. Colin eventually got dressed under the watchful, yet moderately amused eyes of the Sergeant Major. Over in the corner of the tent, the arrival of the disciplinarian quickly wiped the smile of Ernie Poschalk's face.

In one of the more 'mature' tents, Stan Ross was having a little difficulty with a troublesome tooth during his morning ablutions routine. Stan was a 28-year-old conscript who wasn't real keen on being in the Army. The former motor mechanic from Mungindi in New South Wales was eyeing the troubled area of his mouth in the mirror. He'd been to the dentist at the Camp's 59 Australian Dental Unit the previous Wednesday where Dental Officer, Captain Brodie, did some work. For some reason it was still a little tender and it was still irritating him. Not enough though to make him miss the most crucial week of his training thus far. But as Stan finished looking at his troubled tooth, he noticed that he wasn't wearing his identity discs. He generally took them off and hung them up whilst he slept because they often got wrapped around his neck. Normally they were the first thing he put on in the morning. But on this morning he woke with his tooth still troubling him. Distracted from his normal morning routine, he strode out to the Coy first parade leaving his important identity discs hanging by his bedspace.

Ironically, over in Geoff Partridge's tent, he also awoke with a toothache that needed some professional help. Overnight a sensitive tooth had become worse for the young man and any remedial suggestions made by his enthusiastic tent mates didn't help. On morning parade, he'd have to ask Sergeant Tafe if he could miss training and go to the dentist.

Sleeping in the 'cot' next to Jack Nixon in another of the tents, coincidentally, was Stan Emery, the country boy Jack had earlier met at the recruiting office in Paddington. They both knew that they'd end up at Kapooka together, but were dumbfounded when they discovered they were not only in the same tent, but sleeping in the bedspace beside each other. Due to his slightly overweight frame, Jack, the doting new dad, became endearingly known in the tent as 'Porkie'. Carrying a photo of his baby girl, Neryl, in his pocket and talking relentlessly about his angelic daughter, Jack received his novel and initially unwanted nickname from his tent mates—no doubt, which also included new mate Stan. However, nicknames amongst mates became dangerous fodder for others. If expressed outside the confines of the tent nicknames were just the ammunition

Sergeant instructors needed to motivate. Thanks to Stan and the rest of the tent, Porkie's nickname became an unwanted burden. For Jack it mattered little. For him all that mattered was finishing the course and getting back to his young family. Although thirteen years separated the two men, Jack the elder statesman and Stan the larrikin became good mates. I suppose when you sleep next to someone it's inevitable, but for Stan and Jack whether it was sharing a 'Lady Blamey' beer at the canteen or even sneaking off into the authorized 'gambling zone' at the back of the camp, the men forged a solid friendship. Jack thought the world of his wife and daughter and he had no hesitation in telling everyone on a daily basis—especially Stan.

On this beautiful brisk May morning, it was certainly no different. As he prepared to walk out the front of the tent, Jack turned towards Stan, looked at him and with a proud glint in his eye—he barely had time to open his mouth when a slightly agitated Stan piped in ... ' ... I know mate, you miss your wife and daughter and you can't wait to see them again soon.' ' ... How did you know I was about to say that?' came Jack's witty response. ' ... Every bloody day Porkie it's the same.' With that last remark the men acknowledged each other's playful banter with a wry smile. Jack flipped open the entrance of the tent, placed the photo of Neryl in his back pocket, and went off to the parade ground leaving Stan to hurry up for his own first parade. '... See you tonight mate' he yelled as he busied himself in the direction of the parade ground.

Over in the tent where 'Brikie' Hurst and Allan McPaul bunked down, they didn't share the same relationship as Jack and Stan. On this particular morning Allan was off to do underground mining and tunneling lessons with B Coy, whilst Kevin and the other G Coy men were doing demolitions. They did however exchange pleasantries as they departed the tent. Wishing each other a good day, they knew they'd meet up again at the end of the day.

In the 2nd Battalion Junior NCO hut, the gregarious and always impressionable Bill Cousins had just arrived back from his Sunday night ritual with the Titus family. It was also the morning he caught the bus into camp alongside Jack and not surprisingly, Bill was in high spirits. Dropping off his overnight bag and tidying up his bed space for inspection, he decided to talk up a bellyful of stories about himself and his courting of one of the Titus girls. Having the charismatic, smooth-talking Corporal in their hut, it was always a place of raucous and loud behavior. Still, his mates knew better than that. As junior NCOs, their accommodation huts were always full of egos, attitudes, fun times, and most of all tall stories. The most humorous involved descriptions of the conquests of women in town, or even one of the 400-strong AWAS residents at the camp. But they could see right through Bill's attempted ego trip, ' ... the only legs you got to see last night Bill were covered in gravy and tasted delicious!' came a colourful retort from the other side of the hut. Erupting in laughter the NCOs couldn't contain their amusement at Bill's expense. They knew Jack and Alice Titus wouldn't stand for the type of rubbish Bill was sprouting. But every week he went off to the Titus family for dinner, and every week his story got more and more embellished. It was a character trait that defined the tall larrikin. He was, after all, the life of the party. Perhaps jealous of his relationship with the Titus family, and the fact that he was in a position to get away from the camp, only Bill knew the truth about his Sunday night trysts. Being a tall good-looking man, perhaps his stories did have an element of truth. He often left his mates wondering, ' ... was he actually courting one of the Titus girls?'

Further around the range, in more luxurious surrounds than the Corporals' hut, Ron Linthorne greeted the day in the Senior NCO huts with a mix of excitement and nervousness. Ron, a 25-year-old former carpenter, turned soldier in 1942 as a member of the artillery. Like Jack's reliance on his wristwatch, Ron as a Sergeant was no different. He constantly wore his watch. It was the same one presented to him by his new bride, Pauline, prior to undertaking the harrowing five-day journey to Kapooka from Western Australia. Young soldiers turned to Sergeants like Ron for everything, and now he was prepared thanks largely to his wife. Demolitions week was something he was looking forward to, be that though, he had also heard the stories of what 'could' happen if it all went wrong. He'd been in camp now since mid-April when he and his mate from the West, Alf Woods, arrived together. Alf was a little older than Ron. At age 32 Alf was a former hairdresser who was also a married man like Ron. In 1941 Alf and his wife Jean lost their only son and Alf turned to the Army for comfort. It was where he met Ron and the two men became inseparable. The pair managed to join the 1st Battalion's G Coy in early May and together they were going to experience everything that was Kapooka. It's not clearly known whether the inseparable pair bunked down together in the same hut, after all Alf was still only a Corporal, but suspicion suggested that at 32, as the elder statesmen, it wouldn't have stopped Alf enjoying the spoils as a Senior NCO. Tucked away in the corner of the hut, the slightly rebellious yet totally charismatic hairdresser probably offered a free snip to keep the unwary Sergeants on his side and a bed alongside his mate. As experienced soldiers, after two weeks of doing nothing at the camp they were pretty keen to start demolitions week, albeit with many of the inexperienced GR guys who were still all busy over in the Sappers tents getting ready for parade.

G Coy was required at the Company parade ground by 0825 hours for squad allocation. Complete with haversack containing Dixies, with water bottle and bayonet fixed to their belt, their rifle, fanatically cleaned, got a quick dust down, before the complete trainee package exited the tent bound for the parade ground. No doubt tucked away in their trouser pockets and haversacks were the two juicy pieces of fruit that would see them through until lunchtime whilst they were out at the demolitions area.

Chapter Six
Squads

0825—0840 hours

On the edge of the G Company parade ground, Warrant Officer Ed Dodds, the experienced 30-year-old engineer from Sydney, surveyed the next unruly lot undertaking demolitions training with his instructors. 'Milling around'—military jargon for standing around smoking, jabbering over nothing and basically hurrying up to wait a little longer—they were standing on the edge of the parade ground waiting to be called on parade. Normally Ed Dodds was the assistant Key Instructor to Warrant Officer Doug McFarlane who ran the morning parade, but on this particular morning Doug was absent. Having some much deserved time off, he went on leave far away from the daily routine of the camp. Although it was demolitions week for a fresh squad of trainees requiring specific attention, he had no hesitation in handing over his duties and responsibilities to Ed Dodds.

Also taking some time off on this particular Monday morning was the Kapooka Camp Commandant, Brigadier McDonald. Normally arriving unannounced, the Brigadier would visit the demolitions area during the day to 'checkup' on the training and to make sure safety and procedures were being followed. By arriving unannounced, it kept the men, especially the instructors, on their toes. He'd made numerous surprise visits before and not once did he have the need to reprimand any of the instructors for carelessness in handling explosives or disregard for safety—but today he'd be nowhere to be seen. Although today the Brigadier was absent, much to the relief of all, the men would still treat safety and careful handling of explosives as if the 44-year-old Tasmanian Commandant was going to turn up at any minute.

Prior to moving onto the parade ground Ed Dodds retrieved the squad roll books, which had been delivered to Company HQ from Battalion HQ. The Acting Sergeant Major for the day quickly checked the roll book and counted the names of 24 GRs and eleven men, who'd all recently marched in from the Recruit Training Battalion at Cowra. He should've had about forty or so men 'milling' around on the edge of his parade ground waiting for him to step up and start the parade. Calculating in his head a total of 35 men, he allocated them against three of the battalion's seven experienced instructors. He decided that Jack, the bespectacled Roy Tafe, James Conwell, who celebrated his 22nd birthday the Friday before, and their respective assistant instructors, would be tutoring that day. Despite the fact that it had previously been discussed and arranged, Ed Dodds authorised and notarised the instructor appointments in writing in the squad roll books against their potential squads. For James Conwell, he was hoping for a more successful demolitions week than the one he'd experienced earlier in the year. A minor accident with explosives had him a little over-cautious.

Married in September of 1944 aged 21, the former bricklayer's apprentice was used to sustaining minor injuries, especially nicks, and grazes on his hardened hands. But in the January just past, he was a member of the 4th RAE Trg Bn instructing on the demolition of mines when a small explosion occurred. On that occasion, the explosion lacerated a finger on his right hand. Although not permanently incapacitating him, James maintained his cautious and respectful approach to demolitions training. There was no Court of Inquiry called to investigate how the explosion occurred. Meanwhile, unexpected detonations causing minor and more serious injuries were a regular occurrence during demolitions week.

Appreciating his 'troops to tasks' requirements, the temporary Sergeant Major decided that twelve of the GRs would be allocated to Roy Tafe for the day's training. James Conwell would also get twelve of the GRs and Jack would get a manageable eleven ex-Cowra men. Like Jack, Roy Tafe was also happy to be instructing. His wife Mary was an AWAS working at the nearby communications centre and the couple was also living in town. He was relishing the break from active service duties, but, aged 30, he surveyed the next batch of trainees who looked like they could also be his sons. Reassuringly, he spied some older-looking men like portly Jack Nixon, Stan Ross and the charismatic Alf Woods, standing amongst the boys. He also noticed father of two, 34 year-old Norm Dilley, an ammunition factory worker from Rutherford in New South Wales and the oldest man amongst the squad, 36-year-old fettler and married man, William 'Bill' Reid, a conscripted soldier from South Australia also waiting by the parade ground. The older men looked out of place amongst the younger brigade. For Norm, Alf Woods and Jack Nixon, they wanted to be there, but for their more mature counterparts Stan and Bill it was a little different—they'd been conscripted. Now, somewhat against their will, they were entering the most dangerous week of training surrounded by overzealous volunteers who were certainly more energetic than they were.

Meanwhile, assisting Roy that morning was the very competent Colin Kendall. At 31 years of age, Colin first served in the Militia before he joined the 2nd AIF in 1941. He was recently married and as a native New South Welshman, he felt right at home amongst the trainees at Kapooka. He knew the job well and was more than happy as a Lance Sergeant ordering and supervising trainee Sappers in the art of demolitions. Both Roy and Colin respected each other professionally. Roy knew and was confident that Colin could easily take over his lessons as the primary instructor if he was summonsed away for whatever reason.

The Company was allocated the demolitions training area for the week, meaning that every man on the parade ground would need to know the standing orders for the area. Confidently striding onto the parade ground with a copy of the *Standing Orders For the Demolitions Area* and the pre-marked squad roll book allocating troops to instructors in his hands, Ed Dodds fell the squads in. Using the roll book containing his list of 35 names he called the roll into corresponding squads. Amongst the squad of 35 were men were the young and inexperienced Sappers Frank Platt, Kevins Pierce and Hurst, Ernie Poschalk, Stan Morphy, Alf Witt, Denby Grasby, two Colins, Boyd and Hurley, and Joseph Collins who went by the name 'Joe'. Another name called out was Joseph Faull. Joseph was an 18-year-old labourer from Parkes in New South Wales. With no real life experiences, Joseph possessed a few practical skills like being able to drive a car, ride a motorbike, write shorthand and use a typewriter, but the Army, well that was something else. He was the quintessential 'wet behind the ears' trainee. At least that's what he stated on his enlistment paperwork. But he wasn't alone. His new mate, Terence Moore, had also joined up with little experience. 19-year-old Terence was a street-wise city boy who befriended Joseph along the journey of their lives where the young men swore a boyhood 'oath' to look out for each other, no matter what. So, after going through Cowra together, both men stood side-by-side on the edge of the parade ground, waiting for their names to be called out by the Sergeant Major. They were hoping to be in the same squad for the day's training.

For Joseph and Terence, investing in the same dream and vision, of service and sacrifice, was all rolled up into a boyhood adventure. But there was something special about the relationship between two mates wanting to be soldiers and serving side-by-side. In the past, moving stories describing the highs and lows of a war-torn friendships provided fodder for Hollywood scripts; in Australia bush balladeers would write poems and songs about, such but for Joseph and Terence it was just about good mates on the adventure of their lives. Together they were getting valuable life experiences witnessing the highs and lows of Kapooka.

Thankfully, Terence's name was called out that morning alongside Ted Robson, Allan Flood, Les Mather, Geoff Partridge, Paddy Cranswick, Terence and Allan Bartlett. But such was the nature of this roll call much older men like Ivan Merritt, Stan Ross, Norm Dilley, Jack, Porkie' Nixon and Toddy Woods were also listening intently for their named to be called. When their names were read out they were allocated into training squads with their much younger, and energetic, colleagues. For Bill Reid, at age 36 and Norm Dilley 34, the young 18 and 19-year-old members of their squads could pass as their sons.

Suddenly men complete with rifle and equipment were crisscrossing the parade ground to fall into ranks beside their allocated instructor. In less than two minutes, three separate squads were assembled. But confusion reigned supreme. Jack noticed seven men aimlessly drifting around his newly-formed squad. Something had gone wrong with the roll call. The men looked worried and confused; Jack knew their names hadn't been read out because they didn't have a sense of urgency to form up with his squad. Bill Cousins advised the seven men uselessly looking at each other in utter confusion to join in his squad and told them that he'd sort it out when the Sergeant Major had finished with Standing Orders. Unaware of the confusion, perhaps basking in the satisfaction of his allocation, the Acting Sergeant Major, in his clearest and most authoritative voice began:

> Listen in men; these are the *Standing Orders for the Demolitions Area*.' The men responded by ceasing the idle chatter amongst them and paid attention to the normally approachable Warrant Officer who'd somehow morphed into a demon of his former self. Emphasising his position that day as the Sergeant Major, Ed Dodds was trying to exert his authority by using a louder, angrier voice. Bellowing over the top of the busy racket of early morning in camp he began:
>
> *Number one*; personnel in training will at all times carry their rifle to and from the training area.'

Number two; trainees will march to attention to and from the training area. Your rifles are to be carried at the slope or the trail.'

Number three; all instructors will make personnel under their command familiar with all safety precautions prior to the commencement of training, and to see that all trainees adhere to these instructions.'

Jack and Roy looked at each and smiled. In the past safety precautions weren't strictly applied and weren't explicitly read to trainees. On this occasion, the two experienced men smiled in acknowledgement that it was negligent not to do so; especially considering Jack had the squad of men fresh from recruit training. According to him it was not only Joseph Faull who was still a little 'wet behind the ears'; Jack considered them all as being similar in experience to Joseph. Every step of the day needed to be spelt out just as if they were *all* mischievous schoolboys.

Ed Dodds went on:

Number four; dress!' It is every instructor's duty to report to the K/I all personnel who are incorrectly dressed.' (He paused slightly and looked towards the gathered group.) That also applies to any man who is unshaven or thought they could get away with it.'

For half of the men assembled that day, whiskers were barely visible, let alone required to be shaved-off.

Five; instructors must have and exercise complete control of trainees.'

This one caused Jack and Roy a passing wry smile towards each other. They'd heard the Standing Orders read out almost a hundred times but still some trainees didn't listen.

Six; absolute silence on night exercise must be enforced.'

Seven; this is for all you blokes with Chinese bladders ... trainees are only allowed minimum amount of time to visit latrine.'

Number eight; instructors were responsible that all explosives, detonators, fuses etc. were returned on completion of the day's training.'

Nine; when charges have been prepared for night exercises and during the absence of the squad from the training area, the instructor will detail two piquet's from his squad to guard all explosives and stores.'

Last one—ten; the key instructor is to be notified of any breaches of discipline.'

Concluding his formal duties Ed Dodds looked up from his scripted piece of paper and enquired to his audience. 'Are there any questions men?' As no one responded, he assumed the Standing Orders were understood. But as he looked up and surveyed the three squads before him he noticed that Sergeant Pomeroy's squad, which was supposed to be the smallest—was in fact the largest. Counting heads, he noticed a further seven men standing alongside Jack and Bill Cousins. Somehow this squad had a total of eighteen men. With twelve men in Roy Tafe and James Conwell squads, G Company had 42 men this Monday morning. It was a lot more than the 35 names read out by Ed Dodds.

Confused at either his poor mathematical ability, or incorrect details in the roll book, Ed was convinced his schooling hadn't been a waste of time. There had to be a rational explanation. Inspecting the books for clarification, the astute Warrant Officer observed that two pages had been pinned together. He quickly removed the pin and found the seven additional Sapper names that were now staring blankly and confusingly at him from Jack's squad. Satisfied that Jack and Bill Cousins could handle a squad of 18, the acting Key Instructor adjusted the roll to reflect the seven men. Finally he was content. Squad allocation for the day's training on the demolition range was complete. He didn't know it at the time, but the roll book confusion would have serious implications by the end of the day.

Ed's next duty was to announce any messages from Battalion Headquarters before the men marched off to the demolitions area. For Roy Tafe there was some respite in the posts that morning. Nine of his eleven men had been assigned to guard duty and extra duties along with two from James Conwell's squad. One of the fortunate ones to be assigned elsewhere that morning was Paddy Cranswick. On the previous Friday, the 18th of May, Paddy received a letter informing him that his elder sister was ill and was on the verge of passing away. He'd immediately requested to go back to Adelaide to see her. But as she wasn't his official next of kin the Adjutant of G Coy, Captain Archie Smith, declined his request. Paddy was heartbroken and the sympathetic Adjutant could clearly see Paddy's distress at his rejection. He didn't know that coming from such a large family Paddy's sister essentially raised him like a mum. He was extremely close to her, but unempathetic military bureaucracy was in his way. Anyhow, the Adjutant told Paddy that he was to tell Sergeant Tafe that he didn't have to do the training on the Monday. Instead, he advised Paddy that he could be the Company runner at HQ whilst they sorted out the problem of getting him home.

Back on the parade ground Roy Tafe couldn't believe his good fortune. By lunchtime he'd only have three men to worry about, Colin Boyd, Colin Hurley and Stan Ross. His nine men, who included Paddy Cranswick doing runner duties, along with Corporal Conwell's two men, were all scheduled for release from training at 1100 hours. Finally, after all announcements and administration as a collective Company of 42 trainee Sappers and five instructors, the men formed up beside the parade ground ready to march down to the demolitions area. Bill Cousins would be Jack's assistant instructor for the day, Lance Sergeant Colin Kendall was assisting Roy Tafe, and Corporal James Conwell would be working on his own. Bringing the men to attention they were about to set off towards the demolitions area. Roy Tafe took charge. 'Golf Company,' he roared, 'weapons at the slope, QUIIIIIICK MARCH!'

Located north of the headquarters buildings, the demolitions area was sited in one of former landowner Jack Roach's grazing paddocks. Apart from an access road that ran through the area, which would later become San Isadore Road, the area was a barren place. It was an oppressive hotspot in summer, a wind tunnel in the autumn and almost unbearable in the winter chill. Due to the high volume of engineer trainees utilising the camp participating in innumerable demolition type lessons and the absurd weather conditions, the authorities decided to construct a series of four underground dugouts in the designated area during the camp's development. With no real standard of design, the purpose of the underground areas was to provide a workable solution to a number of problems the instructors and staff were experiencing as a result of the high volume of lessons and trainees making their way through weeks four and five.

Whilst their primary purpose was to save time on the high volume of squads safely rotating in and out of the demolition area, the overall safety of trainees was a bonus. More squads could remain in the area within these dugouts without disturbing other squads undergoing training above ground. While careful coordination of the squads ensured no squads were above ground when explosives were being detonated, squads housed underground could continue training and lectures with little disruption due to noise or flying debris. In addition to the strategic necessity, the Wagga Wagga region could be bitingly cold and oppressively hot and dry, therefore, when inclement or hot weather set in, trainees could retreat to the dugout without any disturbance to training schedules.

The dugouts themselves were roughly-built which, according to one trainee ' … looked like they just dug a hole, laid some poles across the top, put some mesh down and then rendered the roof with a concrete before filling in the roof about seven foot on top.' Possibly constructed by Field Company engineers during the early days of the camp's construction, they were always temporary and not designed to be permanent structures or fixtures. Then again, the whole camp was never going to be a permanent fixture. But in any event, some care and attention went into the construction of the dugouts. No one was exactly sure how long the war would go on or even if Kapooka would survive post-war. Overall they were constructed structurally sound and proved very effective over their couple of years of use. They were sturdy, suitable and spacious enough for the training. Their only downfall perhaps lay in the fact that there was only one entrance, which was only wide enough for one man at a time to enter and leave. If there was an emergency, such as a fire or unexpected detonation, getting men out quickly would pose a serious problem and men could be killed.

Each underground dugout was dug to a depth of about 2.1 metres. Pitched at the centre point, they were tall enough for an average height man to comfortably walk around inside. A frame of roughly prepared local timber supported the walls, which measured 6.6 metres long and 5.8 metres wide. No doubt the field engineers sourced some of the rough bush timber from surrounding trees, some mill-cut timbers may have been supplied by Fitzgerald's Construction or Hardy's Timber, both of whom were operating their business out of Baylis Street in Wagga Wagga.

It was expected that the dugouts could accommodate up to 35 people comfortably and although some trainees may have felt being underground in a confined space of that size somewhat claustrophobic, military authorities considered the size was sufficient without any concerns for dangerous overcrowding. To put the size of the dugout into perspective, it was fractionally larger than the service area of an international standard tennis court.

Supplementary to the earthen walls, vertical posts supported approximately fifteen large round horizontal timber poles. The poles provided adequate strength to hold sheets of cement-coated hessian sacks on arc-mesh—steel for the roof. The roof of the dugouts was a few feet above ground level and the mounds of dirt dug to form the shelter were shovelled back into sandbags to provide the edge capping and strength of the roof. Fixed into the construction of the roof was an opening of about 25cm. This gap in the ceiling let in light and fresh air. Once the roof was secured and the aperture constructed with timber and sandbags, the remaining spoil was shovelled back onto the roof to complete the construction. From ground level, the roof of the dugout was about four-feet high. The mound of spoil was tall enough to prominently mark each dugout location on the flat barren demolition area from some distance away. On the same side of the aperture, a series of steps (about five to ten) was cut into the soil to provide access into the dugout. At the base of the steps was a piece of mesh plate, which acted as both a drain, and somewhere for the men to wipe their feet before entering. The earthen base of the dugouts was compacted with roughly three inches of sawdust, providing the men with suitable flooring.

Inside the dugouts a blackboard was placed at one end. Being dimly lit, the blackboard was generally placed at the same end as the access opening, adjacent to the steps in order to rely on the shard of daylight entering the dugout. Instructors used the blackboard for explanation during demolition theory lessons and lectures. The stores for each lesson, which would include training manuals and schedules, training aids and the actual explosives would generally be located next to the blackboard on the immediate right hand side of the instructor. A driver would deliver the explosives for the conduct of the day's lessons to the demolition area every morning. Instructors, facing trainees with their backs to the aperture and blackboard with stores at their immediate right-hand side, could manage most activities taking place in the dugout.

For comfort, a series of wooden ammunition boxes, probably *Dupont Monobel*, which were in abundant supply, and measuring about 45cm long and 21cm in height, were placed tight against the walls end to end to form a hollow square. Facing the instructor, who could also look every man in the face, the setup allowed trainees to pay undivided attention to instruction whilst able to see the blackboard. The men sat on a single box each listening to instruction and were able to lean back on the earthen wall for added comfort during lessons.

One particular dugout, located just off the access road at the base of a prominent box eucalypt tree, became known as 'Number One' dugout. This dugout was Jack's favourite. Its location made it one of the more prominent and popular dugouts, due in part to the sparse shade provided by the eucalypt, which was especially important during lunch breaks on the normally barren field. During lunch breaks instructors could shelter in the nearby ramshackle Demolitions Store shed (often referred to as the 'shop') which was located about 135 metres from the prominent No 1 dugout, as the trainees huddled in the shade of the tree. From the 'shop' instructors could keep an eye on trainees during the lunch period, prepare for future lessons, or just catch up with other instructors or enjoy a well-deserved nap.

Of course authorities considered the issue of the men's hygiene on the range. A rudimetary latrine was located far enough away and upwind from the dugouts so as not to contaminate the area with an unbearable odour. Trainees weren't allowed to linger in the latrines and were ordered to use the minimal time required to do their business.

During construction of the camp in 1942 the demolitions area was safely sited about a mile northeast of the prominent elevated ridgeline. It therefore wouldn't take long to get there when trainees set off on foot from Company parade grounds. On the march down it was a good time to build camaraderie and confidence amongst the squads who weren't all necessarily known to each other. With their .303 rifles carried more formally in the *attention* position ' ... at the slope,' when the men and their instructors were clear of the prying eyes of the camp's senior officers, the instructors gave the orders, ' ... at the trail', allowing the men to carry their heavy weapons in a more informal and comfortable manner. But down on the demolitions area training was becoming a little uncomfortable—especially considering that the rain stopped falling in Wagga Wagga in 1944 and the area, was a dustbowl. Not only that, three years of constant demolitions training combined with a dry summer had decimated the last remaining significant trees and shrubs. The only real area of shade was the lone box eucalypt adjacent to Jack's favourite dugout, but even that was gnarled and suffering the effects of constant demolitions and a dry summer. The drought had turned the range into a dry, uncompromising wasteland and each step raised a small puff of dust which collected in each man's eyes on the march down to the range.

About the same time the men had started the short march to the demolitions area, under the command of Roy Tafe, over in the camp's transport yard Sheila Oehm was 'first parading' her truck, checking oils, lubricants, engine, tyres and electrics. Sheila, like many of the AWAS who'd completed the driver's course, was an excellent driver. Extremely proficient she was also level-headed, an important trait which landed her a job driving the munitions on and off the demolitions area. Sheila had joined the AWAS in April of 1944 and before driving the trucks at Kapooka she'd been busy driving heavy trucks on and off the wharves in Sydney. There she'd tow wrecked vehicles coming back from the frontline to the camp at Ryde; the vehicles would get repaired and sent back to the frontlines. After posting into Wagga Wagga on a cold winter's day, she moved into her accommodation which was an iron hut in the middle of the camp known amongst the women as 'The Bomb Inn'. Sheila filled her straw paillasse alongside the twenty other AWAS calling the iron hut home. At the Engineer Training Centre she did all manner of driving tasks. One of her first tasks was to drive a man around to all the toilets in the camp so that he could clean them. Sheila progressed from carting garbage to the dump to delivering wood to the kitchens, taking dinner and hot meals to the troops around the camp, driving officers into town on business and of course a little later the responsible driving task of delivering ammunition and explosives to the training areas. Sheila enjoyed this job the most because she got an opportunity to flirt with the trainee Sappers. She was tall and statuesque with the poison wit of a comedian, but at the same time had the composure and grace of a princess. Her complete package made her one of the camp's favourite AWAS drivers—especially when she could give as well as she got from the trashy-talking trainees.

On this particular morning, her job was to attend the munitions store, pick up the munitions storeman and his explosive parcels, and issue the parcels to the three instructors, Jack, Roy and James, who were waiting down on the demolition area. Sheila became used to the process of delivering high explosives but never became comfortable or complacent carrying her volatile loads. Although the sight of female truck drivers may have startled some of the junior trainees who were unworldly and had never seen a women drive a vehicle, let alone a large heavy truck, the use of AWAS at Kapooka was becoming so commonplace it was normal. The women were right at home behind the wheel of a Ford Blitz truck. Getting up close and exposed to the engineer training was the kind of job that Sheila loved. Not only that, she loved the interaction and banter with the young trainees.

At the underground munitions store, 40-year-old Sapper Thomas (Tom) John Musto was reconciling his explosives into three equal parcels to be delivered to Jack, Roy and James for the day's training. Tom had been in the Army now for over three years. He'd served in the Middle East and New Guinea, but a stubborn recurrence of a jungle bug meant that his operational service days as an engineer were over. Somehow, for the meticulous former postal worker and returned veteran, he was more than happy issuing munitions to engineering trainees at Kapooka—even if it was a little intimidating riding shotgun with Sheila.

Tom meticulously packed three separate 100 lb parcels of monobel. Ensuring it hadn't expired, he noted that the lot from which he drew his amounts had been received at the camp in late March of 1945. This meant it was still serviceable and able to be issued. Inspecting the packaging for any tampering or opening Tom was satisfied with the explosive compound. He then turned to reconcile the gelignite using the same process.

Using a similar method of inspection, he observed that the lot of gelignite he was drawing his stores from was only received at the camp on the 9th of April. It was free from tampering and hadn't been opened. Like the monobel, the gelignite was also current and therefore serviceable. Tom placed three separate ten lbs piles of gelignite with the monobel. His next task was to allocate thirty electric detonators, thirty safety detonators, ten feet of orange instantaneous fuse, ten feet of prima cord and 25 feet of blue sump fuse into three individual piles for Jack, Roy and James. He ensured they were separate from the explosives. Satisfied not only with the quantity of his three parcels, but also the safety of its packaging for transport, he waited for Sheila to arrive with her truck.

After a short wait, Sheila and her three-ton truck with the high canopy arrived at the munitions store. The two set about safely loading the truck. Climbing into the cabin with his paperwork, ready for issuing to the instructors, the 40-year-old storeman felt a little out of place sitting in the passenger seat next to Sheila. Strikingly tall, Sheila was also slim, but looking dangerous dressed in her uniform. Tom was a little intimidated, if not slightly excited, to be working alongside the good-looking women. But he was a married man. He'd married his childhood sweetheart, Mary Ludwig, in 1929 and to his credit he remained chivalrous and professional towards Sheila. Besides, as a munitions truck driver, Sheila wasn't one to back down from an argument. She often retaliated to the indiscriminate wolf whistles directed at her with her own style. Travelling daily with Sheila, Tom could only laugh at her regal appearance in direct contrast to her colourful and and often poisonous retorts.

Chapter Seven
Morning Delivery

0840—0915 hours

Passing through one dusty paddock, the squads happily marching to the demolitions area soon went through a gate entering an adjoining barren enclosure. This paddock looked like any other, except for a couple of defining features. Four scattered mounds of earth. Their appearance signaled to the men that they'd survived the short march under Roy and had made it to the demolitions area. As they reached their destination the collective spirit amongst the men was high. The morning turned out to be delightfully warm and although Jack's squad was slightly oversized, he too had a good feeling and spirit about the day. After all, he knew the Brigadier was away and wasn't going to show up to scrutinise his instruction and potentially ruin his day and, best of all, it was his 31st birthday. But in the quiet of the morning march Jack's mind couldn't help but drift off. He wondered if a recent requested transfer back to his unit would be successful. And, as the birthday boy, why wasn't he at home celebrating with his family instead of once again trudging out to the range? He spent the last birthday on course surrounded by strangers, and now, once again, strangers on the dusty dry plain of the demolitions area surrounded him. ' ... Not again,' he thought to himself as he made the short but familiar trek.

About 0840hours that morning the squads arrived. With the end of the march initiated by Roy Tafe's loud words of command, 'GOLF COMPANY , HALT,' Jack's mind was soon back on the job. Jack and Bill Cousins' squad, as the largest, peeled off towards the primary dugout under the lone box eucalypt. James Conwell immediately took his smaller squad to an adjacent dugout. Included in James' squad at that time was Bill Rhodes, a 24-year-old trainee Sapper from Box Hill in Victoria. Roy Tafe took his smaller and soon to be released squad to another dugout where his squad set about doing preliminary training—not the full demolitions week training schedule.

Whilst Bill sorted out the squad, Jack went to the nearby demolitions store to get his instructor's box. Although widely experienced, he still kept a copy of the *Military Engineering Volume IV Part I Demolitions* in his trusty stores box. While the manual was a British manual dated 1942, it was the most up to date précis and doctrine Jack could rely on. For the experienced instructors, they often made reference to the manual on a point of technicality if ever questioned by a 'smart-arse' trainee who thought he knew better. On the rare occasion they were questioned, Jack and the other instructors were generally always right. Backed up by the doctrine, the smart arses were quickly put in their place.

Over in Jack's squad, the often loud but always colourful Corporal Cousins took the initiative. At over six feet tall, he was looking every bit like his experienced Sergeant instructor, except for one thing. Unlike Jack, Bill had feet the size of a circus clown. His Army boots looked like they should be attached to a pair of dotted pants stumbling, falling and making kids laugh with a round red nose under the Big Top. Nevertheless, Bill was professional when he needed to be and took control of the trainees in Jack's absence. Forming them up for an administration brief, he spoke with professionalism. In his unique way, prior to Jack's return, Bill advised the men of the location of the latrine, the smoking area, and just where their weapons and equipment would be stored during lessons in the dugout. Most importantly he advised the men what time and where lunch would be served. He concluded his proactive brief with a general layout of the area including the locations where Sergeant Tafe's and Corporal Conwell's men would be working. He also took the initiative of pointing to another squad of trainees about 110 yards away. They weren't from G Coy but had obviously marched onto the demolitions area that day to do other lessons.

The other squad referred to by the vigilant Corporal was under the supervision of Sergeant Francis (Frank) Sim. Sergeant Sim had a squad of about two-dozen 18-year-old trainees who were in week five of training and receiving instruction on mines and booby traps. At 26, Frank Sim was an experienced instructor having served as a 'Rat of Tobruk' with the 9th Division Engineers. He was only instructing at Kapooka because he had contracted dengue fever; however, after being there for a couple of weeks he relished the opportunity of passing on his experience to the young trainee Sappers.

Meanwhile, in amongst the oversized squad of men, Jack returned to the squad and took over by rescuing the trainees from Bill's colourful delivery with more formal briefings. ' ... Thanks Corporal Cousins,' he added with a wry smile. He knew Bill's administration brief was colourful but he could see it worked a treat as the trainees appeared a bit more relaxed. The completion of Bill's brief meant that it was now time for the Sergeant to take over with the compulsory and more important safety brief. Jack was getting organized for the day's activities. As the principal instructor for the day's lessons, safety was Jack's responsibility. He wanted the day to run smoothly. To do so, he needed to lay down the law about what he expected in the area. The Key Instructor, Warrant Officer McFarlane, had previously drawn up a list of safety precautions for the demolitions range. It just so happened that Jack carried a copy of those safety precautions and religiously read them to his squad at the demolitions area before the start of every first period of instruction. They were commonly used for the overall handling of explosives, but Jack believed it important that every trainee knew the procedures. He read on.

'… There is to be no smoking in the presence of explosives … keep explosives away from fires … don't leave explosives in stacks where extreme weather may affect … use underground magazines … don't throw charges … don't run away from charges … don't play jokes … wait half an hour before going to misfires and finally never cut or tamp explosives with steel … does everyone understand?' To the formal list Jack also added, '… under no circumstances were you to smoke in the dugout or in the vicinity of the explosives … I will store the explosives separately and therefore before entering the dugout you will place all your smokes, lighters and matchbooks into the tin located at the rear of the dugout … is that understood?' A collective '… Sarge', sounded out from the gathered men, all acknowledging the order.

Jack continued reading from his scripted safety precautions, '… regarding detonators … keep them away from flames and explosives … don't crush, bite or jump on them … don't throw them to each other … don't hold or roll them … test them before you use them and don't tamper or interfere with them.'

He also added a little extra of course, '… At no time are you to use any metal instruments that would cause a spark and detonate the explosives, Corporal Cousins will therefore inspect your equipment before you enter the dugout … is that understood?' Once again the men let out a collective chorus, '…Sarge!'

'… Whilst in the area you are to obey every instruction given to you by myself, Corporal Cousins or any of the other instructors.'

'… There is to be no running around, playing up or horseplay whilst in the dugout or in the vicinity of explosive stores, is that understood?' Silent nodding of the heads relayed to the Sergeant that the assembled men got the message. '… Are there any questions about what I expect or what I have just read to you?' None were forthcoming. Being a stringent believer in safety Jack got the men to sit down and he proceeded with his own spiel about his experiences and the three principles of demolitions, just so the men got the full picture.

Sitting down on an empty ammunition box, Jack held court with the men sitting before him. ' ... There are three principles of demolition work behavior that you must adhere to.' ' ... First you should be scared through lack of confidence, you shouldn't however be careless through over-confidence and because I've seen what this stuff can do (pointing to the empty monobel box he was sitting on), I am always, always careful with this stuff.'

Hoping his fresh-faced charges got the message about just how he operated, Jack took a look at the men before him and observed an element of fear, excitement, nervousness and apprehension in their eyes. He knew that this was a good sign. ' ... No complacency and smart-arses here', he thought to himself. Understanding that anxiety can also lead to nervousness and accidents, he decided to take a more informal approach in order to ease the tension.

' ... Righto men,' he went on, ' ... I'll just give a run down on how today is going to work.' The anxious, yet quietly enthusiastic group of men sitting before Jack included Kevin Pierce, Frank Platt, Ernie Poschalk, Stan Morphy, Alf Witt, Ivan Merritt, Allan Bartlett, Denby Grasby, Toddy Woods, Joseph Faull and his mate Terry Moore, Kevin 'Brickie' Hurst, Jack 'Porkie' Nixon, Ted Robson, Allan Flood, Joe Collins, Les Mather and Norm Dilley. They all listened intently as they sat at the feet of the Sergeant holding court. Also sitting before him but becoming increasingly fidgety was Geoff Partridge. His annoying toothache was getting the better of him and it needed attention. He was struggling to be attentive. Jack identified his plight and immediately dismissed him from the squad to go to the dentist. From the demolitions area Geoff proceeded back to the HQ alone to let WO2 Dodds know that he wasn't on the range. Immediately after reporting to HQ he went to get treatment. Risking the valuable training opportunity, he'd finally had enough.

Back at dugout No 1, the men listened intently to Jack, still informally explaining how the day would go. ' ... Pretty soon we'll go into the dugout and start lessons ... this morning, up until about lunchtime, it's all about introducing you to the art of demolitions, the safety precautions and the types of fuses and detonators we use and of course the types of explosives we'll be working with. After lunch, we'll get your mind back on the job with some physical training and then back into the dugout where you'll start handling the explosives ... okay ... has anyone got any questions so far about how today is going to run?' The men were taking comfort in the informal approach in which the experienced Sergeant was breaking the day down for them. As none of the men had any questions, Jack went on.

' ... After that we'll spend the afternoon preparing for the night activity which is firing a barrage of these small charges that you'll be preparing.' ' ... After a short night stroll back to camp you'll be at the canteen for a beer before twenty hundred hours.' ' ... If there's no further questions ... oh one other thing ... the transport truck will be delivering the explosives down here shortly so I'd suggest if you want one last smoke to get it in now.'

Scurrying to the designated smoking area pointed out by Bill, the Sergeant gave the men a couple of minutes before they were to start moving back in. As they finished their smoke and started filing into the dugout, Bill Cousins assumed the role of the safety 'enforcer'. At the entrance he relieved the men of their matches and smokes, checked them for metal objects and instructed them to take a seat facing the blackboard. By 0900 hours, eighteen men were seated in an orderly fashion on the wooden boxes around the dugout, whilst the two instructors prepared the first period of instruction. With his instructor's box containing the constantly referenced *Military Engineering, Vol IV (Part 1) Demolitions* manual and lesson notes, Jack started on the first lesson; an introduction to explosives used for demolitions in the field.

Jack had barely gotten fifteen minutes into the first lesson when Tom Musto and Sheila Oehm arrived with the parcels of explosive stores.

The rumbling of the truck alerted the three instructors that their stores had arrived. Like moles out of their holes, Jack, Roy Tafe and James Conwell appeared out of nowhere to greet the storeman and his 'better-looking' driver. In the dugout Bill Cousins instructed two of the trainees to assist the Sergeant unloading the explosive package from the truck. Having a good rapport with Sheila, Jack invited her into the dugout to watch the day's training. He told Sheila that he'd sort it out with her boss and that it would be beneficial, but as she had other pre-arranged tasks that day, she politely thanked the Sergeant for the offer but duty was calling. She was going to be back in the afternoon, so Jack thought he'd try to convince Sheila after lunch to come into the dugout to watch. Inspecting the stores, the three instructors were satisfied that their correct complement had arrived and they signed for their respective stores accordingly.

Chapter Eight
Discretion

0915—1230 hours

As traffic on the demolitions area wasn't restricted just to the training squads, often headquarters staff, catering staff or even transport drivers would be found in the area. For this reason, the security of the explosives was important. They needed to be monitored at all times. It was standard practice amongst instructors that only sufficient amounts of stores for the purposes of the actual demonstration taking place were to be taken into the dugout. In those cases, the remainder of the explosives would be left outside the dugout to be picqueted (guarded). Ordinarily, although there was no formal written instruction to direct otherwise, two Sappers from each squad would picquet their respective explosives being stored outside the dugout. However, policy also gave some discretion to the instructor. He could decide if the weather was inclement that the parcels could all be stored inside the dugout with the squads. It wasn't the preferred practice, but if it were managed appropriately, it would be perfectly safe.

On this Monday morning of 21 May, the weather wasn't necessarily inclement or bitingly cold. Instead, Jack used his discretion and made a senior soldier's decision. He didn't want two of his squad missing out on training. So, flying in the face of the key general safety rule located in the instructors' précis regarding the separation of detonators and explosives, using his discretion afforded him as the instructor, Jack directed the two Sappers to place the complete parcel of explosives inside the dugout alongside the blackboard. He didn't know it at the time, but it was a decision that would have catastrophic results. Following behind the men, who shared the load carrying a medium-sized wooden box containing 100lbs of monobel and 10lbs of gelignite, and additional fuses and detonators in a separate packaging, Jack instructed the men to place the stores separately in the immediate right hand corner of the dugout as they entered. It was an area out of the way of trainees and directly adjacent the blackboard. Stored in their original packaging, which included separate boxes, the explosives would be safe, especially directly next to the right-hand side of the intelligent instructor. From that position, Jack would be responsible for it during the entire lesson and be able to monitor the required separation between detonators and explosives throughout the day.

Roy Tafe and Colin Kendall, having received the exact same parcel of explosives, placed theirs in their respective dugout as well. James Conwell, working in the third dugout with his own smaller squad, also placed his in safety below ground. It was a decision the three experienced instructors agreed upon during the issuing of the munition stores. When they considered that eleven of the men would be taken away later for duty, it was really restricting the option for two remaining men to piquet the explosives all day long. Although storing the explosives in the dugout was unsafe, it was standard procedure at the time and the men were confident that there'd be no problems.

Following the issue and signature for stores, the departure of Sapper Musto and Shelia Oehm soon followed. Now that their stores had arrived, the demolition area was free of any non-essential personnel and the three squad instructors had a couple of hours to instruct before Roy Tafe and James Conwell's men were due to be released for duty. Finally, after all the administration and logistics, the day of instruction was well and truly underway and the three independent squads could get down to business. Jack was aiming to get through the first two period of instruction, which included the safety precautions for the types of fuses they'd be using for the remainder of the week, without any further interruptions.

At 1100 hours, as directed by the Sergeant Major earlier that morning, Roy Tafe and Colin Kendall released nine of their men from training to report to headquarters and get ready to mount guard duty. As they would be on the 24-hour guard, the men weren't expected back until the Wednesday morning. At the same time, James Conwell released his two men. Together the eleven men formed up, and with weapons at the slope, marched back to Company Headquarters to report for guard.

With only three men, Sappers Boyd, Hurley and 28-year-old Stan Ross and his assistant instructor Colin Kendall, Roy Tafe interrupted Jack during a break in his lessons. '…Jack, as I've lost most of my squad for the night can I suggest that I send Colin and the remaining three men to your squad? I'll get them to prepare detonators with your stores for your night exercise and get my men to join in?' ' … Perhaps we can introduce something like an electric circuit with hand charges to your regular night activity of firing a barrage of charges?'

Jack agreed to the recommendation and also suggested that Colin Kendall use his stores of detonators and fuse wire. Before too much longer the two Colins, Boyd and Hurley, packed up their equipment from Roy Tafe's dugout location and moved into Jack's dugout. They got in there just before Jack stared his third period of instruction—types of explosives. It was only fair that they not miss out on any of the training syllabus. Sapper Ross then accompanied Colin Kendall in their dugout preparing the 'electrical circuit' for the night-time activity. Colin and Roy chose Stan because he was a little bit more mature than the two remaining Colins; also, as a motor mechanic, he knew a bit about electrical circuits. Together the two men set about preparing fuses and detonators. Using No 6 electric detonators and blue sump safety fuse, the men continued preparing the equipment for the purpose of using them in Jack's night activity.

After the changes had been made and the squads reconciled, Roy Tafe proceeded back up the hill to Headquarters to inform the Acting Sergeant Major of the changes. It was about 1130 hours when he advised Ed Dodds that Sappers Hurley and Boyd were now in Jack's dugout and that Sapper Ross, who was assisting Colin Kendall in the adjacent dugout preparing electrical circuits and detonators for the night activity, had also been attached to this squad. The practice of reconciling squads when men were taken away for other duties was commonplace. Ed Dodds agreed with the changes and even discussed the fact that Geoff Partridge was at the dentist and would probably return sometime after lunch. Now that Roy Tafe was freed up, Ed Dodds tasked him to work alongside Sergeant McNabb, another member of the Company, who was working on a separate project in the nearby demolition store shed. Being about 150 yards away from Jack's dugout, the shed was close enough that Roy could respond if he was needed to assist Jack or Colin with the instruction or demolitions preparation.

Meanwhile, Jack's squad of 20, which now included Colin Hurley and Colin Boyd, had worked their way through to the end of the third period of instruction with few glitches. Bill Cousins remained his affable self, but the young man was fast becoming ravenous. He couldn't wait for the meal truck to arrive with the regular hearty lunch. Looking at his watch, Jack wrapped up the period of instruction around 1220 hours. That gave the men some time to have a smoke before the field kitchen arrived with their hot lunch.

Chapter Nine
Jack's Growing Squad

1230—1415 hours

Like a scattered mob of cattle, grazing aimlessly on the barren dusty ground of the demolition area, the remaining men from all three squads sat around the shade of the gnarled box eucalypt adjacent to dugout number one. A purpose-built truck, fitted out with its own wood-fired boiler, had earlier delivered hot meals to all the trainees on the range and they were now busy tucking into a hearty lunch of meat and vegetables. The trucks ingenious design kept the water heated, and in turn, the hot water kept the food containers hot. It was just the tonic the men needed. Setting up a server-style dining hall wherever it stopped, the food was laid out on tables allowing the men to file through and receive a nutritious hot meal of meat and greens in their dixies. At the end of the table was a brew point. Cups canteens were soon filled with hot water for a welcoming tea or coffee, or the men filled their cups with ice-cold water in which they mixed cordial crystals for a refreshing cool drink.

Afterwards, grabbing the small available patches of shade, the squad eagerly tucked in enjoying one of the nourishing 'three square meals' provided by the Army. And, supplemented by the generous fruit they'd grabbed at the canteen that morning, bellies were soon full. As the trainees and instructional staff hungrily tucked into the nutritious meal, so too did Sheila Oehm and Tom Musto. The pair stopped in for lunch after making further deliveries on the demolition area and various locations around the camp that morning. Sheila's presence was a welcome distraction for the men. They'd become bored of the familiar sight of the dirty hard-working faces of the colleagues and in Sheila, well; she was a breath of fresh air on the range.

Meanwhile, as always, Jack sat amongst the youngish men finishing his own lunch. Careful not to alert the vulture-like trainees, from his pocket he carefully extracted the small piece of brown paper containing his personal morsel of birthday cake. He'd carefully stashed it away in his haversack before leaving Headquarters that morning. Washing it down with a cup of beautiful English tea, sitting in the shade amongst the young trainees, Jack was in his element. To his utter surprise, out of the corner of his bespectacled eye, Roy Tafe had exited the demolition store shed and walked purposively towards Jack after spotting what he was doing. ' … No thanks Jack,' he quipped. ' … I couldn't possibly fit any more in.' Much to the bemusement of the trainee Sappers, who had no idea what was going on, the men shared a laugh over the birthday cake. As the last morsel of icing and fluffy cake disappeared into Jack's mouth, he savoured the moment by giving his belly a little pat. The trainees, still none the wiser, looked on confused …, ' What'd we miss?'

It hadn't been an over-challenging morning; after all, Jack had delivered the Monday lessons on more than ten occasions before, and for the hungry squad, warm food was a welcome relief and a short distraction from the intense training. The men could enjoy the hour-long break and it was a good opportunity for an informal chat with the instructors. For the instructors who'd decided to stay on the demolitions area, rather than go back to the comfort of Headquarters to eat, lunch was a great lubricator for idle chatter amongst the remaining Sappers; a moment to learn a little more about each other. For someone like Bill Cousins, in amongst the men, he was in his charismatic, charming element. Many of the trainee squad were his age, or younger. At 25, he was young enough to relate to the younger men and their lifestyle, whilst also old enough to understand the elder brigade amongst them. With the added maturity of rank, the gregarious and somewhat 'goofy' Corporal with the oversized military boots, was always a lunchtime hit. Telling stories of his overseas service and his many weekend 'conquests' with the local ladies, to those that had surrounded him, listening to 'Billie's' barrage of tall stories—it was proving difficult to get a word in. The funnyman held court like a seasoned entertainer.

Seizing a break in Bill's volley of stories, Jack was able to establish that many of the men were still living with their parents. About five or so of his squad were married, and, as he suspected, a few of the men had children themselves. Explaining to the gathered Sappers that he was living in Wagga Wagga with his wife and four children himself, it was the opening that 'Porkie' Nixon needed. The new dad, immediately, and without invitation, proudly piped in with some repetitive dialogue about his angel whilst waving a slightly tattered photo of baby Neryl. Now crumpled and almost illegible, having lived in his trouser pocket for months, Jack was excited in his delivery and forgot to mention his wife Marjorie and her part in bringing their bundle of joy into the world.

Ivan Merritt overheard the conversation between Jack and Sapper Nixon, and wasn't able to hold his tongue any longer. Declaring that his lovely wife Joyce was pregnant, he smugly broadcast that the baby was due in the coming August—his beaming smile, a dead give-away at how proud he was. Ivan was hoping engineer training would pass quickly, but if it didn't, he was going to apply for some leave to be at the birth anyway. ' ... Well Sapper Merritt,' Jack responded, ' ... I wish you and your wife all the very best, I know how challenging it can be to be away from your wife whilst she's pregnant, let's hope you get home in time.' Ivan politely thanked the empathetic Sergeant and continued eating his lunch whilst idly chatting to Denby Grasby.

'...Anymore?' asked the Sergeant father figure prompting more voracious dialogue from his nervous charges. Before he knew it, 'Toddy' Woods also joined in on the conversation of fatherhood and mentioned that he and his wife Ettie had a little son Alan. After all the dads in the squad had informed Jack of their children back at home, without prompting Ivan finished eating and spied a football sitting under the table of the server. Walking over to the table, he picked up the ball, fashioned a half decent handpass to Kevin Pierce and declared, ' ... well I'm excited to be a dad for the first time so don't any of your bastards spoil it for me or put me off with all the unpleasant details, now ... who wants to have a kick?' It was the icebreaker the men were waiting for. They all saw the sports equipment under the table, but they didn't have the maturity of Ivan to go over and pick it up and start something for fear of getting in trouble.

Apart from their distinctive Army numbers, it was only when a football was produced that you could establish where in Australia the trainees 'originally' hailed from. The West Australians, Victorians and South Australians kicked the footy like they were on the MCG playing for their State, albeit a little dusty, whilst the New South Welshmen and Queenslanders were passing it to and fro like they were warming up in the tunnel of the SCG for an upcoming interstate clash. There was always some friendly banter about codes and aerial 'ping pong' but it almost never got out of hand. When the dust settled 'literally', they were all there for the same thing and the interstate rivalry could continue over a 'Lady Blamey' ... or two, in the canteen a little later on. For now, football, regardless of the code, was the break from the morning's lesson they all needed.

During the break, the Sappers would alternate between kicking and throwing the footy, having a smoke, standing around in small groups chatting or making another brew. Meanwhile, the issued explosive and demolition stores remained in the respective dugouts. At irregular intervals, the instructors would venture down into their dugouts, or task their assistants to check on the explosives. On this particular day, nothing of an abnormal procedure regarding the security of the explosives was anticipated. It was however perfectly normal standard practice to leave the demolitions stores in the dugout during lunch and physical training, but under no circumstances were they to be left there overnight. To prevent this, transport vehicles with the munitions storeman on board would round up the unused material later in the day. Not that it mattered on this occasion; day one of the fourth week also had a night activity, which the men were also preparing for.

Sadly for the Sappers and instructors, who were all enjoying the freedom of being equals on the footy field, the end of lunch was signaled by the arrival of the physical training instructor. His appearance returned the Sappers to their subordinate status, but for now at least, the atmosphere had become a little more civilized knowing, in some areas at least, the men were all equal.

Finishing his lunch in the camp canteen, Paddy Cranswick was walking back to his tent for a bit of a break before returning back to duty. As he made his way to the Company accommodation area, he noticed Geoff Partridge walking through. The two men knew each other pretty well and on a number of occasions even slipped into Wagga together for a beer and had now stopped for a chat. '... G'day Geoff, what are you doing here?'" asked Paddy, who was a little surprised at his appearance. Geoff told Paddy that he'd been to the dentist. '... Where you going to now?' he asked. '... I'm going back out', came Geoff's reply. '... You're bloody mad, what do you want to go out there for?' asked a somewhat puzzled Paddy. '... Go and tell the Sergeant your tooth's aching and you don't need to go out again.' '... I don't know Paddy.' Geoff didn't listen to Paddy's advice; he just wanted to rejoin the squad. '... I'm going back out mate', came Geoff's reply. Paddy watched as Geoff walked off in the direction of the demolition area. After his short rest Paddy went back to the HQ for further afternoon tasking.

Shortly after lunch at the G Coy HQ, Sergeant Ron Linthorne, Alf Woods, Bill Reid and three other trainees, one which included 18-year-old Sapper Arthur Hanchard from Dubbo, who happened to be a good friend of Stan Emery and 'Porkie' Nixon, reported to Ed Dodds. They'd all been off doing administrative duties but had since been released to return to training. Ed Dodds sent Linthorne, Woods and Reid to join Jack Pomeroy, whilst Hanchard and his two fellow trainees were sent to join Corporal James Conwell's squad. Ed Dodds advised the men that the squads were doing physical training, but they could join in after that in the formal instruction periods. Ably steered by the only Sergeant in the group, Ron Linthorne, the six men marched off with their equipment towards the demolitions area. Ed Dodds followed loosely behind for a quick visit to the demolitions range to check on progress before formal lessons resumed.

Physical training, whilst a necessity to improve fitness, was for many of the men an unnecessary evil, especially standing in the hot Riverina sun. Lining up in ranks, exercises consisted of a series of star jumps, toe touching, stretching and running on the spot. On the dusty barren range it was if a herd of cattle had barged through. It didn't help to whine about the conditions though. Somehow, through the dust, the physical training instructor could single out those not pulling their weight or back chatting. It was a perilous journey for those who'd had enough of the heat, dust, flies and sweat. The instructor's punishment ... of course nothing other than press ups (more commonly referred to today as pushups). Press ups were the military go-to exercise for swift purposeful punishment. No doubt many men on the range that day were perhaps a little boisterous and were handed a few pushups? But it was Bill Cousins who once again stole the limelight. Of course he wasn't just being punished, for Bill it was more about showing off his attributes. As a fit farmhand from the bush, Bill 'showed em all how it's done'. Kissing the soil with every push up, his antics never seemed to cease.

Geoff Partridge had finally arrived back at the demolitions area. After getting his tooth fixed up by the camp dentist and running into Paddy Cranswick, who tried to convince him not to go back down, Geoff just wanted to get involved. The young Sapper approached Ed Dodds who was overseeing the physical training. ' ... Excuse me Sir', ' ... welcome back Partridge, how's the tooth?' came the less than enthusiastic response from the experienced soldier. ' ... Fine thanks Sir', came Geoff's not so honest answer. Ed Dodds continued, ' ... Sergeant Pomeroy is just about to start his lesson on slabs of explosives, you can join him', ' ... Thanks Sir.' As Geoff walked over to join in the rest of the squad Bill Cousins was announcing to anyone who was listening that he was approaching his 30th press up. ' ... That's thirty ladies, any takers?' Geoff had a wry chuckle at the antics of the instructor and slotted unannounced back in with the group—albeit still a little sore in the mouth.

Whilst Bill Cousins was busy showing off, to a less than amused Sergeant Major, Roy Tafe tasked his assistant Colin Kendall to retrieve their parcel of explosives and any fuses that he and Sapper Ross had already assembled. He wanted them moved from their dugout and ordered them to be placed just outside the entrance to Jack's dugout. Kendall and Ross had made up about thirty detonators with safety fuse and match heads and placed them alongside their pile of explosives not far from the entrance to Jack's dugout. Roy Tafe also instructed Colin to grab his two men (Sappers Boyd and Hurley), whom Jack had been looking after, and tasked him to supervise the three men to prepare as many more detonators and safety fuses as they possibly could before Sergeant Pomeroy's large squad arrived back in the dugout after physical training.

Colin therefore placed approximately 95 lbs. of quarry monobel and ten lbs of gelignite in a heap, approximately seven yards from the entrance to Jack's dugout as directed by Roy Tafe. Jack and his men were still going through physical training when Sappers Boyd and Hurley reported back to Lance Sergeant Kendall. Along with Stan Ross, Colin Kendall started demonstrating to the three trainee Sappers how to safely attach blue sump safety fuse to the detonator, crimping it on with the tool, then attaching a match head to the end of the safety fuse to aid ignition. He showed them completed ones from the thirty he'd previously made up with the assistance of Sapper Ross and, at the same time, he showed them how to attach electric leads and bind it all together with used detonator wire.

Studiously the three men paid attention. Having received safety precaution instructions from Jack Pomeroy less than three hours ago, safety was still in the forefront of their mind. It was of particular importance, especially now, as they were working with detonators. The Lance Sergeant explained to the men how to make up one or two detonators and further instructed them to vary the size of the blue sump fuse wire between two inches to approximately eighteen inches in length. When he was satisfied they understood his instruction and he'd observed them for a short period of time make a few up, he decided to place them inside Jack's dugout. In there, they could concentrate on the task out of the hot sun and then later, when Jack came back in to resume lessons, he could also keep an eye on them.

Chapter Ten
A Handful of Detonators

1415—1445 hours

At 1415hours, the Lance Sergeant guided his three helpers, Colin Hurley, Colin Boyd and Stan Ross, down the seven or so entrance steps into the number one dugout. Ordering the men to wait at the entrance, Colin walked over to the northwest corner of the empty dugout. As directed by Roy and Jack, Colin opened up some of the explosive stores that were allocated to Jack which had been placed inside the dugout earlier in the day. Retrieving a quantity of No 6 detonators and the roll of blue sump safety fuse, Colin returned to the men and positioned them just to the left of the entrance. Sitting on old ammunition boxes, the men set about their new task. Being seated near the entrance, they were far enough away that they wouldn't block the path for Jack and his oversized squad, who were due down in the dugout very soon and, as a bonus, they had just enough light to see what they were actually doing. Knowing the volatile parcel of explosives was less than five yards away in the opposite corner, the men remained somewhat comfortable ... but cautious. Quietly and meticulously working away, a little idle chatter broke the otherwise deafening silence the men were experiencing in the emptiness of the dugout.

Meanwhile, above ground a short while later, Roy Tafe approached Ed Dodds before the Sergeant Major returned to the Headquarters. He informed his superior officer once again of the changes within the squads. With the arrival of Ron Linthorne, Alf Woods, Bill Reid and now Norm Dilley onto the range, the squad roll books also reflected the three men from Roy Tafe's squad working just inside the entrance. Roy advised Ed that the roll books had been adjusted accordingly, for Jack though, that meant he now had a squad of 25 men. With himself and Bill Cousins that made 27 and with Colin Kendall coming and going out of the dugout doing lesson preparation, a total of 28 men would be in the small underground lecture room for the majority of the afternoon. It was slightly unusual to have so many men, who, ultimately, were all in Jack's care, but Jack and Bill believed they could handle it. Although capable of sheltering up to 35 men the dugout wasn't designed to have so much going on at any one time—Jack, Bill, and Colin roaming the dugout, three men at the entrance and soon 24 men would be filing in one by one. Understandably it was going to be overcrowded and busy.

After about fifteen minutes Jack and Bill mustered their burgeoning squad. Formed-up in ranks beside Sheila Oehms' truck by Bill, the men were marched to the entrance of dugout. As they marched off, Sheila, accompanied by Tom Musto, returned to her truck and drove off to the magazine to collect an additional box of gelignite, which was supposed to be delivered to the area.

Back outside the entrance to Jack's dugout, the Sergeant instructor was being informally introduced to his new squad members. Surveying the number of men waiting at the entrance, Jack was a little concerned about the size of the squad, especially considering the next period of instruction, lesson five—making up binding and firing single slabs of explosives—required maximum concentration and vigilance. ' ... Hopefully,' he thought to himself, 'it'd be okay.' As he looked around at the men he observed the three chevrons on Ron Linthorne's uniform. Acknowledging that there was another SNCO amongst them, Jack introduced himself and welcomed the senior soldier.

It was all about professional recognition of rank and getting to know each other. Ron introduced Sapper Bill Reid and his mate Alf Woods to the approachable Sergeant instructor. After some lighthearted conversation, the mood of the moment was somewhat hijacked when the best-intentioned Sergeant turned to Ron Linthorne and Alf Woods and asked them the question of fatherhood, ' ... what about you two then, any kids?' asked Jack. Totally unaware that Alf had lost his infant son a few short years ago during childbirth, the charismatic and courageous Fijian-born West Australian replied, ' ... yes Sarge, I had a son but we lost him a few years back now.' ' ... I'm sorry to hear that mate', came the reply from the slightly embarrassed but dignified professional Englishmen, ' ... I apologise for making you uncomfortable.' ' ... That's alright Sarge,' said Alf motioning to his good mate Ron with his head, ' ... if it wasn't for this bloke I'd be a wreck.' Welcoming the two to his squad, Jack looked towards another man who looked much older than himself, ' ... Sapper Norm Dilley Sarge, I'm thirty-four married, my wife's name is Kathleen and we have two kids, one of each, June and Neil.' Not surprised to see an older man as one of his trainees, the slightly younger Sergeant welcomed another 'Dad' into his squad. ' ...Welcome on board Norm, I hope you enjoy the afternoon ... it will be something to tell your kids about when you see them next.'

Along with Bill Reid, Alf Woods, Jack Nixon and now Norm Dilley, Jack and Colin Kendall found comfort in the fact that at least they had a few more men over the age of thirty amongst the dominant younger brigade. For a brief moment, Jack pondered on the diverse nature of this squad. ' ... It's amazing,' he thought to himself, 'how the war has brought so many men from so many different backgrounds and ages together.' After briefly meeting his new squad members and learning a bit about their background, he felt strangely comfortable and was looking forward to a productive afternoon.

Before they entered the dugout Bill Cousins couldn't help himself. Sensing Jack was a little uncomfortable after his conversation with Alf Woods, he let the cat out of the bag about how special the day was for his superior officer. Before Jack knew it, 27 men were wishing him happy birthday. Of course conducting the chorus, Bill broke into an impromptu and somewhat inappropriate and slightly off-key 'Happy Birthday'. Much to the embarrassment of the senior soldier, for a brief moment Jack's guard was down. Appreciating the gesture he turned to his assistant and in full hearing of his squad added, ' ... you're a bloody larrikin Corporal Cousins ... thanks for that, oh and by the way, don't give up your day job.' A collective chuckle rang out amongst the men. The mood was now well and truly upbeat. ' ... come on you lot,' interrupted Jack now motioning towards the entrance of the dugout, ' ... we've got work to do.' On that lighthearted note, one-by-one the men filed into the dugout confident of having a good time during the next few lessons. The time was about 1440hours and the afternoon lessons were soon to begin.

Leading men into his underground 'lecture room', the birthday Sergeant was closely followed by Ron Linthorne. The SNCO often showed glimpses of his leadership qualities. On this occasion Ron set an example of sitting right at the foot of his fellow Sergeant. Having missed the morning's lessons, Ron sat in the northwest corner. It was closest to the explosive stores where Jack would be standing, but being so close he could watch Jack intently. Often when you establish friendships in the military, much like at primary school and to a lesser extent secondary school, soldiers tend to subconsciously gravitate and sit next to someone they know or get along with during formal lessons. Whether it's confidence or a security thing, sitting next to that person creates a feeling of comfort. As the men were only in the fourth week they were still consolidating new friendships. But for men like Joseph Faull and Teddy Moore, who'd been friends for a while, they sat together. It was the same for Denby Grasby and Allan Bartlett; they had mingled together outside the dugout and walked in after each other. Walking straight ahead from the entrance of the two mature Sappers, Merritt and Reid, had struck up a friendship and took up the first positions on the eastern wall close to the southern end of the dugout. Followed by Sappers Mather, Collins, Flood and Dilley, the men occupied the eastern wall from the southern wall to the entrance where Colin Boyd, Colin Hurley, and Stan Ross were busy working making up detonators with varying lengths of fuse wire. Careful not to trip over the three Sappers working on the detonators in the doorway as they entered, the men arrived in the dugout one after another. Stan was seated closest to the entrance and watched intently as the men filed in, careful not to kick or trip over his stores sitting in front of him, they spread out and sat down on the ammunition boxes around the eastern, southern and western wall of the dugout. Jack's oversized squad positioned themselves so they could all see both the blackboard and the Sergeant. Once they were all seated Jack could see they were seated tightly together almost shoulder-to-shoulder.

On the extreme southern wall sat Sappers Pierce, Platt, Partridge, Witt, Nixon, Toddy Woods, Poschalk, Morphy, Joe Faull, Moore and Grasby who had positioned themselves along the breadth of the wall. From here the younger men had a direct view of the Sergeant and his instruction. Sitting down in a comfortable position, resting their heads against the earthen wall, the men crossed their arms, not in defiance but more to relax. On the western wall Sapper Bartlett found himself squashed into the southwest corner beside his South Australian mate Denby Grasby. Denby, who was a little squished on the end of the southern wall, was comfortable enough, and folded his arms whilst stretching out his legs. He'd been looking forward to this lesson and a little uncomfortable seating wasn't an issue. On Allan Bartlett's immediate left was Alf Woods, beside him was Ted Robson, followed by Kevin Hurst and finally Ron Linthorne sat on the edge closest to the front where Jack was standing.

The last men to enter the dugout that afternoon were Lance Sergeant Kendall and Bill Cousins. Bill moved towards the blackboard to assist Jack whilst Colin remained in the vicinity of his three men checking on their progress. At the same time diligently observing the men making up the detonators, Colin could see and hear Jack explain to the squad the purpose of the night activity. Pointing in the direction of Sappers Boyd, Hurley and Ross, cautiously preparing fuses Jack explained what they were doing and gave a quick demonstration about how they were doing it. He did this for two reasons. The first was to reassure Colin Kendall and his three men that they were doing it correctly and secondly so that the entire squads were confident that nothing was going to go 'bang' as a result of the work they were doing with detonators so close to the explosives. After his quick demonstration he got on with the business of the next period of instruction. ' ... Alright men what we're going to do now is get on with the preparation of hand charges.' ' ... First can everyone see?' A chorus of ' ... Sarge' came the collective response in the muffled underground classroom.

Confident that his three inexperienced Sappers making up the detonators and fuse wire at the entrance of the dugout were doing a pretty good job, Colin turned to walk up the steps out of the dugout. As he exited, he ran out approximately seventy yards of electric firing cable from the vicinity of the dugout entrance onto the parched ground outside. Meanwhile, back down in the dugout, Jack started his instruction. Following the same instruction technique and routine George Holdsworth and Stan Emery witnessed him do three weeks ago, for which the Training Centre authorities agreed Jack was doing a fine job, he commenced his fifth period of instruction for the day. As always, the period of instruction was the same: *making-up, fixing, and firing single slabs of explosives.* The only difference between this lesson though, and the nine other occasions Jack completed the fifth period of instruction, three Sappers were sitting in the corner of the dugout making up detonators. They certainly weren't there three weeks ago, two weeks ago or even last week.

However, on Monday the 21st of May 1945, as a result of changes to squad allocation and men being released for duty from other squads, they were there looking in on Jack's important lesson.

Before Jack really got into his period of instruction he heard that familiar sound of the ammunition truck arriving outside. Leaving Bill momentarily in charge of the squad, Jack ventured up the steps of the dugout. As he poked his head out he observed Sheila and Tom Musto standing near their truck. ' ... Would you like to come in this time Sheila and watch what we're doing,' requested the experienced Sergeant, always keen to show others not fortunate enough to experience the exhilarating job of demolitions. ' ... I've just got some more to deliver but on my way back I'd love to come and have a look', responded an obviously excited Sheila. Jack completed the conversation, ' ... Okay I'll see you shortly.' On that notice Sheila and Tom hopped back into the truck and drove off in the direction of the Headquarters. Jack went back down into the dugout to continue the lesson. Mindful that Sheila would be joining them soon, he kept an ear out for her. He also kept an eye on the three men in the corner and after gathering his thoughts, he continued with his lesson.

Above his dugout, busy preparing for the night activity; Colin was inspecting the surrounding area for pieces of wire or shrapnel. Like a hawk looking for prey, he kept scouring the local area because he knew that even the smallest piece of metal might interfere with the electrical circuit he was preparing. If explosives were placed on them, the results would be catastrophic. After a short period of time he decided to return to the dugout to check on his three men. However, as he approached the entrance of the dugout he changed his mind. He turned with his back to the entrance and started walking away.

Meanwhile, inside the dugout, as he has done numerous times before, Jack was standing adjacent the blackboard holding a small 4 oz. plug of quarry monobel in his left hand explaining to the entire squad the correct method for safely handling the explosive compound. He continued to explain that the compound needed detonation to active and that on its own, the monobel was quite safe. ' ... Don't worry men,' he started, ' ... I wouldn't allow you to handle anything capable of doing any damage, what I've got here is perfectly safe.' Again pointing to the three Sappers in the corner, Jack explained that detonators needed to be either in direct contact, inserted into the monobel or gelignite, or be in very close proximity for it to detonate. He stressed to the squad that the detonators the three were preparing in the corner were in fact a safe distance away. In essence, he was explaining the theory of sympathetic detonation to the inexperienced demolition trainees. He didn't call it that, but he did reassure the men that everything was safe.

As Jack was explaining sympathetic detonation theory, Bill Cousins was standing over Colin Boyd, Colin Hurley and Stan Ross inspecting their work as they busily manufactured more detonators for the night activity. Less than three yards away from Jack, the four men kept quiet, even with Bill intimidatingly watching their every move.

Watching and listening to Jack, sitting in the extreme opposite southwest corner, Allan Bartlett could clearly see Bill Cousins standing over the three men inspecting their work. He watched as Bill bent over to receive a handful of completed detonators being handed to him by Stan Ross. Turning to his right to say something to Denby, Allan observed Bill counting the detonators as he was carrying them in his hands whilst he slowly walked the short distance towards the direction where Jack was standing. The Corporal was only a couple of feet away from Jack when Allan again turned to his right to cheekily say something jokingly to Denby about Bill's ability to count.

At the same time, over in town at the Engineer Bridging School in Tarcutta Street, Clarence 'Hank' Keenan was preparing for an upcoming posting to the 14 Australian Works Company in Sydney. Sitting around enjoying a smoke with a couple of mates he was more than three miles away from the camp he referred to as a '... bastard of a place.' For Hank, the Bridging School was barely far enough away for him to forget the terrible time he'd experienced at Kapooka, in particular the very period of instruction Jack was now engaged in.

A lot further away, in Melbourne, Des Surkitt was making his way back to Kapooka via the troop train. He'd finished his Sapper course at Kapooka about three weeks earlier and was sent home to Warrnambool for two week's leave prior to going to New Guinea. He'd be arriving back at Kapooka around midnight on Monday the 21st.

In Company Headquarters, Ed Merry, as the Captain and Officer Commanding the Demolitions Wing of the 1st Battalion, knowing it was one of his responsibilities to visit the demolitions area at least twice a day to check on the training, he was preparing to head to the demolitions area to do just that. After wrapping up some last minute administration, he summonsed his personal driver to bring his staff car around. A short time later, accompanied by his driver, Ed set off for the short distance to the demolitions area to check on Jack and the oversized squad. About the same time, Roy Tafe was in the nearby Demolition Store shed with fellow Sergeant Phill McNabb from the 4th RAE Training Battalion. Both men were working on the project set by Ed Dodds. The afternoon for them was quiet, but Roy was keeping an eye on his assistant Colin who wasn't that far away picking up old pieces of wire near the entrance to Jack's dugout.

For Sheila and Tom, they had also finished their tasks and were soon travelling back towards the dugout in the truck. Sheila was keen to watch what the men were doing and was looking forward to going down into the dugout for the first time. By this time they were about one minute away from arriving at Jack's dugout.

From all accounts, the afternoon of Monday the 21st of May 1945 was a beautiful sunny day and was progressing like normal. But in less than a second, the afternoon calm and normality of the camp would be shattered. The lives of everyone in and around the busy camp and township were about to change forever.

Chapter Eleven
' ... The Dugout's gone!'

'... people were running around everywhere in a panic."

SX30560 Sapper Keith Christopher Kuhn,
2/2 Australian Forestry Company

1445 hours

As Sheila and Tom approached to within 75 feet of Jack's dugout, although slightly deafened by the engine noise of her supply truck, they both felt the earth rumble and heard the unmistakable sound of an explosion. But this eruption was different to the many thousands they'd heard before. This one was eerily unusual. It possessed an unrecognisable gusto about it. Peering through the windscreen down the dirt road, towards what was supposed to be her destination, Sheila carefully navigated her truck along the dusty path. Almost simultaneous with the unfamiliar rumble, they both spied a large dust plume reach high and wide into the Riverina sky. It was coming from their intended destination. Soon it was only dust that consumed the entire view of the windshield. Immediately Sheila stopped her truck. Waiting anxiously for the dust to settle, she got out. Waving away the settling dust from their eyes, what Sheila and Tom witnessed in the spiralling dust and debris would emotionally plague them, especially the gregarious Sheila, for the remainder of their lives.

As the Riverina wind softly blew the dust aside, like peeling back an unwelcome blindfold, a scene of utter disbelief and horror was exposed. Amidst the rubble of sandbags and timber, the extent of the horror unfolded before them. It was a sight for which Sheila's robust AWAS resilience hadn't prepared her. From this point forward, her ill-equipped brain and memory would be horrifically scarred for life. She would never ever forget what she saw that day. In the settling plume, Sheila and Tom anxiously surveyed the horrific scene—a scene soldiers may have witnessed on the front line of a combat zone, but not in rural Wagga Wagga. Strewn amongst the debris, which had earlier rained down and settled in the dust around them, they could clearly see dismembered men everywhere. Some were partly clothed in khaki uniform, some were so mutilated it was hard to distinguish that it was a human body. An eerie silence descended over the scene as Sheila and Tom looked down into the remains of the dugout. Not knowing what to do next, they were frozen in fear.

After silently registering the horror in her consciousness, Sheila awakened and in the immediate distance could faintly make out two men running towards the dugout. Straight away she knew something occurred that was disastrously wrong. The pace and purposiveness at which the two figures ran towards the scene was a true indication that something unimaginable and unexpected had occurred. Sheila was numb, and Tom, regardless of the fact that he was an experienced veteran, was no better. Sheila remembered that less than an hour ago she was having lunch on the demolitions range with a group of men who'd invited her back to watch what they were doing after she'd completed another driving task. Was it the same laconic and inviting bunch of young soldiers whose mortal remains were now scattered around what used to be dugout number one? As she approached the ghastly scene, the first thing she realised was that the super-friendly and accommodating Sergeant Pomeroy was nowhere to be seen. She knew Jack well from the numerous times she'd delivered explosives to him, and it was Jack who'd invited her to come back and watch the demolitions lesson. But where was he? Surveying the unbelievable scenes of death and destruction looking for signs of Jack, and any of the other boys, a brief but frightfully sobering thought caught up with her. In her shocked unaware state she remembered, '... I was meant to be in there.' Sheila then broke down. The long comforting arm of Tom, a man who knew her hard exterior better than most, reached around her shoulders to let her know that he was there, and it was okay to cry.

The first to feel the impact of what was considered to be a supersonic explosion was Colin Kendall. Seven yards away from the entrance to Jack's dugout, the Lance Sergeant had chosen not to re-enter. Instead, thinking about the night activity, he knew he needed some scrap pieces of wire to tie the charges together. Earlier he'd seen some suitable pieces of wire scattered about and decided to forage around near the entrance to the dugout before going back in. Standing close to his parcel of explosives, he was scanning the ground for the elusive wire when, without warning, he heard an explosion and felt a great intense heat engulf his face. Singeing his hair, eyebrows and eyelashes, in a daze he instinctively turned to his own stack of explosives. With his knees buckling beneath him and rapidly collapsing into shock, he spied his own heap still intact.

Falling to his knees, in what he later recalled felt like slow motion, he was soon face down into the dry hot Riverina soil. In the settling dust and debris, his vision was interrupted. Charred limbs were falling from the sky and landing around him. When a dismembered arm landed close to his head on his left side, only then the recently married man began to register the horrible reality of what had occurred. Lying motionless he was conscious, but only just. As the hot Wagga Wagga sun penetrated the dust cloud and beat down on his scorched and painful face resting on the hot earth, his thoughts immediately turned to his wife and family. Mustering any remaining strength not already paralysed by the blast, he craned his neck to the side. Trying to orient himself and gather any sense of reality about what was happening, he was able to look for the origin of the blast. Facing south toward what was the entrance of the dugout, he could just make out the roof of Jack's dugout had caved in on the one side. But deep shock wasn't far away.

As Colin collapsed to the ground, about 150 yards away Roy Tafe was in the Demolition Store shed with 44-year-old Sergeant Phill McNabb. Both men heard a thunderous explosion and felt the earth tremble beneath them. The experienced demolitions men knew something was wrong. The explosion sounded like a crater charge, but it was abnormal for the work being carried out in the demolitions area that day. Phill McNabb went to the doorway of the shed. With fear in his voice and horror etched across his face he turned to Roy, '… the dugout's gone!' Quickly racing to the doorway, Roy looked towards the vicinity of Jack's dugout. What he observed upset him. The largest cloud of smoke and dust he'd ever seen on the demolitions range alerted the two experienced demolitions experts that something catastrophic had occurred in Jack's lesson. They needed to get there quickly. Silently, deep down both men knew what had happened; but they held a faint glimmer of hope that perhaps they were wrong. Hopefully it wasn't as bad as it appeared. Instinctively, the two men raced toward and into the settling dust and debris.

Sitting over in Company Headquarters, Ed Dodds, along with every other life form in the general and wider reaching vicinity, heard the explosion or saw the billowing aftermath. It was the first day of the fourth week of training for the trainee Sappers. As it was Monday afternoon, about 1445hours, that meant his instructors would be well and truly busy instructing groups of trainees on how to make up and fire small 'slabs' of explosives. Knowing very well that they weren't large-scale detonations, he also knew something of a catastrophe had occurred. On foot, he scarpered out of the office and raced as quickly as he could to the demolitions area.

Back on the range, in James Conwell's adjacent dugout, the roof and walls shook violently showering the men inside with a thick coat of dust. James also knew something was horribly wrong. The nature of the explosion had a signature of obvious disaster written across it. He knew the training syllabus well. Therefore he knew what stage Jack would've been at in his fifth period of instruction when the explosion rang out. He also knew that his fellow instructor had an oversized squad in the adjacent dugout. Wisely, James ordered his trainee squad to stay put as he cautiously ventured above ground. Taking the seven or so steps up out of the dugout, James immediately identified a burnt Colin Kendall lying motionless on the ground. In the distance, he could see Roy Tafe and Phill McNabb sprinting towards the destroyed dugout. James lept from his dugout.

Roy arrived first and instantly identified his injured Lance Sergeant lying motionless on the northern side of what used to be the entrance to the dugout. Roy could see that Colin's clothes were still smoldering and he knelt down beside his mate he could see that his face was badly burnt and he was in immense pain. Just as Phill McNabb was crossing the roadway, he looked up to see Captain Merry's car speeding towards the area.

About 800 yards from the demolition area, Ed Merry heard the explosion and saw a column of dust fly up into the air. Over the last couple of days the Battalion had been conducting experiments in the size of artificial caverns (camouflets) using crater charges; he suspected that this explosion was just a crater charge. But something quickly changed his mind. He witnessed Sergeant McNabb race onto the road in an obvious hurry. Sensing something drastic had occurred, the Captain hastened his driver, ' … hurry up, it looks as if something is wrong.' Speeding up he quickly met up with the Sergeant. ' … Get the doctors and ambulances—the dugout's just blown up.' They were the worried words of an experienced demolitions engineer who knew a volatile mix of high explosives and young trainees were in the dugout. In the confusion Sheila Oehm was also ordered by one or all of the men racing to the dugout to return to the hospital, ' ... get as many ambulances as possible.'

Meanwhile, Roy Tafe, and now James Conwell, were tending to Colin's injuries. Observing his scorched hair and smoldering uniform, Roy could see no visible signs of serious injuries such as missing limbs or large shrapnel wounds. He began talking to Colin, in order to gauge his condition, ' ... you alright mate?' Colin unselfishly replied to Roy and James that, although he needed a drink of water, ' ... I'm okay fellas; you'd better look after the others.'

Roy knew many of the men didn't stand a chance. But it didn't stop the urgency with which he needed to get help if there were to be survivors. Captain Merry quickly turned his car around and headed straight for 54th Camp Hospital. Upon arrival he went straight to the casualty section where he ran into 38-year-old Medical Officer Captain Lemual 'Henry' Judd. Explaining to the medical officer that there had been an accident on the demolitions range, the gravity of the situation was evident in Captain Merry's explanation, ' ... you're going to need a few ambulances!' A rescue mission needed to be coordinated immediately.

Unfortunately the Senior Medical Officer and Officer Commanding 54th Camp Hospital, Major Les Tunley was absent from Kapooka camp that afternoon. It was therefore left up to Captain Judd to assume command of the initial medical response and rescue. Over the public address system, which could be heard throughout the entire camp, the shocked Captain advised listeners that there was a terrible accident on the demolitions area. With urgency in his voice he declared, ' ... all available ambulances and doctors, there's been an accident at the demolitions area, report immediately to casualty.' The unusual call for 'all available ambulances' signaled to the entire camp audience that this was a very serious incident. The message continued,

'… Multiple casualties expected.' All available female ambulance drivers and female medical orderlies on duty at 54th Camp Hospital jumped into action.

One person who heard the crackled announcement over the camp PA system was Paddy Cranswick. Paddy was on his way back to G Company when he heard the explosion and turned to see a swirling cloud of smoke above the area he thought was the demolitions range. Straight away he knew something was wrong. '… That's a bit unusual that a large detonation was taking place today,' but when he heard the crackled voice of the Medical Captain over the PA system calling all available medical staff, it instantly dawned on him what had occurred. He thought to himself, '… I wonder if Geoff Partridge went back down there?' '… I hope not, hopefully he decided to have a lie down instead.'

Before leaving 54th Camp Hospital, Captain Merry found the nearest phone. He was trying to call Major Berg, the Assistant Chief Instructor, or Major Macdonnell the Acting Chief Instructor, to dutifully advise his chain of command of the accident. Finally, after a couple of attempts, he reached Major Macdonnell. Devoid of any real information, he informed him of the scant details he knew about the accident. Leaving his chain of command hanging on the phone for more details, the distraught Captain raced to the scene. Immediately after being notified of the situation by Ed Merry, George Macdonnell, the experienced Field Company Engineering Officer of five years, advised his right hand man, Maurice Berg. Together the two experienced men sped off to the scene.

Hank Keenan and his mates at the Engineer Bridging School heard the explosion and felt the earth rumble beneath them even though they were more than nine miles away in Wagga Wagga. Turning towards the direction of the camp, seeing the cloud of dust rise from the Kapooka area, he knew very well it was coming from the vicinity of the demolitions area. A nauseous feeling dropped deep into the pit of Hank's stomach. He had a fair idea of what had just happened.

In another area of the camp, Allan McPaul and his B Coy men were returning to camp when they heard only part of an announcement over the PA system. The announcement that ' ...all afternoon and evening activities had been cancelled.' spelt celebration. But they didn't hear the initial announcement about an explosion. Not knowing the reasons why training had been called off the men let out a raucous cheer. However, when a subsequent announcement was posted that a number of men had been killed on the demolitions range, the joy and cheering turned deathly silent. Deep in thought and sorrow they hoped it wasn't their mates or anyone they knew.

Meanwhile, although residents of Wagga Wagga were treated almost nightly to constant distant flashes coming from Kapooka, it'd become commonplace and a way of life in a garrison town. Often resembling lightning and crashing thunder, the night activity of the Sappers became routine; so routine in fact that it no longer attracted outside interest from residents. But this daytime explosion had the markings of an unwanted earthquake. Shaking the foundations of the Riverina town, this explosion was bigger than anything Kapooka had previously served up and therefore nothing the residents of Wagga wagga had experienced before. When the force of the explosion cracked their plaster walls and shook the dust off light fittings, suddenly, the normally disinterested residents were awakened. Something was wrong out at Kapooka. Whatever it was, it was a blast so powerful that the entire town was on edge.

In Beckwith Street, Dorothy Pomeroy was sweeping the verandah of the small modest Hickson residence when the explosion rang out. She'd recently became frightened of the constant explosions, knowing full well her husband was in danger every day he went out there. When this tremor shook the rail of the verandah, she gripped the broom handle tight in her dusty hands and said a little prayer for her husband. ' ... Lord, keep him safe...not my Jack.'

Chapter Twelve
Faint Hope!

Roy Tafe stood cautiously on the edge of the destroyed number one dugout. Before reacting, he found himself temporarily motionless—mesmerised. His eyes were open, but they weren't necessarily registering what he was looking at. Partial bodies of men and lumps of flesh roughly resembling human tissue, now littered the floor and surrounds. Men still seated on ammunition boxes, although covered in debris were motionless. Surveying the horrific scene, in utter disbelief he found it difficult to take those first tentative steps into what would become his private hell. Only minutes earlier, the area he was entering protected 27 men, all of whom were perhaps focusing attentively on the lone Sergeant instructing them at the front of the dugout. Now, it was eerily silent, his good friend was nowhere to be seen. Only rubble remained from what was once a protective dugout.

At first, his eyes scanned the area seemingly taking in the magnitude. Examining the absolute carnage of rubble and bloodied figures of what he suspected were the men he once knew, including a couple he called mates, the overpowering acrid stench of death, dust and burning flesh overwhelmed his struggling nostrils. Rubbing his temples, he was trying to focus. What do I do first? Steeling himself for the reality of what his eyes were registering, with a rush of adrenalin, he cautiously stepped in. Without any rational process, now with a real sense of urgency, he started sifting through the gruesome and mutilated remains of its former occupants. Lifting twisted timber and mesh, he was desperately searching in faint hope for survivors. With every step, his optimism was shattered. When the adrenalin running through his body finally settled, and he could no longer hear the thumping of his own heartbeat in his ears, his overworked heart felt achingly heavy and cold. He couldn't see Jack or Bill Cousins anywhere. ' ... Where the bloody hell are they?' he thought to himself; ' ... Jack?' he repeatedly mumbled to himself precariously stepping through the human mire, ' ... where are you mate?'

Negotiating the debris, he glanced upon a small number of men eerily looking directly at him from what would have been the western wall of the dugout. Faint hope he thought, ' ... can they hear me?' Even though they were motionless and their fixed gaze had a sense of macabre uneasiness, they appeared to have all body parts intact with no visible signs of trauma. Were they alive? Hopes were refreshed. Quickly moving towards them, Roy started shouting hoping for any sign of response. Soon he was standing before them. Reaching out he shook one of the men's shoulders—nothing. No movement, no reaction, they simply stared right through him. Their empty eyes were still fixed at what were only moments earlier, the front entrance to the dugout. Consumed by a growing grief, Roy began resigning himself to the fact that no man could've survived this horror, not even his larger than life mate Jack Pomeroy. Even though he was beginning to accept the horrific inevitable, he had to keep searching. He had to remain hopeful that, by some miracle, someone had survived.

Continuing his now almost cursory search along the southern wall, he noticed about seven or eight young men sitting slightly upright with their arms folded in the front. They too had no visible signs of injury, ' ... perhaps one of them is alive', he thought to himself. Some of the men had their eyes open, some were closed, and with no obvious physical injuries they looked more like old men sitting on a park bench. Eerily all covered in Riverina dust, they appeared to have been quietly minding their own business when a terrific shock wave immediately fixed them to where they sat. As he approached, once again desperately hoping to find signs of life, they were like ancient stone-faced warrior statues. Not one of them moved. Not even a pulse or faint heartbeat.

Losing all hope, Roy went from frenetic searching to a more controlled approach. Trying to be careful whilst lifting timber and rubble, exposing more grisly disfigured remnants of strong young men, in the far southwestern corner, Roy detected movement. Turning towards the disturbance, he spied one man who appeared to be awakening from a frightful nightmare. The young looking man had an indescribable fear and panic in his eyes. Desperately trying to recover rapidly from an episode of unconsciousness, the young man's scared gaze appeared transfixed on Roy. With his wide eyes indicating a deepening panic, an obvious frantic distressing scream was trying to escape his traumatized lungs via his half-open mouth. The scream and cry for help would have cleared his airways—but on this occasion, nothing came out. His was a painful silence.

Continually gasping for breath, clearly the man was alive but in great distress. His paralysed diaphragm struggled for a lungful of clean air. Suddenly a haunting and most unforgettable scream of horror broke the deathly silence of the eerie dugout. In a half-seated, half-upright position, he was leaning backwards into the wall behind him. Something had forcefully picked him up from his seat and at the same time punched his entire youthful body rearwards and implanted into the earthen wall behind him. Thankfully the youngish-looking man was mildly responsive to Roy's initial questions. He appeared to Roy as if he was trying to stand up as rubble fell around him. Unspeakable fear was heavily glazed in his unforgettably empty stare. Hastily, but with the utmost care trying not to touch the young man, who had obvious burns, Roy started scraping the earthen wall away from the young man's body. All he wanted to do was free him from the mysterious clutch the wall had placed on him. But its earthen grip was tight. Constantly reassuring him that he was going to be alright, Roy carefully started freeing the distressed man. Soon Ed Dodds was in the dugout doing what he could to help. He quickly went to Roy's aid. Slowly the pair liberated the young man. Still gasping for breath, his mouth and nose were full of mucous and thick dirt had packed tightly into his nostrils.

The two rescuers finally pulled the numb and limp body up the side of the dugout wall. Once extracted, they laid him on the rubble making sure he could no longer see the gruesome scene from which he came. An overwhelming look of boyish fear was still clearly visible in his obviously burnt and traumatised face. Unaware of his surroundings and what was happening, the young man instinctively tried to sit up. Fearing internal injuries, one of the orderlies from the first arriving ambulances kept pushing him down. Suddenly, a mysterious female with a calming tone could be heard saying, '…let him sit up if he wants…let him sit.' Even though he was barely able to speak, the medical orderlies cleared his airways and gave him an injection for the obvious pain. They left the deeply-shocked young man in a seated position atop the rubble in the care of medical staff. With a glimmer of hope the two rescuers, Roy Tafe and Ed Merry, re-entered the distressing dugout in the hope of finding more survivors. By then word had got out around the camp that they'd found a survivor. Ed Dodds had earlier raced off on foot to the scene and upon his arrival he was shocked at what he saw. He could recognise some of the men from his Company but a pile of rubble now rested where one of the dugouts once was.

In a third ambulance, now racing to the demolitions area, a pensive Henry Judd sat amongst nervous and visually upset medical orderlies.

Anxious, they were almost there. He knew that as soon as the ambulance doors flung open all the medical training in the world wouldn't prepare him or his inexperienced orderlies for the carnage they'd have to triage. Little did Henry know the majority of the victims were beyond his medical help. For him it was going to be about recovery of corpses and collecting the parts of bodies that littered the gruesome scene. Amongst the small pieces of apples and oranges, the very fruit the men had retrieved from the canteen for lunch that day, fingers, hands, toes, arms, legs, torsos, bone fragments and clumps of skin and bone, some with hair attached, also littered the earth. Still surveying the scene and doing what she could, Sheila Oehm identified one of the men who'd she'd earlier shared lunch with seated atop the rubble. She quietly whispered to herself whilst surveying the indelible horror, ' ... that's one, perhaps they'll find someone else alive.'

About 1455 hours, ten short minutes after the explosion, the ambulances arrived at the dugout with Captain Judd aboard. Exiting the third ambulance with his medical kit he noticed an obviously injured young man sitting on the edge of the rubble; he also observed that Ed Dodds and Roy Tafe were moving amongst the dugout desperately looking for survivors. He immediately moved towards the injured man. Conducting a cursory examination of the young Sapper wearing a burnt uniform and covered in dust, he observed bleeding from the left side of his head. Captain Judd determined that his patient didn't have life-threatening injuries. He did however note the young man was burnt on his face arms and legs, he had a contusion on his back and the doctor wasn't to know at the time, his victim also had two ruptured eardrums. After treating his first patient, the Captain in charge of the medical response jumped into the dugout in the desperate search for more survivors. It was a struggle to find any signs of life, especially in the men whose bodies were still intact and weren't scattered amongst the obvious deceased.

Shortly after Captain Judd's arrival, Ed Merry, the man who raised the initial alarm at the hospital and informed his chain of command, arrived back at the scene. Upon his arrival he noticed that the dugout roof had been totally demolished. A portion of the roof made from sandbags, timber and wire mesh, had been blown to an area outside the dugout whilst a quantity of the roofing material had also collapsed inwards. What he saw next set his mind thinking about what he believed had occurred. It also prompted him to respond. Ed Merry saw the bodies of men he knew from the mornings parade laying motionless in the bottom of the once dugout. He could also see that one young man had been assisted from the rubble and was now being attended to. He knew he needed more medical help than just that of the capable but overwhelmed, Captain Judd. With his driver he jumped back into his car and sped back to the hospital. Upon his arrival he rang the nearby 1st and 4th Engineer Training Battalion Medical Officers. He summoned them immediately to the demolitions area.

The authorities made a number of wise calls during the mayhem of that dark day. The wisest was perhaps the call to ensure that all remaining trainees were to be immediately evacuated from the demolitions area. Only men who've been in combat were authorised to assist with what was known at the time as a 'Kangaroo Hunt' of the scene. The so-called kangaroo hunt was an excruciatingly absurd request, but one which was an unquestionable necessity. A dozen or so experienced combat veterans lined up in a straight line, about one arm's length apart, then, armed with bags, slowly walked through the area scanning the ground around them picking up every piece of flesh and bone they could see. Meticulous and every bit respectful to the victims, in the end, the bags containing bloodied and dirtied unidentified human remains were loaded onto ambulance stretchers or placed on the ground near the ambulances. The kangaroo hunt was all about the methodical, but somehow dignified, recovery of human remains

Meanwhile the Regimental Medical Officer from the 1st Battalion, Captain Victor Bennett, a relatively inexperienced 25-year-old from New South Wales, received a call from Captain Merry to assist Captain Judd.

Accompanied by his Sergeant, the two men leaped into the staff car and straightway drove to the demolitions area. At the 4th Battalion Headquarters, Captain George Miller, the Battalion's 28-year-old Regimental Medical Officer received the call from Ed Merry. The sickening news prompted him to react with the utmost speed. Grabbing the nearest truck he headed straight to the demolitions area.

Like wildfire, the medical fraternity around the camp was advised via telephone or the public address system. Soon the collective medical experience at Kapooka, except Major Tunley, was all gathered at the scene. Even though the camp went into communication blackout, word of the explosion and multiple casualties got out of the camp. At No. 1 RAAF Hospital, near the Showground in Wagga Wagga, Major Tunley was just arriving for a meeting. As he entered the hospital grounds a member of the hospital staff greeted his staff car and informed him of the terrible news. Immediately, the 47-year-old Queensland-born New South Welshman turned his car around and sped directly to the demolitions area.

Arriving at the scene with his Sergeant, the 1st Battalion MO, Captain Bennett, advised his subordinate to stay with the vehicle whilst he proceeded into the dugout. Prior to entering, Captain Bennett watched Roy Tafe and Ed Dodds keep up their valiant, yet fruitless, search for more survivors. A couple of identity discs, no longer hanging around the necks of their owners, could be seen in the bottom of the dugout. Roy bent over to pick them up. Wiping off the blood-moistened dirt on his sleeve, he read the names embossed on the disc. He placed the discs back on the ground in the vicinity of the closest human remains. It wasn't the most scientific form of identification, but what else could he do? The sheer volume of human remains scattered throughout a small area made it almost impossible to accurately identify the body with the name on the identity discs. During the frantic rescue the recovery team managed to salvage all but two sets of identity discs. Some discs were discovered actually attached to the victims following the explosion, but for those who unfortunately felt the full impact of the supersonic detonation, their discs were blown from their torsos and into the general vicinity of where their remains scattered.

Then, just as the rescuers were losing all hope, in the southeastern corner Roy found faint signs of life in an older-looking man. Suffering from obvious serious abdominal injuries, Roy noticed his intestines were protruding through his tattered and burnt clothing. Sitting in a pool of blood and dirt, only very faint signs of life were present. Roy signalled to Captain Judd that he'd found a man alive. They needed to move him out as soon as possible. It was obvious that the man was significantly burnt and barely alive.

Captain Judd quickly examined the man. Noting his critical injuries, he immediately authorised a stretcher be brought into the dugout. He knew the man was only just clinging to life. With the slightest, but careful touches of his rescuers, his charred skin peeled away from those parts of his body exposed to the tremendous heat and gas as he was placed on the stretcher. Accompanied by an orderly, Roy Tafe carried the dying man out of the dugout and into the waiting ambulance. As they slid the stretcher into the rear of the ambulance they all knew it would take a miracle if he were to survive.

At the same time, others also scouring the dugout alerted Captain Judd that the man seated beside the stomach wound victim also showed very faint signs of life. Shouting incessantly in the panic for another stretcher, the Captain watched as this man, also barely clinging to life with horrific burns and life threatening injuries to his limbs and face, was quickly extracted to the waiting ambulance. Henry Judd was left frantically searching for signs of life in all the remaining bodies. Wretchedly, none of the young men were responsive. With time to ponder just exactly where he was and what he was doing, soon came to the realisation that he was standing amongst the remains of more than two-dozen young men. '... How the hell did this happen?'

The young Sapper found embedded against the dugout wall had been attended to by a medical orderly and was finally being loaded into an ambulance. As he was lifted into the darkened ambulance the dazed and confused young man mumbled his name to the orderlies, '... I'm Allan Bartlett.' The South Australian somehow not only miraculously survived being ripped to shreds in a violent explosion; he was only suffering what appeared to be minor non-life threatening injuries. His survival confused the mass of support staff now busily extracting dead bodies. By now they'd given up all remaining hope of finding any more survivors. But with Allan's phenomenal survival they were hoping, praying, for just one or two more miracles. Sadly, it didn't happen. Along with Colin Kendall and the two unidentified men extracted barely clinging to life, the four survivors were quickly rushed to the Camp Hospital.

By 1500 hours that afternoon the demolitions area, specifically the general area of No 1 dugout, was swarming with medical officers and medical orderlies. In addition, 1st Battalion Headquarters staff and now staff from the RAE Training Centre Headquarters were beginning to arrive to survey the grisly scenes. Medical orderlies were still busily removing full and partial bodies. Bags of identified human remains and dismembered bodies were being placed on the ground adjacent to the ambulances before being loaded onto stretchers for removal back to 54th Camp Hospital for formal identification.

In a gust of dust and squealing brakes, Captain Miller, the 4th Battalion Regimental Medical Officer, arrived in his truck. He noticed Captains Judd and Bennett were already hard at work within the dugout. Dodging the flurry of stretcher-bearers carrying tormented torsos and charred human remains in bags, he walked closer to the dugout. He could clearly see more bodies and several others, which had already been removed, were being placed onto the ground outside the dugout. He made a professional mental note, '... These men are clearly victims of a very powerful explosion.' The obvious trauma to their bodies confirmed initial reports he'd received of an unexpected detonation. As he registered the magnitude before him, Captain Judd approached him. The medical officer in charge of the horrendous scene advised him that there was nothing further he could do there. With the horrific scene embedded into his memory he returned to 54th Camp Hospital highly emotional but ready to attend the injured.

In between Captain Merry advising Major Macdonnell of the accident and the latter arriving at the scene, the Chief Instructor also had time to advise 41-year-old Captain Fernleigh Ellis. Fernleigh was the Captain at the RAE Training Centre Headquarters appointed as the Acting Deputy Assistant Adjutant and Quartermaster General (A/DAA/QMG). In essence, Captain Ellis was the Training Centre's Administrative Officer. With his appointment came responsibility. As the coordinator, it was Fernleigh who was responsible for the administration of the aftermath of the greatest tragedy the camp, and country, had witnessed in its short history. He was notified, as the formation headquarters liaison, to coordinate any external support such as police or civilian ambulances. In turn he got a message to his absent Brigadier. Waiting for his transport to arrive it would be some time until the Brigadier was able to personally survey the scene.

Shortly after receiving the call Captain Ellis also sped to the scene in his staff car. Upon his arrival, seeing the carnage and death for himself first-hand he sought out the 1st Battalion Adjutant, Captain Archie Smith, who by now was also on the scene. Fernleigh Ellis instructed Archie Smith to conduct a formal identification of the bodies as soon as possible and that he be informed immediately the gruesome task was complete. Captain Ellis had to report the tragedy to the formation headquarters at Second Australian Army and he needed all the details quickly. In particular, he was reporting to Brigadier Claude Esdaile Prior, the 50-year-old Deputy Adjutant and Quartermaster

General (DAQMG). Amongst the pandemonium of recovery operations, Archie Smith found Ed Merry on the end of a stretcher. He was assisting the orderlies with the removal of yet another body. It was the most horrific of tasks, especially for the female drivers who went about their business just like the men. At times, Ed Merry, who was ably assisted by orderlies and female ambulance drivers with bags, were just scouring the dugout recovering what parts they could. After a brief conversation both men acknowledged that the unnerving and in some cases impossible task of identification of human remains needed to take place as soon as possible. Archie Smith and Ed Merry knew that identification was going to be extremely difficult and in some cases almost impossible.

To aid in the identification, Ed Merry set Corporal Bob Challenor, the Clerk for the demolitions area, an immediate task. He was to go and retrieve the squad roll books in order to crosscheck on the bodies that were present. At the same time he instructed Sergeant McNabb to quickly check the names of the four men about to leave for the hospital in the first ambulance. Phill McNabb already knew Colin Kendall, and he ascertained that the young man who was extracted alive was Allan Bartlett he doubled checked by inspecting his identity disc. Observing that the two seriously injured were still wearing their identity discs he identified them as Ivan Merritt and Bill Reid. They were the two mature trainees who'd first entered the dugout for Jack's fifth period of instruction after lunch and who took up a seat together on the eastern side of the dugout.

By this time Major Tunley had arrived at the scene having turned around at the RAAF Hospital after getting news of the tragedy from Kapooka. He inspected the scene and observed that numerous dead bodies were still being extracted from the dugout by the valiant orderlies, ambulance drivers and staff. He approached Captain Judd who informed him that four men had survived, two were however critical and that the injured had been identified and were now on their way to 54th Camp Hospital. With that news and realising there was nothing he could do at the scene, the Major proceeded immediately to his Camp Hospital to assist.

Ed Merry continued to supervise the recovery of bodies and human remains from the dugout. Unlike modern-day crime scenes that get saturated with photographs of victims where they died in order to capture the grisly scene for subsequent investigations, understandably *no one* had thought to bring a camera. It's understood that no photographs were taken; no sketches were made by anyone who attended the scene. It was all about recovery. It was plain to everyone present and even those attending the scene after the bodies were recovered that an investigation to establish the causes of death was more than likely not needed. Considering the mutilated and dismembered state of the bodies, it's perhaps a blessing that no photos were taken.

Satisfied that all bodies and human remains had been removed from the dugout and surrounding area, Ed Merry conducted a check of the roll. At this early stage it wasn't about names but just numbers. Like piecing together an unimaginable puzzle, the question facing Ed Merry was not so much *who* was recovered but *how many*? Names and identities would be sorted out later, at the moment it was about accounting for all of his men. Ed was however in a position a little later to check the roll against names and supply a list of names of those more than likely killed, but as some of those names were now lifeless and unidentified carcasses and some men still missing, he decided to leave the formal identification to Archie Smith. From the roll books Ed established that a total of 28 men from G Coy had been either killed or injured. Checking the figure with Archie Smith, both men confirmed the figure of 28 was accurate. Fernleigh Ellis was initially advised shortly thereafter that 24 persons had been killed outright; four were seriously injured; two of whom were critical. This was the initial message Captain Ellis transmitted to his Headquarters.

Meanwhile, Adjutant Archie Smith directed the bodies and human remains be conveyed to 54th Camp Hospital where Major Tunley would be waiting. As the remains were being transported to the hospital in the fleet of ambulances, Major Tunley was already at the hospital organizing the response.

With news of the total devastation down on the demolitions area, the camp and town fell into an eerie stillness. Back at the dugout, Roy Tafe assisted Ed Dodds and a host of volunteers to sift through the remains of the twisted wire and splintered timber. With all of the bodies removed, they were now looking for valuable items of evidence. Tirelessly, they inspected every inch of the dugout looking for any trace or indication to explain what happened. The experience of Roy Tafe and Ed Dodds quickly established that *all* of Jack's 110 pounds of explosives stores were gone. So the two men went about trying to establish the origins of the explosion. Familiar with the layout of the dugout before the explosion, both men observed that the collapsed nature suggested the huge explosion originated from the known spot where the instructor's positioned themselves to best instruct trainees in the northwest corner—the same position where Jack should've been during his first lesson after lunch. Foraging around in the vicinity of where the Sergeant instructors normally positioned themselves, they found evidence of an enormous blast scattered amongst the rubble. Pieces of burnt paper, uniforms, splintered timber, mangled wire mesh and shrapnel from objects such as the instructors blackboard, were all visible in the debris. Clearly, whilst sifting through the mess, everyone knew that an explosion of amazing power had caused the enormous death toll and total obliteration of the dugout—but what the whole camp, the small New South Wales town and the entire country would soon want to know was, how did it happen and out of the 8000 or so men who occupied the camp at the time, just who were the victims?

But in the wake of the devastating explosion and the solemn search by the rescuers to find out who the casualties were, there was one man the entire camp was looking for—Jack Pomeroy. Over in G Coy Jack was the soldier's soldier. He was the type of SNCO who earned the trust and respect of subordinates and superiors through his professionalism, dedication, brave actions, instructional ability and importantly his patience in teaching young soldiers the deadly art of demolitions. Young soldiers gravitated towards his calmness and in doing so he became the trusted father figure to many young and mature soldiers alike. Aspiring soldiers wanted to be like him and more importantly superior officers wanted ten more of him. But this was Jack's dugout and it was now spread out over the barren demolitions range. G Coy staff and soldiers knew it was Jack's 31st birthday; it then became especially important to find him. Everyone, especially Sheila, Ed Merry, Ed Dodds and Roy Tafe wanted to know—what happened to Jack?

Chapter Thirteen
A Makeshift Morgue

At the hospital, the unenviable task of formally identifying the deceased men fell upon one man—Captain Archibald William Leslie Smith of the 1st RAE Trg Bn. Archie was a married man from Pascoe Vale in Victoria. He first joined the 26th Light Horse Machine Gun Militia Regiment in 1941, almost five years earlier. He was sent to the RAE Training Centre on one-month probation in November of 1944 to determine if he was suitable for the position as Adjutant. There, convincing his superior officers he was more than competent, he got the job and was subsequently posted to Wagga Wagga in December of 1944. Now, a mere five months later, he was tasked with the hardest job for any soldier or officer—identifying fallen comrades and subordinates. It wasn't a job he necessarily volunteered for when he transferred to the 2nd AIF in July of 1942, but he was the designated officer that day. As the Adjutant—the staff officer assisting the Commanding Officer with the administration of the 1st Battalion, regrettably, it was his solemn duty. When the 30-year-old officer presented for work that morning, he certainly didn't expect to be the man solely responsible for identifying the human remains of men he hardly knew. Not only that, an entire Army and country was relying on him to get it right. With limited experience dealing with heavily mutilated casualties, it was always going to test his resilience, courage and strength.

As Archie made his way to the hospital, the fleet of ambulances that had raced from the scene were now arriving at 54th Camp Hospital. Carrying their distressing cargo, they made a more discreet entrance at the rear of the hospital. Meeting the ambulances, teams of orderlies and other drivers carefully carried the remains of the victims into the small hospital. Major Tunley had already organised the remains to be placed in an unoccupied ward for his inspection. The ward, which normally catered for trainees with the flu and minor illness and injuries, had been transformed into a large makeshift morgue. It was Major Tunley's job to sift through the bags and recovered masses of human flesh, now sitting atop dust-covered stretchers in the unused ward, and prepare them for Archie's identification inspection.

Archie couldn't help but worry about the fact that the mess he was about to confront was all that was left of some of the young men who, only an hour earlier, were promising young soldiers and two of his experienced instructors. Nevertheless, it was his job to sort through the horror. He had to officially put names to the mess and, to do so, he needed to find an unusual strength within himself. Arriving at the hospital, finding the necessary courage deep within his heart, he proceeded to the ward with a plan. It was as simple as it was crude. Inspect a body or mass of flesh, find an identity disc, match it up with the correct corresponding body or human remains, and then record the details of identification.

As he entered the ghastly smelling ward the reality and enormity of his task instantly became apparent. On bed, after bed, after bed, rested the smoldering, dust-covered, charred bodies of men from his battalion. They were waiting to be formally acknowledged by him. Initially, amongst the hundreds of pounds of human flesh, little identification evidence could be seen. For those that were unrecognizable as human forms, Archie would have to sift his way through the mountains of flesh and bone to identify a lifetime amongst the lifeless carnage. Some had distinguishable features representing faces. Some had their faces and heads violently blown from their bodies and for many, only stumps existed where arms and legs used to be. Although some of the men bore remnants of their distinctive uniform, some had their uniform and undergarments blown from their bodies.

According to the roll books, adjusted throughout the day by Ed Dodds and Roy Tafe after men were coming and going from the demolitions area, 28 men were recorded as being in or around the dugout. Of those suspected 28 men, Archie thanked God that four were alive. Although Allan Bartlett and Colin Kendall were both suffering from minor non-life threatening injuries, miraculously they were expected to survive. However, the two men extracted alive, South Australians, Ivan Merritt from Crystal Brook and his mate Bill Reid, from Port Augusta, they were unfortunately desperately clinging to life. The powerful detonation had ripped huge holes through their torsos and burnt their flesh down to the bone. Now in a painful wretched finale, their innocent lives were teetering. They were unresponsive to the frenzied efforts of medical treatment by panicky hospital staff; it appeared as if their traumatised bodies were submitting and shutting down. It was understandable. After suffering significant and unmentionable trauma, the spark of their human spirit was rapidly fading. The two men, who'd found themselves seated together in the ill-fated dugout after lunch, were now silently slipping away together.

On an adjoining bed in the casualty and emergency ward, 18-year-old Allan Bartlett, although still in deep shock, was responsive to the frenzy of medical treatment. And whilst he and Colin Kendall escaped the tragedy with only minor physical injuries, both men had undiagnosed significant psychological trauma.

Archie's task therefore was to account for the remaining 24 dead, injured or missing men. On the initial face of it, his task seemed almost impossible. Archie felt the men had suffered a wretched and pitiable death and perhaps should be laid to rest as soon as possible, but he knew accurate records were needed and, importantly, their families notified quickly.

As a Camp Hospital, nursing at Kapooka was a lot lighter than the General Hospitals. No surgical procedures were carried out and the majority of admitted patients were simply trainees suffering from malaria or a sore throat. However, the dangerous nature of the training program did see other injuries such as the case of someone like Toddy Woods' concussion or the nasty case of Kevin Pierce's tinea. But on this particular day, the normally quiet Camp Hospital experienced the ballistic horrors of war first-hand. It was an inoculation into the depths of hell for some medical staff, and fittingly, it was more than an appropriate decision to shield young orderlies from such horror. Archie wasn't going to do it alone. That afternoon he was attended by a number of nurses and one orderly in her late 30s. The younger hospital orderlies weren't permitted to enter the ward that had been converted into a makeshift morgue. Instead, they were sent to other areas of the hospital to keep the patients occupied so as not to dwell on what had just occurred.

One of those more experienced nurses who was to be exposed to the horror that fateful day was Maida Doubleday. She had only recently announced her engagement to a soldier, Jack Bate, a Corporal who'd returned to Wagga Wagga after four years in New Guinea. So for Maida, her Monday started in high spirits, but her afternoon turned to hell. Her fiancé was at the engineer-bridging site in town when the explosion occurred, and now, with a communication blackout enforced by the authorities, no calls were made in or out of the camp. For Maida and Jack and anyone else who had loved ones at the camp, they weren't able to communicate with each other to see if they were okay. For thousands of expectant family and friends waiting for any type of news to escape from the now deathly quiet camp, it was an agonizing wait.

The time was about 1520 hours, less than an hour since the explosion woke the camp from its Monday afternoon post-lunch slumber, when Archie Smith, the former Victorian storekeeper-turned soldier, tentatively entered the quiet hospital ward to commence his duty. Presented before him was a far cry from the young enthusiastic bunch of young soldiers he welcomed to the camp less than a month ago. Emotionally stunned and momentarily fixated on the overwhelming scene before him, he couldn't help but shake his head in disbelief. The scale and waste of such promising lives was overwhelming. A nervous Archie was shaking from a combination of fear and his own trauma. A brief moment of inner prayer, he quickly gathered his emotions. After all, it was the war years and scenes like this should have been expected. Certainly the scene may have been more appropriate on any of the world's battlefields and easier to accept, but this was a regional New South Wales hospital, and everyone working the makeshift morgue that May afternoon didn't expect this many bodies all at once and in such a state of dismemberment.

Disfigured, decapitated and limbless. What were once strong young men, now laying dead on hospital beds, were presented as lumps of partially dressed, partially charred masses of smoldering human flesh and splintered bones. Stretchers, hospital beds and dark bags containing the clumps of flesh, now coated in a mixture of blood and dirt, provided a macabre contrasting palate against the brightly lit walls and bed sheets of the previously unoccupied ward. Not only that, the unforgettable stench of burnt human flesh and the unforgettable lingering smell of death permeated the nostrils of every man and woman who ventured near the makeshift morgue that afternoon. It would take weeks to wash away the smell from their clothes and a lifetime to forget.

For Archie, his afternoon and early evening work alongside nurses and orderlies that day would forever remain one of the most demanding, upsetting and soul-destroying days of his career. Blankly staring into the makeshift morgue he could only make out a handful of bodies that seemed to be fully 'undamaged'. The remainder displayed the gruesome disfiguring aftermath of being in the path of numerous runaway freight trains. It was a complete mess. Archie knew that 110 pounds of high explosives had perhaps detonated underground, but nothing, not even his own mental preparation, equipped him for what he was looking upon. For a moment he could only stop and stare.

His quivering chin led to tearful eyes and blurred vision trapping him where he stood. In a trance, he didn't want to move. He began doubting his strength and courage and thought to himself, ' ... there is no way I can do this.' He was a shopkeeper, not a forensic pathologist. He doubted that he was emotionally strong enough for such a gruesome task. But, to his amazing strength and credit, the reluctant storekeeper dug deep. Summonsing the type of courage and strength expected of a hardened war-veteran he steeled himself and began looking for anything that could help him tell the world precisely who these men were.

Chapter Fourteen
A Case of Identity

Shortly after entering the makeshift morgue, during the commencement of formal identification procedures, Maida Doubleday was halted in her tracks. There was a familiar face amongst the deceased men. One of the soldiers, who'd recently been a patient in her care at the hospital, was lying motionless atop one of the ward beds. She'd treated the man for concussion and a lacerated face just a few short weeks earlier and knew him well. Straight away he was identified as Sapper Toddy Woods. For Maida it was a personally painful reminder that her job was often too hard. She advised Archie of her agonizing discovery.

Meanwhile, Archie's task was about to be as equally painful and sobering as Maida's discovery. Thankfully, in the majority of cases, it was made a little easier. Resting around the necks of his multiple victims were identity discs. Still attached to many of the bodies, he felt a little at ease that he didn't have to agonizingly linger as to who they were or for that matter rely on individuals who may have known the men to formally identify them. In contrast though, identity discs of six unknown men sat on a spare bench in the ward. But these were not any ordinary discs. These discs were found at the bottom of the dugout. No evidence to connect them with their owner could be established. It was his job, and his assisting medical team, which included Victor Bennett,

George Miller and Major Tunley, to painstakingly match the recovered identity discs with the remaining indistinguishable human remains.

Having placed guards at every entrance to the ward, thereby restricting entry to authorised personnel only, in the presence of his Battalion Orderly Room Sergeant, who'd been instructed to record the details narrated to him, Archie commenced. The plan was simple. He and an entourage of specialist medical advisors would address each body and large mass of flesh one bed at a time. The four men all needed to agree unanimously on the identity of the victim and then relay the information to the Sergeant who could officially record the victims regimental details. With the men's regimental details clearly stamped onto their identity discs, Archie Smith could grab each one hanging around their necks and carefully wipe away the horrors of the explosion so he could read the details aloud. Whilst doing so his Sergeant scribed. After discussing the crude but required plan and addressing the administration, the identification process commenced:

Archie reached in and grabbed the first man's discs,' … WX two seven one six six, Corporal Woods, A E.' Once confirmed and then recorded, Archie's plan was not intrusive and was as brief and simple as that. At the end of the process, Alf Woods, the former hairdresser and mature married man from Western Australia was identified as the first victim. Meanwhile, his inseparable mate, Ron Linthorne, was nowhere to be seen. Archie ensured his Sergeant accurately recorded the details and silently moved to the next bed.

'… SX three four zero five nine, Sapper Grasby, D.E.' Archie had identified Denby Grasby, the 18-year-old South Australian who'd been sitting next to and talking with Allan Bartlett when the explosion occurred. From discs still hanging on his remains, Denby was victim number two.

Proceeding one bed at a time, the Captain momentarily paused at every man as if etching their faces in his memory. He stared into their eyes and joined in with the Padre as the men silently said a little prayer. Quietly mouthing the words, ' … rest in peace soldier', he carefully placed the sheet over the victim's face. As he did, he resigned the soldier's painful face to his memory for eternity. It was the most he could do to maintain his composure as he continued on with the most emotional day of not only his military career, but also his short life.

'... NX two zero five six five two, Sapper Moore, T.R.' The first of the streetwise 18-year-olds from Sydney had been found. Victim number three. Terence's mate Joe Faull had been sitting beside him in the dugout when it exploded. Now the team was trying to find his good friend as well.

'... QX six triple three zero nine, Sapper Platt, F.W.' At just 20 years of age, the clerk's son from Brisbane was identified as the fourth victim. Moving to the next bed, Archie grabbed the identity disc of the man. Wiping the disc clean as it lay on the soldier's chest, the details of the disc identified Sapper Leslie John Mather. '... NX one eight zero two one nine, Sapper Mather, LJ', he read aloud. The 20-year-old moulder from Glebe in Sydney, was victim five.

Following his now familiar ritual, Archie moved to the next bed, '... QX two seven three five five one, Sapper Poschalk, E.F.' At the same time, the Sergeant was also checking the names for spelling and details of the deceased men against the squad roll books. They identified the man as 18-year-old Ernie Poschalk from Townsville, the larrikin nephew of a remarkable Great War veteran, as victim number six. In their sobering procession, almost automatically, the sombre Captain and medical support crew moved to the next bed. For Archie the process was a never-ending nightmare:

'... VX nine six one nine seven, Sapper Pierce, K.F', he read aloud. '... Sir', came the Sergeant's response as he checked names against roll books. The 18-year-old who had spent most of his childhood in a boy's home, was victim number seven. Sadly, his bout of tinea, which required him to swap squads into this unfortunate group of men, coincidentally contributed to his death. With still many beds to get to, the Captain moved on.

' ... NX four eight one double four two, Sapper Hurley, C.L.' Archie as victim number eight identified Colin Hurley, the first man from his family to serve, in either the First or Second World War. As the sheet was pulled up over his remains, the 20-year-old former butter factory worker, from Herons Creek in New South Wales, was farewelled in prayer by two strangers.

The remains of yet another 18-year-old, this time Stan Morphy from Charters Towers in Queensland, were identified by his discs. As Archie read out his service particulars, he became victim number nine. ' ... QX two seven three five six three, Sapper Morphy, S.R.', read the Captain after finding Stan's identity discs underneath his traumatised torso. The fresh-faced son of Wilfred and Elizabeth Morphy was farewelled as a soldier. The padre's prayer was a fitting finale for a life cut horribly short.

Meanwhile, as the Adjutant came upon the next victim, Maida Doubleday moved up beside him. Aided by Maida's previous identification, she confirmed with him that she knew the victim as a recent patient in the hospital. ' ... Sir, this is Sapper Thomas Woods, he was recently admitted for concussion.' ' ... Thank you', came the softly spoken response. ' ... Sergeant ... NX four eight zero eight seven zero, Sapper Woods, T.' The loving husband and father from Brighton Le Sands in Sydney was victim ten.

18-year-old Allan Flood from Sydney was identified as victim 11. ' ... NX two zero five nine six nine, Sapper Flood, A.', read the Captain from Allan's identity discs. The process worker, the youngest of Alfred and Ethel's children, was covered in a sheet and farewelled as a soldier by strangers at his bedside.

' ... NX two zero five eight six three, Sapper Faull, T.W.' The streetwise Sydney-sider and good mate of Terrence Moore had been found and now confirmed deceased. With his dad also serving in the Army, Joe Faull, at aged 19, became victim number 12.

Archie Smith identified 18-year-old Joseph 'James' Collins as victim number 13. He read aloud the details, ' ... NX two zero five nine eight one, Sapper Collins, J.J.' The rabbit trapper with piercing blue eyes and brown hair from Browns Hill was shrouded by a simple white sheet when the Padre and Archie Smith said goodbye with the utmost dignity and respect for the young soldier.

As distressing and painfully heavy on their heart to see so many young men in this condition, the courageous Captain and his medical team continued their task.

The next body was that of Geoff Partridge. '… NX one eight zero two one eight, Sapper Partridge, G.W.', declared Archie whilst the Sergeant scribed. The 18-year-old big brother to Shirley was one of the most unfortunate killed that day. After his dental appointment Paddy Cranswick nearly convinced him to lie down rather than go back to the demolitions range. Geoff's enthusiasm and dedication made him victim 14.

The unmistakable mature appearance of the next victim confirmed that the 35-year-old married man with two children became victim 15. ' … NX four eight one five three six, Sapper Dilley, N.R.J.' Norm Dilley, the reluctant conscript who'd only recently sent a loving letter home to his little angel, just to check on her bunny, was also given a soldier's farewell. Huddled around the bed on which his remains were laid out, Archie, the Padre, and the team of medical staff in the hospital ward, paid their respects. Unfortunately there was no time to dwell and reflect on the life of Norm Dilley. The men completed a prayer and turned the attention to the next bed. There they found yet another mature man.

Archie addressed the bed. Looking upon the remains he reached down to grab the identity discs that had been placed on top of the soldier's remains. Wiping the disc clean, he read out the details, ' … NX two zero five eight three three, Sapper Nixon, J.C.' Again the Sergeant recorded the details. This time it was the details for 31-year-old Jack Nixon from Cobar. Little Neryl Nixon's dad was Archie Smith's 16th identified victim. Sadly, Jack's prized photograph of his daughter was nowhere to be seen.

20-year-old Alf Witt from Western Australia became victim 17. ' ... WX two three one zero one, Sapper Witt, A.G., did you get that Sergeant?' were Archie's sombre words to the Sergeant. ' ... Yes Sir.' The younger brother of two men already serving in the 2nd AIF was blessed by the strangers in a fitting symbol of recognition of their sacrifice as they had done with the previous 16 victims. When the shrouded sheet was pulled over Alf's remains, the team took a well-earned break.

Over at RAETC Headquarters, Captain Fernleigh Ellis, the A/AA/QMG, was busy ringing the New South Wales Police Department at Wagga Wagga to inform them of the accident. He advised them that an explosion had occurred at the Centre resulting in the death of a 'number' of soldiers. Cautious not to raise any unnecessary alarm, he requested a member of the Police Department attend the camp for a more detailed brief. As he hung up the phone, Major Macdonnell, who was now also present in the Headquarters coordinating the response, was briefing Brigadier Prior, the DA at Second Aust Army Headquarters, via telephone, that an accident had occurred resulting in the death of a number of soldiers.

After a brief break, Archie Smith and his team of committed medical experts continued their gruesome, but totally necessary, task. Unfortunately the next body to be identified was presented to the Captain as just human remains. All the staff working the makeshift morgue knew that recognition and positive identity was going to prove difficult from here on in. In many of these cases there were no physical signs of identity. No face, no rank on a stray piece of uniform, nothing, not even a boot size. It was time to draw upon the lonely identity discs sitting on the bench. They were the discs found at the bottom of the dugout indicating that although the owners of the discs were present in the dugout at the time of the explosion, physically, their identities are unknown.

By now, many friends and colleagues of the mounting death toll were starting to assemble at the hospital. Their attendance would actually turn out to be fortuitous for the men conducting the identification. Many in the crowd were cohorts and mates of the victims just waiting for news. They didn't know it at the time, but they would become valuable assets for Archie, Major Tunley and his team. Archie realised that perhaps many of the assembled outside could assist him in identifying human remains that were unfamiliar to him or his team. It was however a last resort. The less people exposed to this the better, he thought.

Like forensic pathologists examining remains looking for cause of death, the brave men commenced their unenviable task. But they weren't trying to establish a cause of death; they were trying to prove identity. A somewhat impossible and hopeless task based solely on what was presented before them. By now they'd eliminated 21 men from the squad roll books. From their calculations, of the remaining possible seven men yet to be identified, the examination team relied on a process of elimination. Built around this knowledge of just who the remaining missing men were, and more than likely amongst the piles of flesh, the Captain and his team commenced sifting for clues.

Before too much later a wristwatch attached to what was the remains of a forearm was located. They surmised that the remains were possibly either that of Sergeant Jack Pomeroy or Sergeant Ron Linthorne. As Senior Non-Commissioned Officers, it was possible that either or both of these men were wearing watches at the time of the explosion. Captain Smith allowed one of the medical officers to remove the wristwatch from the remains. Wiping the watch of debris and skin fragments the medical officer handed the watch to Archie Smith who inspected the reverse and saw the watch had been engraved with the owner's details.

Calling the attention of his Sergeant, he asked him to inspect the squad roll books for any of the unaccounted men with the initials R.I.L. Even though they knew there were only two SNCOs in the dugout when it was destroyed, Archie followed his processes. It was instantly apparent that amongst the human remains they'd found evidence of the gregarious soldier from Western Australia. His identity was confirmed when a gold finger ring was also located amongst the bloody mess of flesh. ' ... Can I have this man's details please?' asked the professional but increasingly emotionally weary officer. His meticulous assistant found the details in the squad roll books. ' ... Sir, WX two five seven nine two, Sergeant Ronald Irwin Linthorne.' With the required professionalism of the moment, the burdened Captain again asked his fellow identifiers if they all agreed that the remains inspected were that of Ronald Irwin Linthorne. The accompanying medical officers nodded their agreement. Although Ron's remains were only partial, his wristwatch and wedding ring was enough evidence for Archie and the concurring party to establish that the recently married Western Australian was the 18th victim. And, until such time as they located any evidence of the only other SNCO in the dugout, Ron was the tragedy's most senior military victim.

Whilst the identification was still ongoing, Sergeant Sherwood of the Wagga Wagga Police Department had managed to get onto the camp. He'd run the gauntlet of commotion coming from family, friends and the press now gathering at the camp front gate. The policeman proceeded directly to the Training Centre Headquarters where Captain Ellis met him. Carrying a copy of the *National Security Regulations* he pointed out in his conversation with the Captain that the police would not be getting involved. Specific sections of the Regulations dispensed with the need to have inquests into the deaths of Army personnel where the occasion warranted it. Both men agreed that this was one such occasion. Sergeant Sherwood further advised the Captain that the Commonwealth Police would therefore also dispense of a Coronial Inquiry into the deaths. Death certificates for the deceased could be produced and signed by the senior medical officer of the camp. There would be no further civilian police involvement.

With Ron Linthorne's formal identification completed, Archie Smith continued his ghastly task as six men continued to be unaccounted for. He moved to the next bed space where he looked upon a clump of flesh supposedly the human remains of another victim, but his concentration was distracted by an overwhelming sense of futility. His brave façade displayed in the makeshift morgue for the last few hours was showing signs of cracking. The magnitude of the horror was now getting to the young Captain. Stopping for a quick break and a glass of water, he excused himself and walked alone outside the hospital walls. Finding himself an isolated corner, as far away as possible from the lingering smell of death, the young Captain's mind was numb. As the afternoon sun began to settle over the shocked and silent camp, the fresh air briefly replaced the acrid smell of death that was permeating his clothes. It seemed to be around every corner and in every thought. With a deep breath the married 31-year-old could no longer control his emotions. Letting the enormity of his task and the indelible history he was creating affect him, blocking out the horror of the ward, his thoughts soon turned to his lovely wife Beryl. With memories of family toying with his emotions, Archie's tears were soon flowing uncontrollably. Allowing himself time to personally grieve at the senseless loss, he soon realised many families were relying on him to identify their loved ones. He knew he had to return. But he just wanted a bit more time to himself. Time to remember his family and conjure up more courage and strength to get the job finished. Already at victim 18, ' ... when,' he thought to himself, ' ... when, will it end?' Composing himself, Archie returned to the ward and the next bed of charred and mutilated flesh.

With eighteen victims identified, two men being treated for life threatening injuries and two men likely to survive, of the 28 men, six men were still unaccounted for. Continuing with a process of elimination, based on the macabre and distressing evidence contained in black bags, some of the valuable identifying clues were becoming harder to find. The next victim especially challenged Archie and the team. The human remains didn't have any identity discs or distinguishing features that immediately alerted the examining officers as to its identity. By now the team was relying on anyone who might know one of the remaining six men. They agreed to call in Ed Dodds. The Sergeant Major inspected the human remains and believed it to be that of 28-year-old Sapper Stan Ross from Mungindi in New South Wales. To confirm, he invited two of Stan's tent mates to review the remains. They moved the remains to an isolated room as the two young Sappers were escorted in. In the room, accompanied by Archie and the team, the two men viewed the body of the man lying on the bed and both agreed that it was Stan, the former motor mechanic who was their tent mate. To endorse his identity Archie requested Stan's dental records. With the assistance of Captain Brodie, the dental officer from 59 Australian Dental Unit who treated Stan about a week ago, the men all agreed that the victim before them was NX91964 Sapper Ross, S.E. His details were formally recorded once again by the accompanying Sergeant. Stan's mates were escorted from the ward, as Stan officially became Kapooka's 19th victim. Now, presumed dead and lying amongst the remaining corpses, only five men were missing.

Confident in his victim processing method, Archie turned to the next body and immediately inspected the clothing. Stripping off a shirt from the remains in, with the assistance of the medical team, he located the fragmented remains of a pair of issued trouser braces. Aiding his task, thankfully the unfortunate owner had inscribed his service details on the braces. Peeling back the braces Archie made out the details and again asked to confirm the victim from the squad roll books, '... NX two zero five nine five one, H...U...R...S...T.' '... Yes sir, that's the details of Sapper Kevin Hurst', came the reply. Ensuring the accompanying men also inspected the braces, the weary Captain asked, '... Gentlemen, once again, do we agree that this is the human remains of Sapper Kevin Alexander Hurst?' The men agreed. The assisting Sergeant, now sickly aware of his process in the unbelievable saga, made the necessary record in the roll books.

Richard and Jessie Hurst's 18-year-old red-headed son from Carathool was victim 20. Now only four men were missing; Jack Pomeroy, his assistant Bill Cousins, 18-year-old Sapper Ted Robson from Liverpool in Sydney and 21-year-old Cowra-born Sapper Colin Francis Boyd from Haberfield in New South Wales.

Moving to the next gruesome pile of remains, Archie's attention was drawn to the fact that the victim was wearing a pair of Army-issued underpants—the notorious 'romance busters'. Amongst charred blood-stained underwear, Archie could make out the details: 'NX 204475'. Turning to his administrative assistant, still holding and checking the squad roll books, he checked the Army Regimental Number against the unfortunate four men still missing. Inspecting the underpants a little further the initial 'C.F.' and name 'Boyd' were identified. Turning to his accompanying medical officers ' … Gentleman do we all agree that this is the human remains of NX two zero four four seven five Sapper Colin Boyd?' The gathered identification team agreed. ' … Thanks,' ' … Sergeant, please make the appropriate mark in the roll book.' Colin Boyd was positively identified as victim 21. Colin's identification left only three of the ill-fated squad to be discovered Jack, Bill, and Ted.

Although the men had been at it now for a couple of hours they weren't going to stop, even when they were only faced with indistinguishable lumps of human flesh and bone. They'd have to sift through the remains and hopefully find something, anything that would identify the remains of any of the three men.

Considering two of the men yet to be identified were Jack and Bill, theories had begun to circulate about what went wrong. With the two instructors' bodies still missing and speculated to have worn the full brunt of 110 pounds of high explosive, it was any wonder that only scant pieces of their human remains were present in the morgue. It did mean that they were perhaps closest to the detonation. But the questions of why and how it happened, what was Jack and Bill's involvement and where were the remains of Ted Robson were questions momentarily unanswered. Focus was still centred on their positive identification amongst the remaining parcels of flesh.

Theories, responsibility, speculation and allegations of blame continued to circulate and puzzle everyone involved in the inquisitive aftermath, but for now Archie Smith still had to find the missing men at the centre of it all. Perhaps their remains, if found, could provide a clue? Sifting through the heavy masses of human flesh, it was difficult to distinguish which part of the body was what. Confident that bone fragments with heavy hair density were scalp, and less hair were arms and legs, the men sifted until familiar anatomy became recognizable.

After a short while there was a breakthrough—a second wristwatch was found. Considering Ron Linthorne had already been identified and was possibly the only other person wearing a wristwatch, the examination team suddenly stopped what they were doing. Was it possible that this new wristwatch belonged to the charismatic instructor who should have been celebrating his birthday? The watch was handed to Archie who, without looking, placed it on a nearby table. At that very moment he defiantly refused to turn it over or inspect it for clues. Knowing full well whose details he'd find engraved on the rear of the watch, he wanted to prolong the faint hope that its owner was about to walk through the door. For what felt like an hour, but was perhaps closer to ten minutes, Archie refused to draw attention to the watch. He didn't want to confirm the identity of its owner. But he acknowledged that he had to prepare himself nonetheless. Finally, as if in slow motion, he turned the wristwatch over. What he read caused him to once again to be overcome with grief and emotion. Refusing to accept the truth, watched by all in the makeshift morgue, he slowly placed the watch back down and moved on to the next victim. He didn't necessarily want the identity to be known just yet. At least that's what he told himself. The remainder of those in the ward looked at each other with a cold sense of realisation. They knew!

After further sifting and having put off the inevitable for long enough, it was time for Archie to turn his attention back to the wristwatch. In doing so he'd not only confirm the owner, more importantly, he'd be able to identify another of the missing men. Sure enough, the watch, although shattered and only slightly resembling its original glory, did bear the details 'J Pomeroy'. Inscribed on the rear, it was what they'd all been hoping 'not' to find. Perhaps yearning, wishing and praying that by some divine miracle the larger than life family man would walk around the corner and wake them all from their nightmare. But sadly that wasn't to be. Jack's wristwatch confirmed not only his identity, but also his existence in the dugout at the time of detonation.

The first words spoken by the shocked Captain amplified the respect they had for the man, '... get word to Ed Merry that we've found Jack's watch'. It confirmed to the many caring colleagues, that amongst the remains of his charges, the man they all looked up to, and trusted with their welfare, had been identified as one of the victims. With them to the very end, the identity team all stopped what they were doing and for a brief moment reflected in the loss. Silence filled the makeshift mortuary. They'd lost a prodigious soldier. Being at the epicenter of the detonation, Jack's remains were few and far between. Death therefore would have been unexpected, swift and thankfully, painless.

At Kapooka on his 31st birthday, the dedicated family man was still training young engineers who not only gravitated towards his experience and hung on his every word, but who also looked up to him as a father figure guiding them through the dangerous training. Suddenly in a blinding instant, they were all gone. Perishing with his men, it was a dignified, noble and soldierly way for Jack to meet his maker. But laying spread out on a bed and table as piles of flesh, the dignity and aura demanded by this larger than life man was certainly being tested. Archie Smith knew this and ceremoniously placed a sheet over what sparingly purported to be Jack's remains. He didn't need to say a word to anyone. Not that anyone was truly listening. Reeling in shock that the latest discovery was a man they all looked up to and knew well, the gravity of the death hit everyone hard. Jack was victim 22.

With Jack's remains identified, it was fast becoming a hopeless and fruitless search for information amongst the human remains for any more names. Not only that, Jack's identification robbed Archie and the remaining staff of motivation. But Jack's assistant instructor Bill Cousins and one of his young charges were still missing. The one key premise of the gathered staff remained their one true motivator—dignity. Two men still needed a dignified conclusion to their life. The easy path would be to resign all effort and accept that Ted and Bill's remains were present amongst the bags of horror, but Archie remained resolute. Jack would have wanted them to finish the task. *All men would be positively identified today.*

Besides, Ted Robson's identity discs had been located in the dugout. That meant his remains had to be somewhere amongst the flesh and bone. Additionally, either Bill Cousins didn't wear his discs on that day, or they too were lost—or worse—blown to pieces, either way, everyone knew he was also there. They witnessed his hijinks during physical training and knew too well that he entered the dugout to assist Jack.

But instead of confirming Ted and Bill's identity with physical evidence, which satisfied all beyond a reasonable doubt, they were now relying on the balance of probabilities—the probabilities being that the size and composition of the flesh and bone matched the size and composition of the two remaining men. With no faces, hair, no distinctive identifiable anatomy, scars, marks, external clues, it was fast becoming frustrating for Archie to identify the remaining two men. Visual examination of the human remains alone wasn't working. They certainly weren't going to be identified by Archie, a less than experienced storekeeper come 'pseudo' forensic pathologist who barely knew the men. It was going to require more than. He needed help. It was time for medical science to take over. Resigned to take a step back and be momentarily relieved of his duties, Archie invited Major Tunley to establish medically that somewhere in the mess of human flesh, the remains of Bill Cousins and Ted Robson could be found.

The experienced Major was faced with identifying the two remaining men based on information at their disposal, albeit scant details found in military enlistment and recruiting records, to describe their physical builds. He didn't have faces to match photos, identity discs, articles of clothing, or even jewelry, just a mangled mess of flesh and bone fragments. Thoroughly examining the gory chaos before him, he knew he had two human trunks. Were these two men slight of build, heavy set, short or tall? He'd have to ascertain which of the two trunks was that of Bill Cousins and which was that of Ted Robson. It was never going to be just a guess. Les Tunley had to get this right.

Calling on all his experience, he established that one trunk was of slight build and the attached leg bone fragments suggested the victim was above average in height. Whilst Bill Cousins was known to be in the region of six feet one inch, and more than six years older than Ted Robson, he wanted to clarify what type of build the older of the two New South Welshman was. Major Tunley sought out two of Bill Cousin's tent mates. Luckily, two of his mates were in the waiting crowd at the hospital hoping to hear good news of the larrikin Corporal. The two men were led away from the crowd and asked to describe Bill to the Major. Confirming that their mate was tall and a slight build, the Major was able to distinguish, with some degree of medical confidence, the difference between the two men and the torsos before him.

The Senior Medical Officer called the now exhausted and distraught Adjutant back into the ward. ' ... Archie,' he said, pointing to the first of the human torsos, ' ... these remains here are that of Corporal Bill Cousins and the other over there is that of Sapper Ted Robson'. ' ... Thanks Sir', came a tired and distressed reply. Calling once again for the roll books, he formally identified the 23rd and 24th victims: ' ... NX one four one seven one five, Corporal Cousins, W.B. and NX two zero five nine three eight, Sapper Robson, E.C.'. The formal identification was now complete. ' ... Thank you Sergeant.'

Meanwhile, back out on the demolitions area, Tom Musto and Sheila Oehm brought the munitions truck to a halt. Doing their regular 1600—1700 hour rounds to collect unused explosives, they came upon a totally different scene that afternoon. The joviality they'd experienced earlier in the day was replaced by the sad overture of men describing, in sad repetitive detail, the events of the afternoon. Collecting the unused explosives from the parcel issued to James Conwell and Roy Tafe, Tom Musto noted that the parcels were complete, except for the fact that Roy Tafe's parcel had thirty detonators made up with varying length of safety fuse with match heads attached. Loading up their stores, Tom and Sheila were in shock. So many men, who only a few short hours ago were joking around with her as she was preparing to join them in the underground dugout, were all dead.

' ...Finally,' thought Archie, his personal pain and discomfort were over. At least that's what he thought. As he checked the Sergeant's diligent work in the roll books his tired eyes surveyed the long list of names. Once again his emotions overpowered his professionalism and he was moved to tears. Welling up in his eyes and flowing down his cheeks were the thoughts of families, of wives, children, parents and friends. With 24 men now confirmed dead he was imagining how the impact of this tragedy would affect those people. But Archie Smith had no time for tears. As the Adjutant of these men he realised his task wasn't finished—it was merely beginning. He couldn't afford to let the gravity of the situation bog him down emotionally, he didn't have the time. He now had families to advise and a large funeral to coordinate.

But word was starting to filter down to the morgue from the casualty department that Ivan Merritt and Bill Reid hadn't made it. Archie Smith's mixed moment of inner peace and anxiety was quickly shattered when the news of the two further fatalities reached him. Medical staff advised him that the two men had sadly lost the fight. With deep sorrow and a weary heart Archie watched as their bodies were wheeled into the makeshift mortuary. With due respect for the men, Archie Smith decided to record their details as soon as possible. Pulling back the sheet on Ivan Merritt's tortured body Archie located the identity discs and recorded the details, ' ... SX three four zero six nine, Sapper Merritt I.W.T.S.'—Ivan was victim 25. Likewise he turned his attention to Ivan's friend who'd just been deposited into the ward alongside him. Pulling back the shroud he reached in, now devoid of emotion, almost angry; Archie remained professional as he grabbed the identity discs and recorded the details, ' ... SX one one five five seven four, Sapper Reid, W.' Pulling up the shroud to cover the troubled face of Bill Reid, '...Is that victim twenty-six Sergeant?' asked the emotionless Captain of his assistant, ' ... Yes Sir', came the relieved reply. Bill Reid was the 26th and final victim.

At the start of his task Archie Smith promised himself that all of the the fatalities of this inescapable hell would be handled with professionalism, respect and above all dignity. The brave and courageous Adjutant achieved all that and more. Every man identified was given a soldier's farewell.

It didn't take long before Fernleigh Ellis was advised that the two critically injured, Ivan and Bill, had unfortunately also died from severe injuries. It forced him to transmit another disturbing signal message to his Headquarters. Identification was completed by about 2000 hours that same evening. But for Fernleigh his job was far from over for the night. He had a throng of Press representatives at the front gate requesting permission to enter the camp and conduct interviews. Of course he denied them access.

But the Press soon learnt that the wife of at least one of the victims was living in Wagga. Fernleigh Ellis became made aware of this and instructed the padre and a number of men to go directly to Jack's house and look after Dorothy and his kids.

Unscrupulous as always, somehow the local Press found out that Dorothy was at the Hickson's, before anyone else in the country. They were waiting at the house when the padre arrived to deliver the news of the tragedy.

For Dorothy Pomeroy, standing on the verandah of the Hickson home, the appearance of media and an unexpected visit by the unit padre and senior officers of the Engineering Training Centre that evening was not the birthday surprise she was expecting. As a soldier's wife she knew that the appearance of the padre could only mean one thing. The young mother and wife went numb. Surrounded by her children and in the arms of Mary Hickson they delivered the most horrific of news that her husband was one of the confirmed killed. The padre consoled the distraught wife and mother who collapsed to the floor. Placing an arm of comfort around the mother of four, the padre advised Dorothy that Jack died doing what he loved doing, instructing young men. He quietly told her that Jack's death was instant and he didn't suffer. Leaving her with the lasting memory of her powerfully strong husband and loving father of her children, the padre reassured her that not a mark was found on his body. He convinced her that viewing the remains of her husband was not necessary. Many colleagues, who respected him as an instructor and genuinely nice bloke, had already identified him. The news caused her to go into shock. Inconsolable, and if not for the support of the Hicksons, and her own loving children, her world may have ended there on the verandah of the family home. She refused to speak to any of the gathered media.

The final process for the professional but burdened staff at the camp hospital was to formally certify the victims as deceased. 26 Death Certificates needed to be signed reflecting the carnage and the ignoble circumstances of the instantaneous death of these men. After enduring the most horrific few hours of his life, Archie Smith was now in a position to formally brief Captain Ellis. As directed, he provided him with a nominal roll containing the accurate details of all the deceased men. He also briefly mentioned his preliminary thoughts about the funeral. ' ... Sooner rather than later', he declared.

In his own office, away from the sobering sights of the makeshift morgue, Major Tunley briefly sat and rested his head in his hands. In his 46 years, he'd seen his fair share of death and destruction, but what he just witnessed left an indelible scar on his memory. Later that evening, having had the opportunity to shower, he washed the stain of death from his trembling hands. The Major knew he had an important task to perform.

With identification established, using a combination of medical officers and Captain Smith agreeing sufficiently beyond a reasonable doubt for the majority of the fatalities and on the balance of probabilities for the more difficult cases, the cause of death was left up to the Officer Commanding the 54th Army Camp Hospital. Confident in the work of Archie Smith and the team of novice forensic pathologists, Fernleigh Ellis requested Major Tunley formalise proceedings. Considering the Wagga Police had agreed with the *National Security Regulations* that any coronial inquiry would be dispensed with, responsibility fell to the experienced Medical Officer to compile Medical Certification as to the cause of death for all 26 victims. But were his findings going to be a decision that would ultimately satisfy the families of the 26 men considering that no individual autopsies or coronial inquiry was going to be held? Establishing the cause of death without coronial inquiries and autopsies left the Major with the responsibility of certifying death to the best of his knowledge and belief. Based on all the evidence and eyewitness accounts by the two survivors, he didn't categorize individual deaths according to blast injury classes of either: primary, secondary, tertiary, or quaternary. He knew his words on any subsequent Death Certificate would be scrutinized over time immemorial, especially in a tragedy of this magnitude. He kept his findings realistic.

Although all the men were in an enclosed area they all would've been subject to flying debris and bodies, however, many of the deceased weren't displaying any signs of external injuries. This would suggest that many died as a result of primary injuries, or multiple organ rupture owing to over-pressurization from the blast. For the unfortunate few such as Jack Pomeroy, Ted Robson, Stan Ross and Bill Cousins, the magnitude of their injuries suggested that they were perhaps closest to detonation and therefore suffered a combination of all four classes of injuries, simultaneously. Considering one pound of high explosive can cut a hole through one inch of steel, for these four men they unfortunately wore the brunt of a detonation that could've leveled two standard brick homes. They had utterly no hope of survival.

For the remainder of the deceased, traumatic amputations of limbs and torsos suggested that they also were subjected to a combination of blast injury classifications. In a confined space, no larger than a modern-sized Australian living room, arguably they also sustained primary blast injuries due to the supersonic over-pressurization blast caused by the detonation of 100 pounds of high explosives. Secondary injuries from debris flying at more than 3000 metres per second ripped flesh open and penetrated bones. In addition, tertiary injuries were sustained through having been thrown around inside the dugout like rag dolls against each other and the walls due to the displacement of air. Finally the men would've sustained quaternary injuries—injuries sustained through horrific gaseous burns, amputations resulting in sudden loss of blood and crush injuries from more than five feet of concrete and earth falling down upon them. With due respect to the men killed and their families, apart from Ivan Merritt and Bill Reid, who remained barely alive for nearly two hours post-explosion, thankfully death was instant and therefore painless.

As macabre and insensitive as it sounds, Major Tunley's certification of death was made easier by the fact that all 26 died in the same circumstance. In the absence of formal autopsies and medical terminology confusing the cause of death with blast injury hypotheses, 26 pre-prepared Death Certificates were presented to the OC of the hospital along with the nominal roll of the deceased. All the experienced MO had to do was write the regimental particulars of the deceased in his handwriting and sign the form; all the other details had been typed up by his administrative staff.

Captain Ellis had earlier briefed Major Tunley that the men were involved in demolitions training and were, at the time, enclosed in an underground dugout surrounded by more than 110 pounds of high explosives. Even though Major Tunley knew of the practice of working in the dugouts in harm's way having the explosives down there, he still winced at the thought of unexpected detonation of 110 pounds of high explosive virtually in the shirt pocket of every man down there. Suffice to say, and, in the absence of a Coronial Inquiry, Major Tunley was left with no other option other than to record the cause of death on all 26 Certificates as: *Blast Injuries.*

About 2020 hours that evening, Archie Smith, carrying the personal details of the victims in the nominal rolls he so studiously and cautiously compiled to ensure accuracy, arrived at the Engineer Training Centre Headquarters. Fernleigh Ellis was eagerly awaiting his arrival so that he could communicate the details of the deceased to his chain of command. At 2030 hours, Fernleigh contacted the New South Wales Echelon and Records Casualty Section in Sydney via telephone and advised them of the full regimental details of the lost soldiers. But like Archie, his duty was far from done. He drafted a signal message to confirm the telephone conversation with the details of the deceased and a copy of the signal message was also sent to HQ Second Aust Army.

At the hospital, Les Tunley was completing the unenviable task of recording the 26 names onto standard Army Death Certificates. Then, having completed them all, he needed to rest not only his head—but also his heart. It had been an emotional afternoon. He didn't seek out the ultra-busy Captain Ellis to hand over the certificates; instead he placed them in a pile on his desk, locked his office door, and walked down to the makeshift morgue. He was well aware his actual morgue at the hospital would not be large enough to store all 26 bodies. Making sure the human remains were now all contained in body bags, he checked to make sure guards remained on the ward entrances to prevent the event becoming a circus in his hospital.

For reasons perhaps only known to the men responsible, the Death Certificates later supplied to Captain Ellis weren't the ones formally signed by Major Tunley. Instead he received a typed copy with no original signature by the Senior Medical Officer, just the hand-written details of the victims and Major Tunley's signature block with his details typewritten. Perhaps the Major, overcome with the scale of the tragedy and the senseless loss of life, simply typed his name on each of the certificates and let someone else fill in the men's details. They were later provided to allow Captain Ellis to communicate the details of the men to the chain of command. That being the case, perhaps the Major left them in his office and retired for the night as nothing further could be done. He'd need a fresh start in the morning organizing the bodies, so perhaps he just simply went home and decided to get orderly staff to type up the official certificates containing the details of the men, leaving a space for him to sign at the bottom first up the following morning. Whatever happened, the official Death Certificates were signed by Major Tunley the following day; 22 May 1945.

In light of the day's events, Captain Ellis called for a *complete* communication blackout. That meant no telephones or telegrams in or out of the camp until the names of *all* the unfortunate casualties had been determined. The last thing he wanted were the families of 8000 or so men across Australia panicking. He neither had the time or staff to address such a catastrophe.

Armed with the knowledge the Army was responsible for handling the notification of death, he was a little more comfortable that he could control the details. That night, on the advice of Fernleigh Ellis, military authorities released a statement for the press, it read:

> 'An accident occurred today at an engineering training centre in New South Wales, when 28 men were involved. Unfortunately 24 of them were killed immediately and four seriously injured. No further details were available at present.'

Meanwhile Sapper Des Surkitt arrived in camp later in the evening; around midnight, of the 21st. He'd been on two week's leave in Warrnambool prior to going to fight in New Guinea. On the Monday night he was returning to Kapooka to rejoin his unit. He'd caught the troop train from Melbourne, which dropped him and his colleagues off at the Kapooka Loop stop. They knew nothing of the days events as they entered the front gate. The remaining press representatives still loitering at the camp entrance did alert him that something was amiss. Passing through the front gate, little was said about the explosion from the guards. They were too busy keeping 'mum' so as not to give the press a snippet of details or names of the deceased. It wasn't until the following day, when Des paraded with his unit, that he found out the extent of the situation. He was informed that his good mate, Toddy Woods, along with the instructor he most looked up to for his professionalism, who had taught him only a few weeks prior, Jack Pomeroy, perished along with 24 other men. It was the most shocking news Des could've received before going overseas. It was a day that he never ever forgot.

Regardless of the communication blackout, the initial media release of the accident catapulted an unprecedented saturation of the tragedy via news outlets around the country. However, as no names were made public, and more than 8000 men were running around training at the RAE Training Centre, a well-informed population knew only too well that the only heavily-concentrated engineer training location was Kapooka. The public statement threw the country, especially families who knew loved ones were still at the bustling camp, into unparalleled panic. No names were being released. It therefore left 8000 Australian families desperately seeking any information about their sons, husbands, brothers and fathers. In the camp, hundreds of soldiers were trying to send telegrams to loved ones advising them that they were okay. But, with a communication blackout and therefore a deep unknown, many families throughout Australia were left praying for an answer to the one burning question—was their loved ones one of the unfortunate death toll?

Chapter Fifteen
Front Page News

Military signal traffic at the best of times operates at a slow pace. In 1945, courtesy of telegrams and teleprinters, it was a snail's pace. The slowness at which the signal traffic was transmitted and received, following the explosion, meant that it wasn't until the following day, Tuesday the 22nd of May 1945, when Headquarters Second Aust Army sent their first signal message to Landforces Headquarters informing them of the explosion. Urgent telephone calls were made, but the Army relied on signaling traffic as official confirmation and notice to initiate actions—especially to the government. On this occasion, Captain Ellis based the information in his signal on the earlier reports he'd received. He sent the report even though he knew Ivan Merritt and Bill Reid had earlier succumbed to their injuries. At 0857 hours that day, the following *'Important Restricted'* message was dispatched:

> Regret advice accident 1 rae trg centre 21 may.
> 24 Pers killed 4 seriously injured by explosion.
> Court of inquiry convened and despatched todays.

In less than two hours, an additional message from Captain Ellis was sent. His second *'Important Restricted'* signal message was transmitted at 1005 hours. It read:

> Accident at 1 rae trg centre.
> 2 Injured personnel since died making number deaths 26.'

At Victoria Barracks in Melbourne, the teleprinter in the office of the Secretary of the Department of Army, Mr. F.R. Sinclair, was running hot with incoming distressing messages. Later that day in the House of Representatives in Canberra, a question was asked by New South Wales Labor and Federal Member for Hume (Wagga Wagga's electorate) Mr. Fuller. In light of the tragedy becoming parliamentary knowledge he requested if the Minister for War was in a position to make a statement on behalf of the Government? Mr. Fuller took the opportunity to publicly offer his condolences. ' ... As this tragedy occurred in my electorate, I offer my deepest sympathy to all who suffered bereavements and to those who were injured.' Mr. John Dedman, the Minister for Post-War Reconstruction in the House of Representatives responded:

> ' ... The government learned of the accident during the weekend with deep regret and it also tenders its deep sympathy to relatives of the deceased. I am not yet in a position to give to the members the details of the accident. The matter is being investigated and no doubt considerable delay will occur before full particulars can be made known owing to the fact that there were only two survivors and they were probably suffering severely from shock. As soon as the details were available I shall communicate them to members.'

Australia's media, especially newspaper print, went into overdrive. Fuelled by the Minister's empathetic response in the House of Parliament, the story quickly circulated throughout the country. Nonetheless, the calamity at Kapooka wasn't always the lead story of the day in the headlines the following day. The headlines were reserved for the negotiation between Yugoslavia and Italy over a matter of the sovereignty of the northeastern Italian seaport of Trieste. However, in most of the popular newspapers in each Australian State and Territory, the disaster appeared as front-page news alongside the problematic Yugoslav situation. Regrettably, some periodicals ran their stories before the two critically injured Sappers—Merritt and Reid—had died. This caused widespread confusion around the country about the inaccuracy in total deaths being reported. Who were the dead and just who were these survivors? Sadly, the rate at which the news spread around Australia didn't have the sense of urgency in reaching fellow troops serving overseas. It would take days for the news to filter through of the tragedy. Regrettably, the fathers, brothers and loved ones of the victims still fighting in New Guinea and the Middle East were some of the last to know.

In Sydney on the afternoon of 22 May 1945, the front page of the late edition of the *Sydney Morning Herald* ran an article titled: '*24 DIE IN ARMY CAMP EXPLOSION*'. Although not mentioning any names of the deceased or dangerously injured, the article reported the explosion had occurred at the Engineer Training Centre in New South Wales. It further reported the victims were, ' ... involved in bomb disposal work at Kapooka camp when a terrible explosion occurred.' For Sydneysiders taking in their daily war news from home and abroad, it was horrendous and unbelievable news. Nonetheless, the demands for details of the tragedy were insatiable. Concerned and distraught Australian families caused interest to quickly escalate and spread to other major, subsidiary and regional newspapers over the next few days.

In Brisbane, the front page of the *Courier Mail* ran a similar article with the headlines: '*Explosion Kills 26 Soldiers*'. Much like the Sydney paper, the article didn't contain any names and details. It did report that the military authorities were in charge of the impending inquiry. *The Argus*, Melbourne's premier newspaper, ran the front-page story: '*24 KILLED IN ARMY CAMP Demolition Explosion*'. Reporting that four men were seriously injured suggests that the information was relatively fresh after the explosion. It notes further that the survivors were so shocked that they were 'not able to relate the circumstances of the accident'. On page four of the *West Australian*, Western

Australia's number one newspaper, the story ran: '*TERRIFIC EXPLOSION—N.S.W. CAMP TRAGEDY—24 MEN KILLED—FOUR SERIOUSLY INJURED*'. Relying on the information gathered from Wagga on the Monday, much like the Melbourne reports, it was word for word the same article from *The Argus*. Meanwhile, that same day, the *Daily News* in Perth ran a more up to date story: '*TWO MORE EXPLOSION VICTIMS DIE*'. Relying on the second round of information about the explosion, it reported that no casualty list will be released until next of kin had been notified and that funerals would be taking place in the military section of the Wagga Cemetery tomorrow (Wednesday). The article also added that military authorities were awaiting advice from relatives whether they wished to or were able to visit Wagga for the funeral. According to the article, it was so far decided to hold funerals for more than half of the 26 who'd perished.

The front page of the *Adelaide Advertiser* for 22 May 1945 reported: *24 SOLDIERS KILLED-Training Mishap In New South Wales camp*. Whilst not the lead story, it was prominently positioned adjacent the lead story for all to see. North Western Tasmania's *The Advocate* ran a story on page five titled: '*EXPLOSION IN CAMP KILLS 26 SOLDIERS*'.

The horde of press representatives eager for more news, who'd earlier been camped outside the front gate of Kapooka after being refused entry, shortly began filtering the snippets of information to national and regional subsidiary news carriers. Before too long remote newsprint locations such as: the *Border Watch* in Mount Gambier, South Australia which accurately reported on page one: '*24 Soldiers Killed In Explosion at Wagga*', the *Barrier Miner* in Broken Hill reported: '*24 Men Killed And 4 Injured in camp Explosion At Wagga*'; *Geralton Guardian and Express* and the *Mirror* in Western Australia; *Morning Bulletin* Rockhampton reported '*Personnel Killed In Kapooka camp Explosion*'; the *Cairns Post*; Townsville Daily Bulletin *Army News* in Darwin; the *Canberra Times*; *Singleton Argus*; the *Narandera Argus and Riverina Advertiser* ; the *Newcastle Morning Herald & Miners'*

Advocate and the *Northern Star of Lismore* which ran the headline: *'Explosion Kills 24: Discovery Delay'*, delivered the tragic news to unsuspecting family and friends. Inaccurately, adding more unnecessary anxiety to waiting families, the *Northern Star* incorrectly reported that a Corporal McKinnon was one of the victims. He was not.

With so much media scrutiny and public interest in the aftermath of the tragedy, the RAE Training Centre hierarchy and Army Headquarters authorities, were under immense pressure to contact the registered next of kin. Until such time the heartbreaking task was complete, it was decided that no official release of names would be made. Withholding names at least gave the authorities more time to contact family direct as a matter of courtesy, rather than them finding out via rumour or reading their loved one's name in an insensitive newspaper. It was a decision which was later criticised, and became the subject of a parliamentary investigation requested by New South Wales Liberal politician Sir Frederick Stewart. It was, however, a frivolous attempt by Sir Frederick to discredit the Army and the Minister for the Army by criticizing what he believed was an 'undue delay' in releasing the names. Apparently, the anxiety of thousands of people was intensified by the delay. Not realising the magnitude of the task that faced authorities, the Army replied that there was no undue delay. It assured the public and any next-of-kin that, following a similar accident of this nature, if no advice had been received from the Army, then there was no cause for anxiety. It was a fitting reply.

All 26 deceased Sappers came from the far reaches of the country. It was always going to be a monumental task to contact families with a speed quicker than the press. This at a time when not everyone had a telephone, were incapable of receiving a telegram or were living on the land in remote cattle stations and state forests. Spare a thought also for the families of Western Australian's Ron Linthorne, Alf Woods and Alf Witt. Thankfully Pauline Linthorne was in Melbourne, but what of Ron's mum, or Alf Woods' wife Jean and Alf Witt's father Frederick, all of who were thousands of miles away in the West? It took their brave family members more than five days on troop trains to get to New South Wales, how long therefore was it going to take them to first be informed of the tragedy, and then make their way to Kapooka in two days?

With the media releasing more details, more than 200 telegrams started flooding into Wagga Wagga Post Office from anxious parents and wives from the far reaches. The pressure was well and truly being applied to the Army to release the names and ease the angst being applied to more than 8000 families waiting for confirmation. But the Army remained resolute in their respect for the *actual* victims and their families. A flurry of regretful telegrams and phone calls was soon leaving the Engineer Centre to 26 families as early as 0700 hours of the Tuesday morning.

That same afternoon in Lorne, Marjorie Dilley was handed the small Postmaster-General's Department Telegram. It was folded multiple times. As she opened each fold, her heart got heavier: It could only mean bad news:

> It is with deep regret that i have to inform you that NX481536 SPR Norman Rourke John Dilley was accidently killed on 21st may 1945 and desire to convey to you the profound sympathy of the minister for the army,
>
> Minister for the army

By the end of the day, it was a similar story for 25 remaining families. Kevin Francis' aunt Mary, was advised in Collingwood of her nephew's death; Frank Platt's mother, Ernest Poschalk's father and Stan Morphy's mum had all been informed of the accidental death of their sons. Ivan's pregnant wife Joyce Merritt, Eric Grasby, Denby's father and Bill Reid's wife Ruby, had been advised via telegram and telephone that their loved ones had been involved in an explosive accident during training and killed instantly. In Semaphore, South Australia, Allan Bartlett's parents were advised that Allan had been badly injured in an explosion but he was alive. It was a welcome relief for his parents Ivy and Alf. As soon as Ivy got the news she sent a telegram to her other military son Alf, that his little brother had been badly injured in an explosion. As you'll soon discover, the news caused Alf to go AWOL and embark on an unauthorised journey to Wagga Wagga to be with his 'mate'.

In Mungindi, Stan Ross's mother got the distressing telegram and when Jack Nixon's wife Alice got the news, it became the most saddening moment of her life trying to explain to baby Neryl that Daddy wasn't coming home from the war. And, without warning, Ettie Woods and baby Alan, much like Alice and Neryl Nixon, would have to consider a life growing up without a husband and father. It was after all the unbearable reality of many Australian families during the war and the price for being a military family.

Colin Boyd's father and many other victims' fathers such as Roy Faull, Richard Hurst, William Cousins, Jack Robson, Stan Partridge and Ted Moore, all received word that their young soldier sons had been killed. For Joe Collins and Les Mather, their mothers were delivered the cruelest of news. Meanwhile, Allan Flood's elder sister Ellen, already the inconsolable wife of a soldier husband missing in action, was delivered even more distressing news. Her much loved little brother had died. Suddenly the two most important men in her life were both gone. Her husband was missing presumed dead, and now, the one who should have been the safest, was confirmed dead in an accident.

Over in Herons Creek, when one of *those* telegrams arrived for Mark Hurley, word quickly spread throughout the village. After passing unopened through several sets of hands, it finally landed in the hand of Colin's father. Mark fearfully opened the message to reveal the heartrending news from the Minister for the Army of his young son's death. Mark and Elsie were distraught; their youngest child was gone.

By the night of the 22nd of May, all next of kin had been informed. But for many families the notification of burial arrangements was too short. The Army was planning a large funeral for the following day (Wednesday the 23rd of May), despite the fact that Army Headquarters hadn't yet received all the replies of acknowledgements from the families. It had been widely reported in newspapers that the funerals were going ahead on the following day (Wednesday) and already major planning was occurring behind the scenes to have the men buried as quickly as possible. The reality was that 54th Army Camp Hospital's makeshift morgue was over-crowded. Without a suitable storage facility for their remains, the dead soldiers needed a swift, but dignified, farewell. But it appears families were a minor distraction to the even bigger idea of a memorable funeral. The Army had released details to the public that in just two days, they were planning to bury all 26 men at the same time in the military section of the Wagga Wagga Cemetery. For many of the families this was impossible. In some cases, it left them only one or two days' notice to get to Wagga Wagga. It just wasn't enough time. Their sons were about to be buried without them having the chance to say goodbye in person. It was a ludicrous plan, but the Army juggled their empathetic responsibility to the family and the need to deal with the biohazard of the remains appropriately.

In Bonegilla, Victoria, at the time of the accident, Alf 'Brick' Bartlett was attached to a transport unit working the busy Victorian military camps when he received his mother's telegram about Allan. Due to the seriousness of the disaster, he feared the worst and desperately needed to be by his younger brother's side. He approached his Sergeant Major and demanded to take some leave to comfort his younger sibling. Stubbornly, the serious senior soldier refused his request. Nevertheless, for a determined big brother, no one, not even his bully Sergeant Major, was going to stop him getting to his family. Allan was lying seriously injured, and alone, in an unfriendly military hospital a long way from home and family, and Alf knew he needed to be there. So, like any soldier would, he did the only thing he could think of to be by his brothers side in a time when he needed family—he went AWOL.

He asked his mate Jimmy Foster to run him into Albury where he jumped a train to Wagga Wagga. Meanwhile, his unsympathetic Sergeant Major reported his AWOL subordinate to authorities. He reported that more than likely the distraught Alf was on his way to Wagga Wagga to get to his brother. As Alf rolled into the station at Wagga Wagga, an unfriendly welcoming committee greeted him. He stepped off the train and into the waiting arms of two Army Police who immediately grabbed him and placed him under arrest. Fortunately for Alf an Army Padre was at the station watching the antics. As the two unmovable Provosts listened to Alf's explanation of who he was and that he needed to see his brother, the Padre jumped to his aide. The concerned man of the cloth approached the three men and asked the obvious suspect his name. When he replied to the Padre that he was Alf Bartlett, the soldier's pastoral caregiver took command of the situation. '… Hang on a minute,' he said, '… I think it's your brother who is still alive?' A sense of emotional relief came over Alf; he had to get to the camp, quickly. He knew then the Padre was his divine ticket. Every soldier gets a chance or two to thank god for one reason or another, and for Alf, this was it. The Padre rushed into the busy railway station office and took out a piece of railway telegram paper. He not only wrote an impromptu leave pass for Alf, he ordered the two police to escort him out to the camp immediately!

Arriving at the camp Alf was quickly ushered to the Camp Hospital where he looked in on his badly burnt, battered and bruised sibling. He could see his chest was seriously black-and-blue and as he got closer he saw the substantial and distressing burns on his childhood mate. With tears welling in his eyes, Alf was thankful to be there, but he couldn't stand to see the sight of his brother and best mate in such horrific condition. Although Allan was resting, Alf sat down beside him and kissed his burnt and bandaged forehead, '... I'm here now mate', he quietly whispered. Even though Allan's punctured eardrums weren't receiving noise, with his brother's touch his eyes opened. Allan's eyes met Alf's. He could see the emotion in his big brother's eyes. Upon seeing Alf for the first time in a very long time, a welcome smile caressed Allan's burnt and painful face. Through painful tears, their parallel smiles were the type of non-verbal communication the two close brothers needed for reassurance and comfort. In his brother's eyes Alf could see the painful story of how he came to be there, but in his smile he knew Allan was telling his concerned brother he was going to be okay.

The hospital staff graciously set Alf up in a bed beside his little brother. As he watched nurses bathe Allan in oil for the next four or five days, Alf couldn't erase the memory of watching nurses peel burnt dead skin off Allan's, traumatised body. Allan's cries of pain were a haunting reminder of the suffering he was enduring. Almost every day of his stay Alf witnessed Senior Officer's parade before Allan quizzing him about the accident. Everyday he told them the same story—he couldn't remember a thing. Throughout Allan's time in hospital Alf remained his loving brother's moral support through the relentless search for the cause of the blast and the unwelcomed intrusion of repetitive questioning. In the end, the fruitless questioning, in between Allan's around the clock care, resulted in frustrated authorities succumbing to the realisation that the distraught young soldier, although lucky to be alive, should be left in peace to recover. Physically and more importantly, mentally, Allan and Alf needed to look to the future.

In that hospital ward the two Bartlett men formed a bond that surpassed just siblings and friends during Allan's painful recovery. Alf's journey of support to be by his brother's side was a moving and loving example of his dedication, sacrifice and compassion, not just for his brother and a soldier, but also for another human in trouble. The two Bartlett men didn't go to the large military funeral to bury the victims on the Wednesday, and it would take Allan almost fifty years of recovery to get up the courage to do so. Of course he was supported by Alf when he eventually did.

Chapter Sixteen
Court of Inquiry

'… I have no idea how the accident occurred. The instructors were so experienced and well trained that I can only think that it must have been some unaccountable force. The lack of any other eye witnesses to the tragedy creates great difficulty."

Colonel W.D. McDonald,
Kapooka Camp Commandant.

Described as a 'monumental error' by unknown person/s involved in the ill-fated demolitions lesson, every Australian citizen who followed the voluminous media coverage, including every war-time politician, current serving sailors, soldiers, and airmen and every parent—especially those still frozen in grief—wanted to know the answer to the burning question, '… What went wrong at Kapooka?' Mr. Dedman, the Minister for Post War Reconstruction, relayed in his address to Parliament, '… an investigation was underway'. It was to take the form of an Australian Military Forces *Court of Inquiry*. Ordered by Brigadier Prior, the DA&QMG of the Second Aust Army, the purpose of the inquiry was to *'inquire into and report upon the deaths of and injuries to Army personnel by reason of accident which occurred at Wagga on 21 May 1945.'*

Brigadier Prior appointed Brigadier Alexander Forbes, the highly decorated Commander of Fixed Defences, New South Wales, to preside as President of the Court. Born in St Kilda, Victoria on 19 May 1892, Brigadier Forbes' military career commenced when he entered as one of the first ever cadets of the newly formed Royal Military College at Duntroon on 22 June 1911. He was later awarded Duntroon's first Sword of Honour for 'exemplary conduct and performance of duties'. In August 1914, following the outbreak of the Great War he was graduated early for active service with the Australian Imperial Force, and in 1917 as a Captain, he was awarded the Military Cross for gallantry. In 1918, having reached the rank of Major, he was Mentioned in Dispatches by Field Marshal Sir Douglas Haig. He returned to Duntroon following the end of the war and was appointed Adjutant of RMC. In 1938 he was the organiser of the unveiling ceremony of the Villers Bretonneux War Memorial by His Majesty King George VI and was appointed as a Member of the Royal Victorian Order for service to the Crown. He was a well-credited senior Army officer and hopefully the right man to get to the root cause of the unbelievable disaster.

In May of 1945, in his position as Commander of Fixed Defences, the experienced Brigadier was ideally suited to inquire impartially into the Kapooka accident; he did however need support from *specialist* officers. He needed advisers to explain and inform him on certain technical matters, especially when it came to engineering doctrine, the practice and procedures for conducting field demolitions, and he importantly needed advice on matters of law. For the engineering speciality, 33-year-old Major Wallis Dunphy, an experienced Field Company Engineer from Land Headquarters was appointed. For legal matters, a 5-year Major, Clive Slade, a member of the Australian Army Legal Corps from HQ Second Aust Army, filled the other important Court of Inquiry position.

Without hesitation, the Court members travelled from their respective headquarters in Sydney on Tuesday May 22, and convened at Kapooka Camp at 0900 hours the following day to start proceedings. The Court, pursuant to the order of Brigadier Prior under the *terms of reference*, was to take evidence from all available witnesses, conduct whatever experiments or re-creations as necessary and finally, make a report. The intention of the Court was to spend three to four days at Kapooka to undertake the important interviews and record all relevant statements. As it was a *Court of Inquiry*, it wasn't therefore a criminal investigation. They weren't looking for suspects to prosecute, there was no lawyers representing the families of the dead men, its main function was to conduct examinations with a view to make recommendations and arrive at well-informed opinions as to the responsibility for blame.

The first task for Brigadier Forbes and his Court was to inspect the scene and following that, extensive interviewing of all available witnesses. Prior to any formal proceedings though, as a matter of professional courtesy, the Training Centre Commander, Brigadier McDonald, met with Brigadier Forbes on the evening of the 23rd of May. The two men discussed the sadness of the tragedy and the finer details of what was to follow. Brigadier McDonald of course offered as much support as he could to the Court President, including unrestricted access to all necessary documents in order to formalise his upcoming proceedings. Contained in the requested documents was the training syllabus for the fourth week of demolitions training and of course the critical Monday afternoon instructional lessons.

The influential Officers also discussed the importance of Allan Bartlett and Colin Kendall's eyewitness testimony. Therefore, before official proceedings commenced, both men paid a courtesy visit to the two survivors in the camp hospital. Like the hundreds of thousands of everyday Australians who'd been following the tragedy via media reports, the two influential Officers were especially keen to hear what Allan and Colin had to say. However, as Allan was still receiving around the clock medical attention in the care of nurses and his compassionate big brother, he wouldn't be capable of attending the formal proceedings to be questioned.

Instead, Corporal Grace Anne Carroll, a 35 year old AWAS working at 54th Camp Hospital, recorded, via shorthand, an interview with Allan where she asked him a series of questions which had been set by the Court. Allan responded to the questions as best he could which were later presented to the Court.

When the Court formally convened indoors to interview the first witness on Wednesday morning, using military protocol, they called the Training Centre Commander, Brigadier McDonald as their first witness. Being duly sworn in to give evidence, the Brigadier explained that he had no personal knowledge of the explosion purely because he wasn't present at the camp that day. He did tell the Court that during his previous visits to the demolitions area he hadn't had occasion to reprimand any of the instructors for carelessness in the handling of explosives. Nor did he have to speak to them about the disregard of usual safety precautions. He added that he had never observed anything occurring on the demolitions area at any time that he considered to be dangerous to any personnel. Brigadier McDonald did add that there had only been one very minor accident on the demolition area since he took up the position in April of 1944. He wasn't able to elaborate on the specifics of that particular accident but stressed that the amount of explosives actually used on the training area in more recent times had in fact become quite heavy. Neither the President, nor the members of the Court, asked the Brigadier any more formal questions. The court recorded a typed statement, which was read back to Brigadier McDonald. He took the time to read through his testimony carefully; initialed five grammar corrections and he signed it accordingly as his true and accurate record. Proceedings were well and truly underway as the first important Commissioned Officer was dismissed from the Court.

Maintaining the military protocol of interviewing senior officers first, the next witness called was the Acting Chief Instructor, Major Macdonnell. After introducing himself and his duties, the Major proceeded to explain to the court that at 1450 hours Monday the 21st of May, he became aware after a conversation with Captain Merry that an accident had occurred. Then, accompanied by Major Berg, he proceeded to demolitions area. The Major explained to the Court the process of how and why Sergeant Pomeroy had a squad of 25 men that afternoon. The Court then presented George Macdonnell with the *Training Syllabus*, which had earlier been presented to the Court by Brigadier Mcdonald, along with other important documents. The Acting Chief Instructor explained to the Court that he had written up the syllabus and pointed out that the accident occurred in the fifth period on the Monday. The syllabus was tendered to the Court as an exhibit and marked A. He further presented the *Demolition Area Standing Orders*, which were respectively tendered as exhibit B. The Court then commenced questioning the Major regarding his actions immediately following the explosion. Satisfied with his answers, George Macdonnell produced the *RAE T.C. Précis of Safety Precautions Demolitions, Fuses and Detonators*. He told the Court that the Training Centre issues the précis to instructors as an aid only. Directing the Members of the Court to the appropriate area at the end of the précis, he pointed out the area regarding the three safety-first rules when working with explosive, in particular rule 3. Rule 3 states '...keep detonators separate'. The précis was tendered as exhibit C. During his testimony the Major also produced a plan of the dugout in which the accident occurred. Having satisfied his important actions after the explosion, and the production of necessary doctrine, the line of questioning changed and matters of technical knowledge were referred to the experienced Major. Majors Dunphy and Slade commenced probing the experienced engineering officer looking for details of how the accident might have been caused:

> Question (Q.) '... Is the dugout in which the accident occurred of a standard design?'
>
> His reply (A.) '... I know of no standard design for that type of dugout.'

Q. '... Is there another similar dugout in the area in question?'

A. '... Yes there were two including the one which I conducted the Court to this morning.'

Q. '... Will you tell me what the purpose of the dugout is?'

A. '... They were designed as lecture rooms to enable the continuity of lectures to be unbroken by practical demolitions in the area. Trainees can sit in these dugouts in safety from flying fragments.'

Q. '... Do you consider the design adequate and quite safe for the occupants during demolitions in the area?'

A. '... Yes.'

Q. '... Explosives were never left in these dugouts apart from training sessions?'

A. '... They may be left there during lunch hour and period for P.T. They would never be left there from the close of the day's work.'

Q. '... Do you know why three instructors were with the squad at the time of the accident?'

A. '... Yes. There were only two actually with the squad. The third instructor concerned, Lance Sergeant Kendall was engaged in practical work connected with the night exercise for which the squad was being trained by the other two instructors. It is standard procedure to have two instructors per squad.'

Q. '... Was it the intention when the Syllabus was drawn up by you that the trainees should handle the explosives during the period in question?'

A. '... Yes Sir.'

Following his admission, that in his experience he believed that trainees should handle the explosives on day one training according to the syllabus he designed, he signed his typed transcript of statement and was dismissed from the Court. Strangely, his assistant, Major Berg, was not called to give evidence.

Captain Ed Merry, the Commissioned Officer who perhaps wore the weight of the world on his shoulders as the Officer Commanding the fourth week of training at G Coy, was the next witness called. Being responsible for training and safety, he was expecting his questioning by the Court to be brutal, direct and constant, but equally fair and just. He proceeded by explaining to the Court his position as the Staff Officer Royal Engineers Grade III and Officer Commanding 'G' Company and Captain Instructor in charge of fourth and fifth weeks of training. He advised the Court that he'd been in the position for the last four months. The 40-year-old Captain relayed to the Court his version of events as they unfolded just two days prior. Being fresh in his memory he gave a good account of where and when he was informed of the explosion and his actions up until the evening of Monday the 21st of May.

Exhibits A, B and C were presented to the Captain and he provided evidence that he knew the documents well and he was able to comment as to their validity and use. Once the exhibits were removed Captain Merry advised the Court that orders regarding safety precautions were based on *Military Engineering Volume IV Part I Demolitions 1942*.

'... After the accident,' he advised, '... the remains of a copy of the above textbook were found amongst the debris in the dugout. I saw such remains there myself.'

As expected his thorough examination continued:

Q. '... Who was in charge as instructors of the squad at the time of the accident?'

A. '... Sergeant Pomeroy was in charge and Corporal Cousins was his assistant.'

Q. '... What experience had each of these instructors had in handling explosives and were they in your opinion thoroughly dependable?'

It was the one question that Ed Merry had been expecting and fearing; but his answer came from the heart.

A. ' ... Sergeant Pomeroy had been in the 2/8th Field Company for approximately three and a half years. In civil life he had been a miner and I understand that he had worked in munitions on the explosive side before joining the Army. He was a thoroughly reliable and conscientious instructor, and was the most solid instructor in fourth week. He was the last man I would have expected to have an accident with explosives. Corporal Cousin was a good and reliable instructor.

Both these NCOs were instructors under my command for approximately four months.'

Q. ' ... Do you consider the existing circumstances desirable to carry out instruction in the handling of explosives in a dugout rather than in the open?'

For Ed it was a difficult question to answer considering he had just told the court that Sergeant Jack Pomeroy was the most conscientious instructor and the person least likely to have an accident.

A. ' ... All things being equal it is always desirable to handle explosives in the open rather than in a dugout but owing to the necessity of protecting troops under instruction from weather and from other demolitions being conducted in the area the use of the dugouts as lecture rooms is essential.'

Q. ' ... During this period what size charges were supposed to be made up by the instructors and the recruits?'

A. ' ... A one four-ounce plug charge of monobel, together with detonator and safety fuse would be made up by the instructor. Normally all charges prepared by the recruits were made up above ground and not down below.'

Q. ' ... Do you know how much explosive is normally issued to the squad for this exercise?'

A. ' ... One hundred pounds monobel and ten pounds gelignite with the necessary fuses and detonators. A complete list is kept by the storeman.'

Q. ' ... Was an extensive search made for any explosives in the dugout or in the immediate vicinity?'

A. '... Yes an extensive search was made for explosives both in the dugout and the surrounding areas.'

Q. '... Do you consider that all the explosives issued to Sergeant Pomeroy was destroyed in the explosion?'

A. '... Yes but not necessarily all detonated, some of the monobel may have burnt without detonation.'

Q. '... Could five pounds only of explosive have caused the damage that was caused?'

A. '... No.'

Q. '... Why were there two instructors with each squad?'

A. '... One instructor gives the actual instruction and the other is used to supervise the squad as it is considered that an instructor cannot instruct and supervise at the same time in demolitions.'

Captain Merry's time in the witness box was over with the final line of questioning regarding why two instructors were underground; in this case Jack Pomeroy and his assistant Corporal Bill Cousins. Meticulously Ed read his three-page typed transcript. He knew his testimony would be one of the ones heavily scrutinised not only by the Court, but also by the Minister for the Army. Being a staff officer, he corrected some typographical errors before signing the document.

By now the first day of the inquiry was drawing to a close. However, the President was determined to hear the testimony of Colin Kendall and Allan Bartlett whilst the testimony of Ed Merry was still fresh in their minds. Their principal evidence was perhaps going to explain in some way the possible cause. Before calling a recess to the proceedings until the following morning, he called for Colin Kendall.

Colin Kendall's time in the witness box was in no doubt by any of the Court personnel to be of far greater importance than Ed Merry's. Colin was one of the closest to the explosion and was present in the dugout just moments before it imploded entombing the 26 men. Having given his evidence and therefore dismissed by the Court, Ed Merry took up a seat in the gallery. He also needed to hear the testimony of the only experienced survivor. The Lance Sergeant opened his testimony by giving an accurate appreciation of where he was, what he was doing at the time and what led him to place three of his men just to the left hand side of the entrance of the dugout. ' ... These men,' he advised, ' ... were making up detonators with varying lengths of safety fuse with match-heads attached along with electric leads and the whole thing bound together with used detonator wire.' He told the Court how he was only seven yards away from the entrance to the dugout, in the vicinity of his own parcel of explosives, when he felt an intense blast, which singed his hair and eyebrows. Graphically, he depicted to the Court how a portion of a man's body rained down on him as he lay near the entrance to the collapsed dugout. Colin continued to explain the post-explosion treatment given to him by Ed Dodds and Roy Tafe and then being placed into the ambulance. The two Majors and the Brigadier commenced a cross-examination of Colin's facts:

> Q. ' ... Before you left the dugout did you notice whether Sergeant Pomeroy, his assistant or any of his squad had taken explosives from their containers or were handling them?'
>
> A. ' ... No I did not notice.'
>
> Q. ' ... Was Sapper Bartlett one of your three men?'
>
> A. ' ... I could not say. It was the first day of training and I had not up until then ascertained the names of my squad. Sergeant Tafe had called the roll of the squad that morning.'
>
> Q. ' ... If one of your three men had accidently exploded a detonator do you consider it possible for sympathetic detonation to have been set up in any explosives which might have been amongst Sergeant Pomeroy's heap of training stores?'
>
> A. ' ... No. They were too far away.'
>
> Q. ' ... From your knowledge of Sergeant Pomeroy and Corporal

Cousin do you consider that they were dependable men who could always be relied on to observe all safety precautions?'

A. '… Yes.'

During Colin's examination, the Court had received word that Allan Bartlett's testimony, as recorded by Grace Carroll, had been transcribed and read to Allan in his hospital bed. Brigadier Forbes momentarily dismissed Colin and asked him to leave the room but advised him he still had further questions to be asked. The importance of Allan's testimony was paramount. The President wanted to hear the testimony of the lone survivor from actually inside the dugout when the explosion occurred. After much anticipation, Allan's testimony was read to the Court in a question and answer format:

Q. '… Were you a member of Sergeant Pomeroy's squad?'

A. '… I was.'

Q. '… Can you remember whether just before the explosion occurred Sergeant Pomeroy was handling any explosives?'

A. '… He was showing us how to handle it correctly.'

Q. '… Do you remember if he had any explosives in his hands?'

A. '… I don't remember.'

Q. '… Do you remember if Corporal Cousins had any explosive in his hands?'

A. '… He had his hand full of detonators with blue fuse.'

Q. '… Was he demonstrating with these?'

A. '… No. He was getting ready for the stunt on Monday night. He was counting them.'

Q. '… Had any of the squad been handling any of the explosives up till then?'

A. '… No. He might have passed a few around. He would not give us any that would be capable of doing damage.'

Q. '… You have no idea what caused the explosion?'

A. '… No. The last thing I remember the Corporal had these things in his hand. I turned to say something to my mate and the next I remember I was being dragged out of the hole.'

> Q. '... When you saw Corporal Cousins with the detonators in his hands was he standing close to the heap of explosives and training stores?'
>
> A. '... Was he the tall Corporal?—no he was right on the other side.'

Regrettably, 18-year-old Allan's memory could only capture the period of time where he observed Bill Cousins standing with a handful of detonators in the dugout and then, awakening from unconsciousness and being dragged from the dugout. It wasn't what the Court was hoping for. His testimony didn't answer the burning question of—how? His testimony was tendered to the Court in his absence in the form of a statement. As far as the Court was concerned they were now in receipt of the only evidence from within the dugout and called for the return of Colin Kendall to the witness stand. Completing Colin's questioning would signal the end of the first day.

> Q. '... While you and your three men were working in the corner of the dugout making up fuses and detonators with match heads were the materials you were using part of the explosives which had been drawn for Sergeant Tafe's squad or were they part of those drawn for Sergeant Pomeroy's squad?'
>
> A. '... They were part of the explosives drawn for Sergeant Pomeroy's squad.'
>
> Q. '... Before you left the dugout were you at any time standing in the dugout with a handful of detonators and blue fuse?'
>
> A. '... No.'
>
> Q. '... When you left the dugout were your three men continuing their work under the supervision of Sergeant Pomeroy and/or Corporal Cousins?'
>
> A. '... Yes.'

The Lance Sergeant completed his testimony by firstly having his three pages of typed transcript read back to him. In return, he read the documents and signed them accordingly once he was satisfied it was a true and correct reflection of his oral testimony. Dismissed from the witness box, day one of the inquiry concluded in the shadows of the key testimony from two survivors. At the end of day one the Court was none the wiser as to the cause.

What the Court did establish after day one though, especially following the evidence of Allan Bartlett and Colin Kendall, was that the actions of assistant, Corporal Bill Cousins, when handling detonators, might have been a contributing factor in the catastrophic explosion. It was now speculated that if what Allan Bartlett observed just before losing his memory was an important piece of circumstantial evidence. According to Allan, before he lost consciousness, witnessed 25-year-old Bill Cousins standing to the right of Jack handling the detonators with blue fuses, which were being made by Colin Kendall's men in the corner of the dugout. Colin Kendall denied that it was he who was standing in the corner of the dugout holding the detonators but did confirm the three men working under the supervision of Jack and Bill were members of his smaller squad. The Court was left to ponder overnight the testimony of the day and would need to focus on whether or not the actions of Bill Cousins caused the detonation or was it something that the experienced demolitions Sergeant had done wrong!

Thursday the 24th of May started early for the Court; it was expected to be an important day, very much like day one. Both Major Slade and Dunphy were looking forward to questioning the remaining witnesses, especially Roy Tafe as the first man on the scene. His observations of where Allan Bartlett, Ivan Merritt and Bill Reid were found and, his observations as to the condition of the dugout, might give some indication as to where the explosion originated. Was it in the southwest corner where Jack was standing or somewhere else? What of Allan Bartlett's testimony that he observed Bill Cousins carrying the only real source of ignition of Jack's volatile parcel of explosives?

Apart from Roy Tafe they also had more than a dozen witnesses to interview including Ed Dodds, Archie Smith, Doug MacFarlane and of course Fernleigh Ellis. On top of that they still had the matter of victim identification to address. And finally, all the medical officers involved in the recovery of bodies and treatment of injured were also to be interviewed.

The first interviewed on day two was Warrant Officer Class Two Doug MacFarlane. He advised the Court that he was not present on the day and had no direct evidence concerning the explosion. However, as the Key Instructor and Sergeant Pomeroy's direct superior officer in the chain of command, he told the Court as much as he could including his knowledge of the exhibits and how they applied to him and his position. When shown the *Training Centre Précis*, (Exhibit C) he advised the Court in his opinion all instructors were familiar with it. He proceeded to tender notes that were copied by Sergeant Pomeroy, which were notes on safety precautions in the demolitions area that weren't laid down in the Military Engineering Volume, but notes he relied upon in lectures. They were tendered as exhibit F. He wasn't sure if Bill Cousins was aware of the notes, but was almost certain that Jack knew of these additional safety precautions. Following his tendering of the document formal examination by the Court began. The two Court Majors were specifically looking to validate the testimony of Allan Bartlett and Colin Kendall:

> Q. '... Sapper Bartlett in his evidence refers to a tall Corporal instructor standing with a handful of detonators and blue safety fuse which were prepared for the night exercise. It is normal for the squad to have prepared these fuses and to have done so at this stage?'
>
> A. '... It is normal for the squad to prepare hand charges for the night exercise but not at that stage. In my opinion the instructor had prepared the detonators and fuses referred to. It is quite possible that one instructor had made up these fuses in the morning when he was not required by his co-instructors.'

After a series of questions establishing that Bill Cousins was the tall Corporal and at six foot one inch was the same height as Jack Pomeroy, further probing questions of a technical nature were asked of the Warrant Officer:

Q. ' ... Sapper Bartlett states the tall Corporal holding the detonators and fuses was standing in the dugout on the side opposite to that against which the explosives and the training stores were stacked. Do you consider that if by accident these detonators had exploded would it have been possible to detonate the ten pounds of gelignite known to be amongst the training stores?'

A. ' ... At that distance, no.'

Q. ' ... If the tall Corporal had have been standing within a few feet of the gelignite and the detonators had exploded could the gelignite have been detonated?'

A. ' ... It is possible but extremely unlikely.'

Q. ' ... Is it normal for such a large amount of explosive to be in the dugout during the lectures?'

A. ' ... It was normal except in the case of adverse weather in the discretion of the instructor.'

Q. ' ... What do you consider would have been the effect if only ten pounds of gelignite had been in the dugout and had exploded?'

A. ' ... I do not consider that the ten pounds of gelignite would have had the same effect but would not like to express a definite opinion.'

Having told the Court that he has been a member of the RAE for over five years, that he has extensive experience with explosives, and the fact that he'd been at the Centre for seven months, the experienced Warrant Officer wasn't prepared to make an informed opinion on the nature and effects of ten pounds of gelignite. The interviewing officers did not probe him any further on this issue but if he had agreed that ten pounds of gelignite could have caused the same destruction and, according to Captain Merry's testimony that the monobel may have burnt not detonated, it may have been possible to isolate the cause of the explosion to the gelignite, not the four ounce plug of monobel Jack had in his hand at the time of the explosion. With his mute opinion tucked away on record, the transcribed evidence was read back to the experienced demolitions Warrant Officer. He dutifully read and signed his transcript at which time the President summarily dismissed him.

The Court members discussed the issue of what explosive compound detonated and what compound may have burnt and soon arrived at the decision that the only way to replicate the effects of both ten pounds and 100 pounds of high explosives was to conduct similar explosions. Based on the urgency at which the Court of Inquiry was operating, the Court requested the RAE TC set down Friday the 25th of May to conduct the tests. The experimental detonations would become known as the 'two experiments'. For now though, examinations continued.

Ed Dodds was the next to be interviewed. He too preliminarily advised the Court of his actions on the day and the confusion with the squad roll books when he called the roll on the Coy parade ground on the Monday morning. Ed explained the issue of releasing nine men from Roy Tafe's squad and two from Corporal Conwell's from the day's training, which subsequently placed all the men in Jack's dugout. He explained to the Court that about 1130 hours on the day of the explosion, Roy Tafe advised him that as he only had three men he placed them in Jack's squad so that they could get the benefit of the morning's training. He admitted that he was the one responsible for allocating more men to Sergeant Pomeroy's ill-fated squad. When six extra men reported to him after lunch he sent three to James Conwell's squad and three to Jack's to help Colin Kendall with preparing the night charges. When asked what the usual numbers of men were in a training squad the Assistant Instructor replied, '... the largest squad I have known is 35 and they vary below that'. The line of questioning by the two Majors towards Ed Dodds primarily concerned the fact that Sergeant Pomeroy was working with explosives and detonators together underground:

Q. '... Is it a usual practice for work with detonators and fuses in preparation for a night exercise to be carried on in a dugout while a lecture is in progress in that dugout?'

A. '... No it is not the usual practice it is usually done outside.'

Ed Dodds had managed to throw a snippet of confusion into the proceedings. Although he was correct in his summary, that it was usual practice for work preparation of the night charges to be made up above ground, there was no pursuance by the two questioning Majors of the reasons why experienced Sergeants like Jack Pomeroy and Roy Tafe agreed for this to occur on this occasion. Perhaps missing vital questioning to find out why, Ed Dodds was dismissed by the Court after signing his statement.

Tom Musto replaced Ed Dodds in the witness box. Although his only involvement on the fateful day was to issue the munitions to the three squad instructors at 0915 hours on the Monday morning and ordinarily tasked to collect the munitions at the end of the day, he did have important information to add about the stability of the explosives. Tom advised the Court that the gelignite had been received at the camp on the 9th of April 1945 and the monobel was received a little earlier on the 26th of March. He further added that the explosives were in good condition, they hadn't been opened, and he was the storeman responsible for issuing to the men on the demolitions area that morning as he always did. Following the munitions storeman, a pensive Archie Smith was due to give evidence next.

Having endured a couple of the hardest days of his life, the last thing Archie Smith wanted was to be grilled by a number of senior officers about what he did and how he identified the victims. It was a case of having to relive a nightmare over and over again—26 times. Archie gave his evidence that at approximately 1520 hours on 21 May 1945 he attended the Camp Hospital and inspected the bodies of men from his battalion. Only one man, Corporal Cousins, wasn't from his battalion. He added that nineteen of the men:

Alf Woods, Denby Grasby, Terry Moore, Frank Platt, Les Mather, Ernie Poschalk, Kevin Pierce, Colin Hurley, Stan Morphy, Toddy Woods, Allan Flood, Joe Faull, Joe Collins, Geoff Partridge, Norn Dilley, Jack Nixon, Alf Witt, Ivan Merritt and Bill Reid were still wearing their identity discs which confirmed their identity. Of the remaining seven men, he told the Court that a wristlet watch bearing the initials 'R.L' identified Sergeant Ron Linthorne and upon the discovery of a gold finger ring bearing the initials 'R.L' it confirmed his identity beyond a reasonable doubt.

Archie Smith explained how Sapper Stan Ross was identified by two of his tent mates, which was further confirmed through dental records for treatment a few short weeks prior to the accident. A pair of trouser braces bearing his details, which were subsequently destroyed for hygiene reasons and not shown to the Court, identified Sapper Hurst. A pair of issued underpants identified Colin Boyd, again which were destroyed and not shown to the Court. The Captain proceeded to inform the Brigadier and his two Majors that the instructor, Sergeant Jack Pomeroy, was identified by a wristlet watch bearing the details 'J. Pomeroy' which was shown to the Court.

Of the final two victims, tent mates identified Bill Cousin's remains. And although Sapper Ted Robson's body was not formally identified, his identity discs were found in the dugout. Considering he was known to be in the area at the time by virtue of his name in the roll books he was identified as one of the casualties. With no further examination or questioning, a relieved and emotional Archie Smith signed his transcript and left the witness box.

That afternoon Major Tunley and Captains Judd, Bennett and Miller, corroborated the testimony of identification made by Archie Smith and added their own story to the Court records about where they were and what they saw that terrible afternoon. Major Tunley added to his testimony that he was the Medical Officer responsible for signing the 26 Death Certificates.

Captain Fernleigh Ellis briefly appeared to give his evidence. In particular, he gave testimony of his action and interactions with the civil police. He stated to the Court that the Wagga Wagga police were notified at 1535 hours on the Monday, less than an hour after the explosion, where he informed them that an accident had occurred at the camp resulting in the death of a number of soldiers. He added that approximately 25 minutes later, at 1600 hours, Sergeant Sherwood of the police arrived at the camp. He further advised the court that Sergeant Sherwood produced a copy of the *National Security Regulations*, which dealt with the dispensing of Coronial Inquests of Army personnel where the occasion warranted it. It was the opinion of Captain Ellis and Sergeant Sherwood that this was one such occasion, and he advised the Court of the reasoning for his decision. Concluding his testimony, the Captain presented the Court with copies of the 26 Death Certificates compiled by Major Tunley. They were later tendered as exhibit G.

The final witness on day two was in fact the most anticipated—Sergeant Roy Tafe. Opening his testimony with a detailed account of the day, starting in the afternoon working alongside Phill McNabb in the Demolitions Shed, he concluded his summary with his actions of removing bodies from the dugout. He told the President that he was responsible for locating Allan Bartlett and carefully extracting him from the dugout. As anticipated the questions were being asked of him about his observations of the dugout and the nature of the damage. The Court was trying to establish through Roy's testimony any additional information about the origins of the detonation. Throughout the day little evidence was received which could assist them in establishing a cause of the explosion and they hoped his detailed observations would help:

Q. '... Describe the state of the dugout after the explosion.'

A. '... It seemed to me that the roof and all wall timbers from the centre line of the dugout to the western side had been lifted right out of the dugout. From this centre line to the eastern side all timber members of the structure were shattered but portions of them were lying in the dugout.'

Roy was shown a copy of the interior of the dugout; exhibit D, which was Major Macdonnell's plan of the type of dugout in which the accident occurred. Roy was asked to comment on the plan and what he observed in comparison:

> A. '... In my opinion the demolished dugout is a replica of this plan.'
>
> Q. '... From your observations of the conditions of the dugout at what approximate point of the dugout do you think the explosion occurred?'
>
> A. '... From the condition of the floor and western wall I would say that the explosion occurred near the western wall and towards the northwest corner.'

Roy's opinion as to the origin of the detonation, in the northwest corner, was the same location in which Jack would've been standing beside his parcel of explosives and detonators. Roy's testimony concluded with the Court asking him to explain what his squad was supposed to have been doing during the period of instruction in which the accident occurred. He completed the day's testimony by adding that he left Colin Kendall to supervise three men in removing his explosive stores from the dugout and placing them outside Jack's dugout. They were to get as much work done preparing detonators before Jack's squad returned to the dugout to continue lessons after lunch. He added that he advised Ed Dodds of *all* the changes.

Roy Tafe's testimony ended the day for all members of the Court and the remaining anxious witnesses. Roy signed his transcript statement of testimony and was dismissed; however, the President advised him that he would also be required in the demolitions area tomorrow (Friday) to watch two experimental explosions. The President of the Court also advised Captain Merry, as the other key witness to the explosion, to also make himself available on Friday to observe the experiments. Interestingly, it's not clear why Phill McNabb, Sheila Oehm or any other instructors, such as Frank Sim, or any of his two-dozen or so 18-year-old trainees he was instructing on mines and booby traps in an adjoining area that day, weren't called to give evidence.

The following morning, Friday the 25th of May, the President and his Court Members accompanied a host of eyewitnesses including Roy Tafe, Ed Merry, key RAE Training Centre staff armed with cameras, and demolitions experts carrying the equivalent amounts of identical explosives from Monday's disaster, onto the demolitions area. The task of the RAE demolitions staff for the morning was to replicate two particular underground explosions for the purpose of the Court. Having inspected the actual dugout in which the tragic explosion took place on the Tuesday, the Court was interested to observe not only the power and effects of the blast, but examine just how Allan Bartlett survived whilst 25 men around Allan were killed instantly. In some cases, the men seated around him were reduced to small parts and were extracted from the dugout in bags, yet puzzlingly, he escaped with minor non-life threatening injuries. Perhaps they hoped their findings would support the hypotheses that they might already be opining about the cause of the tragedy and therefore an explanation of just who or what saved Allan's life.

Using two of the remaining dugouts, two separate detonations were to occur. The first was to determine how much damage would've occurred if only ten pounds of gelignite exploded and 100 pounds of monobel just burnt away, as suggested by Ed Merry in his testimony. The intention of the second demonstration was then to replicate the volatile demolition by detonating 100 pounds of monobel and ten pounds of gelignite in a structurally similar dugout. Again, it was so the President and the Court could observe the effects and ponder the question of whether or not Jack's entire parcel of explosives all detonated at once, or was the amazing casualty rate the result of just ten pounds of gelignite? As the dugouts were all roughly the same dimension and construction, the Court wanted to observe firsthand the aftermath of both explosions and understand which of the two, or both, could've caused the alarming number of deaths.

As their questioning of witnesses sought to discover, especially that of Ed Merry, was a 10-pound gelignite explosion sufficient to cause the types of injuries and disfigurement of the dugout occupants? The detonation experiment of 10-pounds of gelignite was created to replicate the conditions inside the dugout. To do so, ammunition boxes, similar to the ones the men were sitting on during the fatal explosion were placed around the dugout in the same 'hollow-square' formation. The gelignite was placed in the northwest corner according to the evidence of Colin Kendall and Allan Bartlett; the identical location where Jack Pomeroy stood over his large parcel that Monday afternoon.

From a safe distance witnesses observed as the first of the two experiments was detonated. The Court and accompanying witnesses watched as the roof of the dugout, partially enveloped in dust, seemed to lift a few inches before settling back in its original position. Once the dust had settled, and the area declared safe, the President and his two members walked down the rear steps into the dugout to inspect the interior scene. What they observed stunned them a little.

The bottom two feet of one of the upright timbers was missing. The empty boxes, representing the soldier's seats around the inside walls, were shattered. But the most alarming observation was reserved for the height of damage clearly visible on the inner earthen walls. At a height of approximately twelve to eighteen inches from the floor, a distinct indentation, approximately six inches deep and twelve inches wide, appeared around all four walls. Major Slade called for another ammunition box to be put in the dugout. When the box was brought in he placed it against a wall and sat down. Sitting atop the timber box, he observed the height at which the damage was sustained to the wall, it corresponded approximately to his stomach region. It was a startling, but not necessarily alarming, outcome. The Court members reached the same conclusion. There was no doubt the injuries sustained, in just a 10-pound detonation, would've been fatal. The Brigadier, his two Majors and the entourage of witnesses and assistants, exited the dugout. It was time for experiment two.

Experiment two, being a larger detonation, was expected to be both noisy and violent. To replicate the location of witnesses at the time of the fatal explosion, Roy Tafe was positioned at the Demolition Store shed just as he was on Monday afternoon. Ed Merry was placed on the road about 800 yards away from the dugout, just as he too was positioned on Monday as he drove to inspect Jack Pomeroy's training of the oversized squad. Thankfully, this time, Colin Kendall was as far away from the danger as possible. During experiment two he was standing between Brigadier Forbes and Majors Slade and Dunphy watching it all from a distance. In Wagga Wagga, thanks to heightened media interest in the deadly blast, the town was well aware that the Australian Military Forces' Court of Inquiry was in town investigating. Almost hourly, for the last day or so, the town had been bracing itself for yet another large underground rumble. They knew all too well an experimental earthquake was coming.

With cameras fixed on the last remaining dugout, demolition experts prepared the charges. Their aim was to detonate 100 pounds of monobel and ten pounds of gelignite. It was the very same parcel of explosives that Jack was standing next to as he gave his instruction to his 22 attentive trainee combat engineers and three of Roy Tafe's trainees working in the corner of the dugout.

The charge was set off.

Like a jet of hot earth, the detonation spewed Riverina soil high into the air. It was an impressive yet equally disturbing sight. The impatient Brigadier Forbes wanted to inspect the area as soon as possible, but a large cloud of dust hung around the site like an unwanted fog. When it eventually cleared, the President waited for the 'all clear' from the engineering staff and hurriedly walked in to the obliterated dugout to inspect the damage.

What every onlooker initially understood, after witnessing the highly volatile explosion, was that a cloud of dust would've enveloped Colin Kendall and Allan Bartlett for a significant period of time. It also showed that Roy Tafe, regardless of whether a secondary explosion was possible, heroically ran blindly into an unforgiving dust storm to help his comrades. The Court duly noted his bravery.

Walking across the shaken earth towards the dugout, the extent of damage was clearly and instantly observable. The results were devastating. The top cover of soil, which was raised several feet high and gave the dugout overhead protection, had been leveled. Upon closer inspection the Court President and his members conceded that the damage was remarkably, almost identical, with the remains of the No 1 dugout which claimed the men's lives. After spending an emotional and reflective time inspecting the remains of the dugout, Brigadier Forbes motioned that he'd like to re-interview Roy Tafe and Ed Merry. The two men were also standing at the precipice of the demolished dugout and were amazed at the similarity in damage; only this time the smell of charred and burning flesh and the organized chaos of the recovery mission was missing. The two men were visibly shocked. Standing in an eerie silence in comparison to the mayhem of four days ago, both men had lived a horrible existence since. But in the short term there was no relief from their private nightmare and the pain of reminiscence; the President wanted them to be re-interviewed the following day.

Roy Tafe took to the stand early Saturday morning the 26th of May. It was now five days after the initial explosion:

> Q. '… Where were you when the charge of one hundred pounds of quarry monobel and ten pounds of gelignite was exploded?'
>
> A. '… In the same vicinity from which I witnessed the explosion on the day of the accident.'
>
> Q. '… Would you compare the two explosions?'
>
> A. '… In my opinion the appearance and sound of this charge was identical with that of the former one, on twenty-one of May.

'The details of similarity that struck me was the mushrooming effect, height and appearance of falling debris at first glance. On moving over to the site of the experimental explosion I noticed that similar to the original charge the roof of the dugout had completely blown-out and the effects of the explosion generally on the dugout were similar to those caused by the explosion which occurred on twenty-one May.'

Finally, after the longest five days of his life, Roy Tafe's questioning and 'reliving' of the tragedy was completed. His final testimony, after witnessing the two experiments, gave enormous strength to the theory that *all* of the explosives in Jack's dugout, not just the ten pounds of gelignite, had detonated at once. For the Sergeant the finding was a cold comfort for his conscious. The detonation was unexpected, frightenly quick, and for the victims, painless. Signing his testimony he left the witness box for the last time. Roy's departure did signal Ed Merry's arrival as the final physical witness for the Court. Ed Merry's observation of the two experiments solidified the Court's theory as to how much of the high explosives actually detonated:

> Q. '... You were present at the demonstration held on Friday twenty-five May forty-five in which a dugout almost identical to the dugout in which the accident occurred on twenty-one May forty-five was demolished. The first demonstration was the exploding of ten pounds of gelignite in the dugout. From your inspection after this explosion would you consider that if there had been occupants seated in the dugout would there have been any chance of survival?'
>
> A. '... In my opinion there could have been survivors but the injuries received would have been so severe the majority could not have survived for long.'
>
> Q. '... In the second demonstration one hundred pounds of quarry monobel and ten pounds of gelignite were exploded in the dugout. Did you witness this explosion from the same place as that from which you witnessed the explosion on May twenty-first?'
>
> A. '... Yes.'

Q. '... Would you compare the effects of the two explosions?'

A. '... The sound appeared to be the same and the earth was thrown up to the same height by both explosions. I then proceeded to the scene of the explosion and inspected the result. The dugout appeared to be in the same condition as the dugout which was blown on the twenty-first May.'

At 1100 hours on Saturday morning, Brigadier Forbes and his two Majors, packed up their Court at Kapooka. Shortly thereafter, after exchanging pleasantries with the RAETC Headquarter Officers and staff, returned to Sydney to finalise their Inquiry. Considering all the evidence and the outcomes of the experiments, the intention of the Court was to have the final report, complete with causal theories, hypotheses and opinions, signed off within the next seven days. Upon completion it was to be delivered to Major General Lloyd at HQ Second Aust Army who was to forward it to Land Headquarters. Then finally, after Land Headquarters had analysed the findings in order to make training improvement recommendations, the report would end up in the hands of the Secretary of the Department of Army for the Minister's Office to make further comment in the House of Representatives.

Soon, the whole of Australia would learn exactly what happened that day and just how 26 men were killed instantly during a routine training activity.

Above: The war office turned every available building into recruiting depots. This building is somewhere in Melbourne circa 1939.

General Sir Thomas A. Blamey GBE KCB CMG DSO ED, Commander-in-Chief, Australian Military Forces. Wagga Wagga's influential WWII leader who chose his hometown as the ideal location for engineer training.

Major General Sir Clive Selwyn Steele KBE, DSO, MC, VD. Kapooka's 'founder'.

Six man tents, similar to the above, complete with floor boards and rifle racks, dotted the Kapooka landscape and was home to more than 8000 RAE trainees.
AWM P01105.009.

Physical training- Military style. After a hearty lunchtime meal on the demolitions area at the RAE Training Centre, trainees were required to parade in the afternoon sun for even more compulsory physical exercises. Star jumps, press-ups and running on the spot were popular exercises of instructors.
State Library of Victoria image H98 105/3832.

'Realistic Training'. On dangerous assault courses Sappers moved through barbed wire obstacles as bombs explode around them.
State Library of Victoria Image H98.105/3877.

'... make your body a bloody bridge and get over it'. The RAE Training Centre obstacle course 'The Mad Mile' was legendry. Sergeants encouraged the men to throw themselves over the barbed wire entanglements whilst fellow trainees negotiated the obstacle by stepping on their backs and heads. Meanwhile, explosions were detonated just mere feet away. It was both a frightening and exhilarating part of the engineer training program.
State Library of Victoria Image H98.105/3876.

Sappers in training. Climbing and throwing themselves over a rope net was only one of the many treacherous obstacles at Kapooka Camp.
State Library of Victoria image H98.105/3869.

More realistic training. Trainees at the RAE Training Centre cross a cable bridge whilst under attack from a low flying aircraft.
State Library of Victoria image H98.105/3830.

Trainee Sappers at the RAE Training Centre experienced realistic training in order to best prepare them for war. Trainees advance on the enemy as demolitions are exploded before them.
State Library of Victoria image H98.105/3871.

Sappers in training practice attacks under cover of wood.
State Library of Victoria image H98 105/3831.

At the RAE Training Centre, engineers were instructed in the use of volatile explosives. Engineers take cover as a demolitions exercise blows debris high into the air.
State Library of Victoria image H98.105/4644.

Moving across a field, RAE Training Centre Sappers are confronted by heavy "enemy fire" and an exploding minefield.
State Library of Victoria image H98.105/3872.

The lure of service in the 2nd AIF convinced Herbert John 'Jack' Pomeroy, and his five brothers to make the transition from civilian into soldier. The excitement and danger of military engineering saw Jack enlist into the 2/8 Fd Coy in June of 1941.
Courtesy of Jack's daughter, Maureen Raunic (nee Pomeroy).

Above left: *Apart from his adored family, one of Jack's great loves was cycling.*
Above right: *Sapper Jack Pomeroy during his operational duty with 2/8 Fd Coy, somewhere in the Middle East. Photo taken circa 1942.*
Courtesy of Maureen Raunic (nee Pomeroy).

Above: *Jack and Dorothy with baby Maureen circa 1945. Maureen was just 10 months old when the explosion rocked the Wagga Wagga Township in the afternoon of 21 May 1945.*

Courtesy of Maureen Raunic (nee Pomeroy).

Above: *A timber box of Du Pont Monobel explosives similar to the one trainees sat on around the wall of the dugout during demolitions instruction.*
Courtesy of google images.

View of the entrance and front of the underground dugouts on the RAE Training Centre Demolitions Area. Note the steps leading down underground and the small aperture allowing light and fresh air into the dugout. This particular dugout was later blown up during the Military Court of Inquiry.
NAA MT885/1 51/1/209 page 5.

View of the actual dugout in which the detonation of 110 pounds of high explosive claimed the lives of 25 Sappers and Sergeant Jack Pomeroy on the RAE demolitions area, Kapooka Camp Wagga Wagga. The remains of the men were removed (some in plastic bags) by RAE Training Centre Staff, which included female AWAS soldiers who were as ambulance drivers and medical orderlies at the 54th Camp Hospital.
NAA MT885/1 51/1/209 page 27.

Above & Below: *More views of the ill fated dugout in which 27 men sat around the inner walls listening to the experienced instructor Jack Pomeroy talk about demolitions. The force of the explosion lifted the roof of the underground dugout on one side and collapsed it upon the men on the opposite side.*
NAA MT885/1 51/1/209 page 17 above and page 25 below.

The end result of 110 pounds of high explosives detonating underground with deadly force.
NAA MT885/1 51/1/209 page 23.

The ground view of the actual dugout showing the box eucalypt in the background of the demolished dugout in which 26 men were killed instantly. Note the location of the box eucalypt tree which provided shade for the trainees during lunch breaks. Today, the very same tree provides a reference point poignantly marking the location of the ill-fated dugout.
NAA MT885/1 51/1/209 page 21.

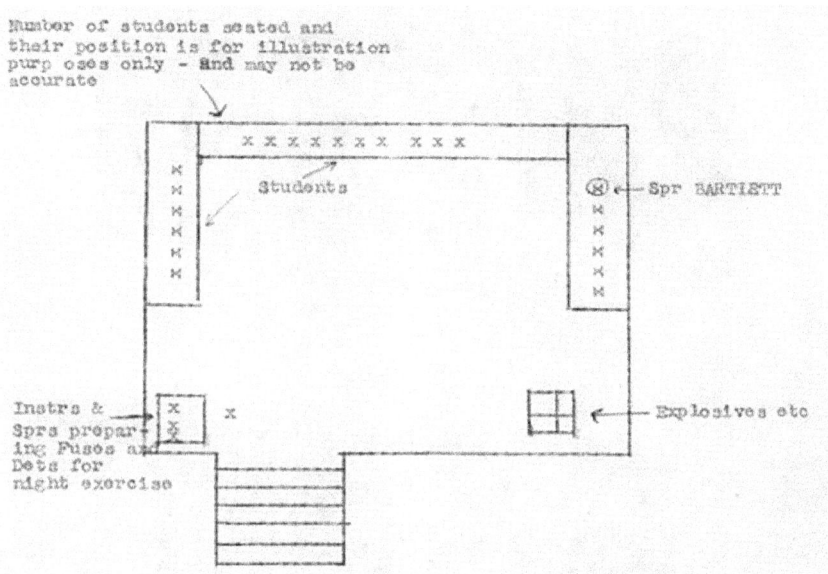

Above: A sketch of the interior of the dugout just moments before the fatal explosion. The sketch was compiled from information provided by Sapper Bartlett to the Australian Military Forces Court of Inquiry. Note the position of Sapper Bartlett who was the sole survivor of the tragedy. Perhaps shielded from the blast by his mates. Sappers Merritt and Reid, who survived the initial blast, were seated on the left side wall.
NAA MT885/1 51/1/209 page 113.

Above: *During the Court of Inquiry, held just days after the tragic explosion, two experiments were conducted in adjacent dugouts on the RAE Training Centre demolitions area. The experiments were designed to compare and assess the survivability and the damage caused by 10 pounds or 110 pounds of high explosive. Debris can be seen being thrown high into the air. According to witnesses '... the earth grumbled angrily beneath them'. Note the concrete buttresses in the foreground. They were important landmarks to identify the scene of the explosion almost 50 years later.*
NAA MT885/1 51/1/209 page 11 top, and page 12 above.

A close up view of the damage caused by the experimental detonation of 110 pounds of high explosives in the underground dugout. Note: This view depicts the same front entrance where the steps and aperture were shown in earlier photos.
NAA MT885/1 51/1/209 page 19.

A studio portrait of NX12713 Edward Owen Merry circa 1941 shortly after his promotion to commissioned officer. Ed Merry was OC of the ill-fated squad from 'Golf' Company 1st RAE Training Battalion, Kapooka Camp. The veteran soldier was on his way to the demolitions area when he observed a plume of dust and smoke reach high into the air. He knew something was drastically wrong. As one of the first on the scene he raced to the Camp Hospital to alert ambulances and proceeded back to the scene to coordinate the rescue/recovery of 'his men'.
Courtesy of Granddaughter, Fiona Bichel.

Above: *On the dangerously ill list - Sapper Allan Bartlett (right). Photo taken upon Allan's enlistment in 1945, just a few short months before the horrific explosion which led to his medical report (left).* Alan's Photo: Courtesy of the Bartlett family, AAF D.11 AWM. The report contains Allan's statement of his injury and a report by the treating Doctor who diagnosed, amongst other injuries, ruptured eardrums.
NAA MT885/1 51/1/209 page 142.

MEDICAL CERTIFICATE OF CAUSE OF DEATH

NAME OF DECEASED   VX57880 A/Sgt Herbert John POMEROY

DATE OF DEATH   21 May 1945

PLACE OF DEATH   KAPOOKA...NSW

CAUSE OF DEATH   Blast Injuries

I HEREBY CERTIFY that (I viewed the body of the above Deceased after death) and that the particulars and cause of death above written are true to the best of my knowledge and belief. I FURTHER CERTIFY that the death of the above Deceased was caused whilst he was engaged in Military duties.

(Signature) _____ MAJOR

(Qualifications) MB Ch M

(Date) 22 May 45

(Residence) 54 A C H

Above: *The Death Certificate of one of Ed Merry's most respected demolitions instructors – Sergeant Herbert John 'Jack' Pomeroy. Major Les Tunley, the Officer Commanding of the Kapooka Camp Hospital had the difficult task of making and signing 25 similar Death Certificates following the tragedy. He later handed the 26 Certificates to the Military Authorities who notified the Government of the official names of the deceased Kapooka Tragedy victims.*

Photo of SGT Pomeroy: Courtesy of Maureen Raunic (nee Pomeroy).

Death Certificate: NAA MT885/1 51/1/209 page 134.

Above & Below: *Wednesday 23rd of May 1945. Scenes of the funeral procession making its way through the streets of Wagga Wagga on its way to the War Cemetery. Leading the procession of coffin-laden trucks is an escort truck adorned in floral tributes to the 26 men. 7000 people lined the streets of Wagga Wagga during the largest Military funeral in the countries history. Men removed their hats and bowed their heads as the trucks silently rolled by. Many men were reduced to tears, whilst some women in the crowd fainted at the sight of such sadness. Behind the last truck, more than 100 cars conveyed distraught family members.*

Courtesy of Mr Peter Wyatt.

Military Funeral For Victims

AT WAGGA TO-DAY

Official Inquiry Opened

What is stated to be the largest military funeral ever to take place in Australia will be held in Wagga this afternoon when the 26 victims of the unfortunate demolition explosion at Kapooka Engineering Training Camp on Monday afternoon are to be buried in the military cemetery.

Above: *The funeral procession arrives at the WaggaWagga War cemetery. Soon the bodies would be removed from the trucks by six comrades and escorted by family members to their final resting place watched by thousands of mourners.*
Supplied courtesy of Mr Peter Wyatt.

Left: *News reports of the day announcing the funeral and the opening of the Court of Inquiry.* Daily Advertiser (Wagga Wagga, NSW: 1911 - 1954), p. 2. Retrieved July 5, 2014, from http://nla.gov.au/nla.news-article144968786.

Following Pages: *More sombre scenes at the Wagga Wagga War Cemetery as 26 men are simultaneously lowered into the ground. It was Australia's largest Military funeral.*
Supplied courtesy of Mr Peter Wyatt.

Above left: *Norm and Marjorie Dilley and right, their two children June and baby Neil seated in the pram, circa 1944.* Bottom: *June, left and Neil, right, stand either side of former AWAS and munitions truck driver Sheila Oehm at the 2000 Kapooka Tragedy Memorial Service. Their father Norm was killed in the tragedy.*
Courtesy of Neil and Susan Dilley.

Above left: *Sheila and the Commandant of the 1st Recruit Training Battalion, Colonel Gordon Hurford (circa 1992) during an AWAS Reunion at the Kapooka Camp.*
Above Right: *Sheila Oehm (nee Sly), the young AWAS munitions truck driver at the RAE Training Centre Kapooka Camp, Wagga Wagga.*
Courtesy of nephew Malcolm Tapscott.

Above: *Shirley Booth's most prized memory of her big brother, Sapper Geoff Partridge. Geoff sent this 'thrupence' bracelet to his little sister for her 13th birthday just days before his young life was tragically cut short. He was just 18 years old.*
Courtesy of Mrs Shirley Booth.

Above: *An aerial view of modern day Kapooka Camp and the location of the 1945 RAE Training Centre demolitions range including the site of the ill-fated dugout (labeled as the 'original bunker site).*
Below: *Aerial image of site (circa 2013) depicting view of demolitions area. The lone box eucalypt directly adjacent the farming shed on the right marks the location of the fatal dugout. The new Kapooka Tragedy memorial site can be seen opposite the site on Kapooka Road.*
Courtesy of Google Maps.

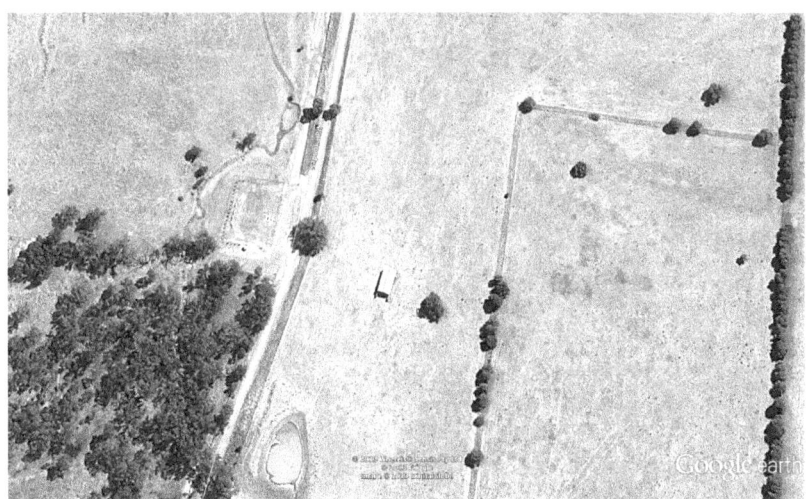

Blast survivor relives horror

By KEN GRIMSON

ALLAN Bartlett has no words to describe adequately how he feels about being the only survivor among 28 soldiers blown up during a military training exercise that went tragically wrong at Kapooka 50 years ago yesterday.

Back in Wagga yesterday as an honoured guest at a memorial service to mark the half-century of the disaster, 68-year-old Mr Bartlett, a young Sapper training in demolition in 1945, said he could still shed no light on the cause of the blast because he did not see it happen.

"It never occurred to me when they lifted me out of the dug-out what had happened," Mr Bartlett said, explaining the miracle of his survival.

"We were being lectured — the gelignite was still in the boxes," he said.

Some people have explained that Mr Bartlett survived the blast because his mouth was open at the time of the explosion. Apparently, many of the 26 victims died from the concussion of the explosion and having his mouth open may have saved Mr Bartlett from that cause of death.

Yesterday, he said he regained consciousness in the bunker standing up with his back against an earthern wall that had been tilted to a 45 degree angle by the blast.

"My first thought was

Continued Page 10

SOLE SURVIVOR: For 50 years, Allan Bartlett has known little about an explosion at Kapooka that official records say killed 26 soldiers even though he was the only survivor. Yesterday, he and about 150 people, including family members of the fallen soldiers, attended a memorial service to mark the anniversary of the tragedy. With the help of a photo supplied by Max Howison (below), Mr Bartlett returned to the scene of his miraculous escape. (Note tree in both pictures) More KEN GRIMSON stories and PETER TREMAIN photos Pages 10 and 11.

1995 marked the 50 th Anniversary of the tragedy. That year, the only survivor, Sapper Allan Bartlett, then aged 68 journeyed to Wagga Wagga and relived the horror of 50 years ago when he lost 26 comrades and mates.

Supplied courtesy of the Bartlett family.

Allan Bartlett seen standing in front of the original memorial plaque affixed to one of the concrete buttresses on the former demolitions area. The original plaque was placed on the concrete buttress in 1992 following the work of the 1st Recruit Training Battalion. Being on private property, the plaque was moved in 2007 to a more public accessible position on the roadside adjacent the former demolitions area, now a grazing paddock.
Supplied courtesy of the Bartlett family.

Above: *The original site of the demolitions area. Note the concrete buttresses adjacent the graziers sheds and the same box eucalypt some 60 years on. It was in the shade of this very tree where the Sappers lost their lives in the 1945 explosion. It stands proud as a living monument to the men.*
Supplied courtesy of Mr Peter Wyatt.

Below: *2010 Commemorative Service. Maureen Raunic (left), daughter of Jack Pomeroy and Neryl Hogan (right) daughter of Jack Nixon, stand in front of the newly unveiled Commemorative Plaque. Both women were just 10 months old in 1945 when their fathers were killed instantly by the blast*
Supplied courtesy of Mr Peter Wyatt.

Above: *The 2010 Commemorative Plaque at the new site.*
Below: *Final resting place. The men are buried together with other Military victims at the Wagga Wagga War Cemetery.*
Supplied courtesy of Mr Peter Wyatt.

VX57880
Herbert John 'Jack' Pomeroy,
aged 31

WX25792
Ronald Irwin Linthorne,
aged 25

WX27166
Alfred 'Alf' Edward Woods,
aged 32

NX141715
William 'Bill' Barclay
Cousins, aged 24

N480870
Thomas 'Toddy' Woods,
aged 25

VX96197
Kevin Francis Pierce,
aged 18

S115574
William 'Bill' Reid,
aged 36

NX180545
Colin Leslie Hurley,
aged 20

WX 23101
Alfred George Witt,
aged 20

N481536
Norman 'Norm' Rourke
John Dilley, aged 34

NX204475
Colin Francis Boyd,
aged 21

NX205652
Terence Ronald Moore,
aged 19

NX205938
Edward Charles 'Ted'
Robson, aged 18

NX205951
Kevin 'Kev' Alexander
Hurst, aged 18

NX205969
Allan Flood,
aged 18

NX180218
Geoffrey Wilton
Partridge, aged 18

NX81964
Stanley 'Stan' Ernest Ross,
aged 28

SX34059
Denby Eric Grasby,
aged 18

SX34069
Ivan Walter Thomas
Merritt, aged 26

Q273551
Ernest 'Ernie' Frederick
Poschalk, aged 18

QX63309
Frank Wilfred Platt,
aged 20

Q273563
Stanley 'Stan' Robert
Morphy, aged 18

NX205833
Jack Clinton Nixon,
aged 30

In loving memory of SX34062 Allan Raymond Bartlett (1927-2005).

The miracle soldier, father and grandfather, was the sole survivor of the horrific blast that killed 26 of his mates. Sadly Allan passed away just one year after this photo was taken. He never forgot the day his mates fell. In his passing, after 60 years of waiting, finally the men were reunited once and for all. Only now can they 'all' truly rest in peace.
Photo: Courtesy of Bruce and Di Bartlett.

Chapter Seventeen
A Terrible Procession

When Brigadier Forbes drove through the front gate of Kapooka on Tuesday the 22nd of May to commence his Court of Inquiry, he would've paid little attention to the convoy of trucks passing in the opposite direction. If he did, he would have seen more than ninety or so trainee engineers, all weighed down with picks and shovels, their 'tools of the trade', on their way to the military section of the Wagga Wagga cemetery. 26 graves was a monumental task for the local gravedigger to accomplish in just two days, but it was nothing for the men of the RAETC. When the call went out over the PA for volunteers to dig graves, there was an overwhelming response. For fellow trainees, it was the least they could do for their fallen mates.

For many local Wagga folk, the sight of Army trucks laden with troops was a common sight in town. But the unmistakable reality of the scale of the tragedy became evident when the number of trucks stopped at the cemetery that Tuesday morning. And, when the force of volunteer graveDiggers piled out of overloaded trucks and set about digging fresh holes in the sparse ground, locals looked on in disbelief. Quickly the news spread that 'hundreds' of soldiers were digging graves over at the cemetery, by then the town knew a historic funeral wasn't far away. In fact, the men were digging at fever pitch as the funeral had been arranged for the following day, Wednesday May the 23rd. For the townsfolk, the high profile funeral would be their one opportunity to pay their respect to the fallen. For now though, they weren't allowed to pitch in and help. It was left up to the Army to take control of the mass burial at the cemetery. What no one knew at that time was just how hard Archie Smith, Ed Dodds, Ed Merry and Fernleigh Ellis were working behind the scenes organizing the largest military funeral ever witnessed, not just in Wagga Wagga, but the entire history of the country. Normally they had about a week to organize a soldier's farewell, this time they had just one day! Killed on the Monday afternoon, authorities wanted all 26 men positively identified and buried by the Wednesday morning.

14 Army and 41 Air Force personnel were already interred in the military section of the cemetery. Located on the aptly titled Cemetery Road (later Kooringal Road), the dedicated military section of the existing cemetery was established at the start of the war. Authorities knew that the latest military casualties of WWII needed a resting place amongst friends and military colleagues. At Wagga they'd be surrounded by locally-interred soldiers, airmen and airwomen already killed in combat or those who lost their lives during training. One of those waiting for the men was 24-year-old Air Force Warrant Officer Lloyd Loftus. After an aircraft accident at nearby Uranquinty claimed his life, Lloyd had been laid to rest in the cemetery less than a month earlier. Alongside Lloyd was 20-year-old Aircraftswoman Gladys Hardey. Gladys was the daughter of a Great War veteran and was herself a member of the Women's Auxiliary Australian Air Force. She was laid to rest in July of 1943 after succumbing to illness. The demolitions accident added more than two dozen Sappers names to the Cemetery Honour Roll, which already recorded the names of eight Sappers previously interred at the cemetery. One of those original interred included William Watson. William was a 61 year-old former member of the Engineer Training Centre staff who was run over by a car in Baylis Street

Wagga Wagga in August of 1943. Sadly, he died at the RAAF Hospital that same day and was buried in the War Cemetery alongside his former fellow engineers.

Over at 54th Camp Hospital, 26 coffins were delivered to the makeshift morgue. Carefully, and with the dignity becoming their status as fallen soldiers, their remains were placed inside the coffins in preparation for Wednesday's funeral procession. Unlike the feverish activity at the Cemetery, the mood and tempo at the hospital was calm and calculated. By mid-Tuesday, as the soldiers were finishing their backbreaking duties out at the cemetery, official dignitaries were beginning to arrive at Kapooka. The dignitaries included 62-year-old Major General Lloyd. As GOC Second Aust Army, he'd only taken up the position two weeks prior and now found himself, in one of his first official appointments, presiding over one of the Army's darkest days. Representing the highest uniformed member in the victims' chain of command, Major General Lloyd would be positioned in one of the lead official cars during the procession of caskets through the proud military township.

Over in town, incredible events, those you wont find scribbled in history books, were occurring. During those critical defining moments following the tragedy, the reaction of a stunned community allowed Wagga Wagga's garrison identity to be set in stone. Kapooka wasn't just an Army camp; it was Wagga's camp and the town responded to the tragedy with compassion, empathy and the valuable human spirit of selflessness.

At the Town Hall, Wagga Wagga Mayoress, Mrs. Doyle, offered her sympathies to the families who'd assembled there after being met by Army Officers at the Kapooka Loop train station. With the arrival of distraught families, a solemn atmosphere appeared to engulf the town. But this was a proud garrison town. They weren't going to just to sit back and watch grieving military families struggle in their town under unimaginable personal pain of loss and the stigma of being stared at as 'one of those poor families'. Instead, they began rallying support for the grief-stricken families. When an appeal was later broadcast from the Town Hall, via the Wagga Wagga Chamber of Commerce, for taxi drivers and car owners to make their cars available to convey relatives and friends from the train station out to Kapooka, and later for the funeral, it was clear just what type of impact this tragedy had on the locals. The townsfolk responded in droves. Openly welcoming them into their oft-times restrained part of the world; the negative atmosphere was quickly replaced by an immeasurable gift of community spirit. Although falsely reported in some Sydney press at the time that police were actually commandeering cars, it was a fatuous proclamation about a very selfless town. In fact, the number of cars offered by the community far outweighed the number of cars actually required. Inspired by an avalanche of volunteer support, it wasn't only cars that were forthcoming. Townsfolk offered up their homes and were cooking meals in a moving tribute to the unfortunate suffering families. The response from the entire Wagga Wagga community spoke volumes of just how much the town was supporting the distraught families. They treated them as if they were their own. Just how the tragedy moved this small community was both inspirational and memorable. Their reaction to such a sorrowful and tragic time was a true example of Australia's proud community spirit of mateship and helping a friend in need.

The coordination of the funeral may have been rushed for Archie Smith and the men from the 1st RAE Training Battalion and RAETC HQ, but his intention was always to provide a truly remarkable and fitting tribute. Delivering a touching and heartfelt farewell, his planning and forethought for the proceedings were always respectful, dignified and in the finest military traditions of honouring their war dead.

Scheduled to leave 54th Camp Hospital at 1400 hours on Wednesday the 23rd of May, the route for the large funeral procession would allow many townsfolk to observe and pay respects. The route followed the Sturt Highway from the Kapooka Camp front gate to Edward Street in town. It would pass along the normally busy thoroughfare to Tarcutta Street and a final turn into Cemetery Road. The procession would include a police escort and road blockages to allow five military tenders an uninterrupted thoroughfare. Of the five tenders one would be carrying the large volume of wreaths and flowers, which would then be followed by four open-bed trucks laden with 26 coffins each draped in an Australian flag. It was expected to be a highly emotional sight for all who summonsed up the courage to witness.

In the following motorcade, the relatives would be assembled and driven to the cemetery for the funeral service. At the cemetery, individual religious ceremonies would be performed at the corresponding gravesite. Then, on a given signal, the men would be simultaneously lowered into the ground to the haunting notes of the Last Post.

What the families didn't expect to see was the sheer level of support shown by the town. Although Wednesday afternoons were usually a half-day holiday in Wagga, the Mayoress asked all banks and business houses to be closed for the afternoon. For essential workers, those working in jobs vital to the war effort, special leave had been granted to enable them to line the tree-lined streets of the route of the funeral cortege. Flags were to fly at half-mast at all government and private facilities and at all shops and private houses. The Mayoress announced that Municipal offices were to be closed between 1400 hours and 1500 hours. She asked the returned soldiers of the district to assemble at the intersection of Baylis and Edward Streets at 1400 hours to form a Guard of Honour. In addition, more than 3000 pupils from State and Catholic schools were organised to line a portion of the route in Edward Street opposite the South Wagga School. Farmers and graziers from miles around the district organised transport so that they too could pay their respects. Effectively, Wagga Wagga was expected to stop functioning as a town for half of the day. Contributing to the largest ever outpouring of public tribute to 26 men, most of whom relatively unknown to them, it was the least they could do as the Camp's supporting garrison town.

Until then, unless they witnessed the activity at the cemetery, for many of the townsfolk the tragedy had all been talk about what that had read or overheard. Nonetheless, it was feared the sight of 26 coffins, draped in the inspirational symbol of national pride and identity, might be too emotional for some locals. Organisers were preparing for an avalanche of distressed civilians requiring medical attention. They would soon find out how accurate their assumptions were. Hopefully they wouldn't be required, but the Wagga District Ambulance Service was organized to be in attendance.

Back at camp, hundreds of floral tributes had begun arriving from family, friends and relatives. More than twenty floral tributes alone were received from Parkes for the deeply-missed Corporal Bill Cousins. On behalf of the RAE Training Centre, the Commanding Officer supplied a wreath, as did many of the victims' colleagues who also provided individual tributes to their mates. The Commander-in-Chief, Sir Thomas Blamey, Major General Lloyd and other military dignitaries, also supplied floral tributes and garlands of commemorative flower arrangements.

Long before the scheduled arrival time of the motorcade, crowds had already begun to form at intersections along Edward Street. Along a one-mile stretch of Edward Street between Bolton Street and Bolton Park, the crowd of mourners was building two and three deep. Big local industrial firms and businesses permitted staff to attend the funeral procession. Due to the magnitude of the service and the expected number of onlookers, the Military didn't permit the general population of Wagga to attend the service at the cemetery so instead they turned up in droves along the route.

At 1400 hours, as scheduled, the leading police motorcycle escort led the first truck overloaded with blooms and floral tributes out of the Kapooka

Camp front gate. Four trucks followed it. The first two were bearing seven flag-draped coffins and the following two a further six coffins each containing the remains of the men. Driving one of the following vehicles was Sheila Oehm. Having witnessed the horror first-hand, Sheila wasn't allowed to grieve; instead she was back on duty tasked with driving the Officer-in-Charge of the funeral in the procession. Slowly the procession made its way down the highway towards the western end of Edward Street. Behind, more than twenty civilian cars conveying relatives, followed by over 100 trucks, jeeps and other cars filled with servicemen and friends completed one of the largest organized funeral processions in Australia's history.

The approaching police motorcycle escorts signaled to the gathered crowd that the cortege was on the way. As the first tender turned the corner, the waiting crowd moved closer to the edge of the road. Then, as the first coffin-laden tender came into view, the large gathered crowd fell silent. Families of the dead men travelling in the following vehicles could see a crowd ahead. Estimated to be more than 7000 of the registered 14000 population of Wagga, it appeared as if the whole town had turned out to pay their respects. Although the victims may have been total strangers, the emotional outpouring of grief on the faces of the gathered masses meant a lot to the victim's families. Upon witnessing the extraordinary turn out of townsfolk paying their respects, only then did it really hit them just how much these coffins containing their sons, fathers, dads and friends meant to this small town. For all who witnessed the spectacle first-hand that day, it was a journey of unreserved emotion; a spectacle they would never forget.

Meanwhile, the eerie silence of the afternoon was replaced by the slow murmur of military trucks in the distance. As the leading truck passed the edge of the burgeoning crowd, grieving families in the motorcade watched as lines of well-wishers came to attention. Men removed their hats and bowed their heads. Using shirtsleeves and handkerchiefs to wipe away tears, the significance of the moment brought even the strongest and bravest to publicly show emotion. Constantly, women pressed handkerchiefs to dab away the continual flow of emotion from their eyes. Needless to say, many men, women and children witnessing the never-ending motorcade of grief were themselves overcome. The magnitude of their grief and sympathy, not just for the relatives, but also for the many officers and men of *their* camp, was in plain view for all to see. As the 45-minute procession passed by, numerous bystanders were totally overcome with sorrow. Their hearts and sentiment was going out to the relatives who'd been so suddenly and violently robbed of their loved ones. It may have been painful to watch, but they didn't want to be anywhere else. It was a matter of respect.

In front of the Wagga Wagga Base Hospital, amongst the bowed heads of a forlorn township, nurses stood at the roadside in their striking red-caped blue and white uniforms, also paying their respects. As the cortege made its way between Baylis and Peter Streets towards the main town intersection of Baylis and Edward Streets, as directed, more than 200 members of the Wagga Sub Branch of the Returned Soldiers League formed an emotional guard of honour to fallen comrades.

As the first vehicle in the procession turned into the cemetery, the last vehicle was still more than a mile back. It was an understandably long funeral procession. Apart from Major-General Lloyd, among other dignitaries attending the funeral were Major General Clive Steel, Brigadier Prior representing the Engineer-In-Chief Colonel McGowan, Brigadier McDonald, George Macdonnell and representing the RAAF Wing Commander Chambers, Squadron Leader Jensen-Muir and Chaplain Collins. Sadly, no doubt due to the accelerated nature with which the military wanted the men buried, only forty relatives representing the 26 fallen were able to attend the ceremony. Many family and friends were still travelling to Wagga Wagga as the men were being conveyed to their final resting place. For one or two families, they were always going to be too late.

Waiting at the cemetery for the procession to arrive were more than 150 men from the Training Centre. They were the fortunate trainees and friends given the honour of being pallbearers. One of those was Allan McPaul. As a tent mate to 'Brikie' Hurst, alongside his B Coy colleagues, they were given the honour of carrying mates to their final resting grounds. Also preparing themselves for the solemn services at the cemetery were thousands of comrades and other grieving family members who didn't join the official procession. As the men came from varying religious upbringings, also standing by at the cemetery were Church of England, Anglican, Roman Catholic, Methodist and Presbyterian ministers all preparing themselves to conduct the appropriate religious denomination services.

The funeral service, as expected, was a well-coordinated affair, executed with the polish and professionalism of military procedure and precision, thanks largely to Archie Smith and the administrative team from the 1st RAE Trg Bn. As each procession car carrying families and loved ones arrived at the cemetery, they were assisted to an adjacent area beside the truck carrying their loved ones. For some it was the first time they'd been in the presence of their son, husband, father or brother for a long time. As the procession came to a halt, teams of pall-bearers mustered themselves in a position to accept the coffins from the tender and, as the coffin was lifted from the rear of the truck, the name of the victim inside the coffin was announced to waiting relatives. As Kevin 'Brickie' Hurst's coffin was unloaded off the truck, the strong arms of Allan McPaul were there to carry his tent mate to his final resting place.

Unfortunately, the first portion of the military area at the Wagga Wagga War Cemetery had been filled with little white crosses displaying the horrors of war. This caused another new section to be opened to bury the latest victims on the southern side. The pallbearers were therefore required to carry the coffins in a particular way and many were cautious that they didn't drop the casket as they negotiated the long journey from the roadside to the new burial site. Gathered families then accompanied their loved one as the party of six pall bearers were, ably led to the appropriate gravesite by an attending Sergeant who coordinated the men's slow and mournful steps.

Passing through a guard of honour formed by hundreds of officers and soldiers, sadly many of the coffins weren't trailed by loved ones. For some of the young men, with no relatives to shed a tear, they were alone at the end. Each coffin was carefully negotiated on its final journey with not only military exactness, but with the dignity of a fallen warrior. A wreath bearer at the rear of each party made for a moving spectacle. The families were then able to follow their loved one to his gravesite. The waiting ministers and priests met the incoming processions and took up their positions in the allocated sections of the cemetery. Being militarily precise, which perhaps detracted from the spiritual significance of burying the dead, pallbearers and officials often clumsily matched the coffins against their intended final resting place. However, in the end, thanks to Archie Smith, the appropriate religious priest or minister buried the right man in the correct location.

Standing tall and proud, like strong beacons alongside distraught families, were also hundreds of fellow soldiers. Taking in the gravity of the service, experienced soldiers and inexperienced trainees stood side by side. However, the significance of the moment also placed a strain on the normally resilient and brave façade of the soldiers. The scene was emotionally intense. It caused many of the Army's normally stoic members to faint in grief alongside extremely distraught civilian onlookers.

In the presence of representatives from almost every public body and military organisation in Wagga Wagga, the military chaplains and ministers commenced their impersonal, yet supremely unique, unattached multi-denominational services. When the separate funerals were completed, and final prayers recited, pallbearers readied themselves to lower coffins simultaneously into their final resting place. Initiated by a bugle call, the remains of the young men were gently lowered into the ground. At the flagpole, Warrant Officer Swift and his six assembled buglers sounded the 'Last Post'. With military reverence, the bugles echoed off the nearby hills reducing many to uncontrollable tears as the coffins were carefully lowered into the ground and out of sight for the final time.

As 'Reveille' played, at the base of the flagpole Major General Lloyd brought his officers to attention. Ordering his men to face south, the senior mourner paused in reflection and saluted to the south, respectfully marking the men laid to rest in the south. With a right about-turn, he brought his officers to attention again. This time the party was facing north. And, after another reflective pause, Major General Lloyd saluted the fallen in the north.

As the 'Last Post' and 'Reveille' brought an end to the ceremony and the thousands of mourners departed the cemetery grounds, their departure left the distraught families to say their private final goodbyes in peace. Standing beside the small mound of dirt, which would soon entomb their loved one, it was a quiet moment of reflection about a life well-lived, but ended too short.

At the conclusion of all ceremonies the families and relatives of the men were finally escorted from the cemetery to the Soldiers Hall where members of the Woman's Auxiliary of the Returned Soldiers League supplied afternoon tea. Back at the cemetery, in the lengthening shadows of a highly emotional day, the plots had been filled in. 26 small white crosses were placed at the head of 26 freshly-dug graves. At the end of the moving service, 26 souls were able to finally rest in peace allowing 26 families to start a lifetime of remembering.

The *Daily Advertiser* in Wagga Wagga reported on the funeral in its Friday May the 25th edition. Graciously, the Commanding Officer of the RAE Training Centre, Brigadier McDonald wrote a letter of appreciation to the editor:

> ' ... Sir, I desire to express my deep appreciation and thanks to the following public bodies and persons for the invaluable assistance and cooperation extended to myself and staff in connection with the funeral arrangements being made respecting the internment of the 26 soldiers who accidentally met their deaths at Kapooka on May 21

1945:—Wagga Police. Mayor and Municipal Council Staff. Chamber of Commerce. Returned Soldiers League. Railway Department.

... It is desired to specially thank Inspector Randall, Det-Sergt Cloke, Sergt Sherwood and Mr. C. A. Bender, Inspector of Telephones for their unstinted cooperation and all car owners who made transport available for the relatives. May I also thank the people of Wagga for the sincere and silent tribute expressed by their presence at the roadside while the funeral procession passed.

W.D. McDONALD.'

Chapter Eighteen
Just Theories

The question of 'how' and 'why' this life shattering event occurred no doubt remained in the forethought of families for many years following the funeral. Adding to the conspiracy theories emanating from various people over time trying to explain the cause and its solemn aftermath, I puzzled as to why the military and the government decided to bury the men in a mass funeral just 48 hours later? When I've asked the question of others whom I've managed to track down who were there, including relatives of the victims—it certainly raised some concerns and resulted in an anti-Army sentiment by virtue of a supposed lack of compassion. With many calling the decision by the Army as both 'secret' and 'hasty', when you review the circumstances in context, it's difficult to add a voice of support to the skeptical views of the detractors. Mind you though, their views are appropriately and understandably guided by grief and perhaps anger.

Despite the fact that Sapper Murphy, a young trainee killed by drowning in the Murrumbidgee River, was sent home via rail to be buried, many of the victims of the Kapooka tragedy were mutilated beyond human recognition. It would have been logistically, and more importantly unethical and insensitive, to send a bag of unidentified human remains back to families to be buried in separate private ceremonies in faraway places such as Perth or Mungindi.

For that reason, I believe the Army made the *right* decision. Additionally the 54th Camp Hospital was a small medical facility. It had no dedicated mortuary and therefore was unable to maintain the human remains of 26 deceased men for an extended period of time. It would have been especially concerning if families demanded autopsies and coronial inquiries looking to blame someone. Although the remains could have been transferred to a larger facility if this was the case, the decision to dispense with a civilian Coronial Inquiry, under the provisions of the *National Security Regulations*, left the Army with limited responsibilities other than to confirm death, issue death certificates and organize a fitting burial. As this was the most appropriate response based on all the deciding factors, the men were buried two days later.

As for the concerns raised by many I spoke to about a cloak of secrecy concerning the tragedy, that the Army was somewhat hiding the disaster from the public, I'm sure difficulties were experienced in contacting the relatives of the 26 before the names were published in national newspapers. Men like Ron Linthorne, whose next of kin was his new wife in Melbourne and his mother in Perth; in 1945 they might as well have been in another country. The same could be said for any of the families living in towns and cities that were more than a day's travel away via train. I believe the Army was endeavoring to do the right thing but was perhaps pressured and persuaded by the government and authorities to limit the burial arrangements. But trying to advise the authorised next of kin as soon as possible can be problematic for a variety of reasons. It's the very reason why soldiers were and are asked to nominate a next-of-kin on enlistment to enable them to be informed first-hand of tragedy. Who is advised beyond the next-of-kin is not the military's responsibility.

During my research I discovered an underlying premise from many that a perceived lack of information about the funeral meant the tragic events of that day and subsequent funeral was a secret. What I discovered was the complete opposite. Information about the funeral was not withheld from the public; it was carefully disseminated to appropriate channels to lessen the impact upon both families of the victims and those who were anxious to discover if their loved ones had also perished. It was however unfortunate that 8000 Australian families were virtually 'left in the dark' about the names of the victims and were forced to anxiously wait for a telegram advising them to travel to Wagga Wagga for a funeral.

The bottom line was that the funeral was no secret. It was broadcast in national newspapers and on radio that Wagga Wagga was preparing for the biggest mass funeral in the country's history. Regrettably though, and not the fault of the best-intentioned Army, some men were buried without any family presence at all. It was perhaps one of the Army's deepest regrets that I'm sure no doubt played on the mind of Archie Smith, Ed Dodds, Ed Merry and Fernleigh Ellis when they planned the funeral service. Although in hindsight it's perhaps unacceptable, I'm sure military authorities felt the utmost sympathy for those men, and especially for their families stuck on trains. But I believe the men making this hard decision, to bury soldiers without a next-of-kin or family member present, was indicative of their moral strength during such a tumultuous and moving time of their lives. I dare say they all experienced a combination of regret and pain on that Wednesday afternoon as the soil was being pushed back into the 26 plots. You could say though, that in the absence of family members, none of the men were of course buried alone. 7000 members of *their* Wagga Wagga family and thousands of soldiers from their military family gave them a most honourable and fitting farewell. It was something Archie Smith made sure of!

Turning now to the important theory of the cause. Enormous speculation began circulating amongst seven million Australians as to just how a catastrophe of this magnitude could occur on home soil. Notwithstanding the incessant Government pressure on Brigadier Forbes to complete all formal investigations in accelerated time, especially now following the well-publicised mass funeral which effectively shut down an entire town, Friday the 01st of June 1945 was the day an entire country had been waiting for. Finally some answers to their burning speculation about what went wrong at Kapooka. Exactly eleven days after the fatal explosion, the Court President and his two experienced Majors signed off on the Australian Military Forces' document detailing their findings. The report, briefly titled: *'Report of the Court'* although much anticipated by the Government, was written and submitted like all other formal military investigation documentation of the time. Not surprisingly, there were no startling revelations or 'smoke and mirror' theories outside the information gleaned from the Court from their earlier investigation and experiments at Kapooka. The dossier contained all the necessary paperwork—including typed statements, death certificates, copies of exhibits from proceedings and details of medical treatment to Allan Bartlett and Colin Kendall. With adherence to procedural bureaucracy of the time, the report was 'thorough and concise'.

The first nine pages of Brigadier Forbes' findings contained compulsory formalities. The key formalities and details included key headings and administrative details, including opening introductions and paragraphs covering the summary of the accident; the system of command for training purposes; an overview of the syllabus for the fourth week of training; safety precautions; information concerning the dugout; composition of the squad; the type and quantities of explosives issued to the squad and the facts of the accident, as brought out in evidence.

Addressing the finer details of the tragic aftermath, the final report covered additional areas such as the actions taken to clear the killed and injured; the administration and logistics of processing 26 deceased soldiers; Archie Smith's painstaking identification of the human remains and any further available medical evidence.

The cause of death of every single one of the victims was neither an earth-shattering finding, nor a verdict that wasn't predictable. The report simply amplified what Major Tunley recorded on the Death Certificates:

> '... The Court is of the opinion that the medical evidence establishes the fact that all the deceased were killed instantly by, or subsequently died from injuries received in the explosion."

Then, much like the findings as to the causes of death, the reported cause of the accident was not shrouded in controversy; it didn't disclose any unexpected or alarming results. Relying on scant information from a frightened and unawares Allan Bartlett, the Court based their theories of cause on Allan's last two seconds of memory. The following explanation provided via '*The Courts Theory as to the Cause of the Accident*' is extracted from the actual Final Report itself. The extract has been re-written by the author:

> '... The Court is of the opinion that NO definite decision can ever be arrived at as to the cause of the accident. After a very careful study of the evidence, particularly as regards the disposition of personnel and materials in the dugout and based largely on Spr BARTLETT'S evidence that shortly before the accident, the tall Corporal (Cpl Cousins was 6ft 1 in.) was standing on the opposite side of the dugout to the heap of explosives with his hands full of detonators with blue fuse, the Court has formulated a theory which appears more feasible than any other.
>
> At the time when Cpl COUSINS was seen by Spr BARTLETT he had moved across to the north east corner of the dugout to see how L/Sgt KENDALL's three men were progressing with their task of preparing detonators and fuses with match heads for the night's exercises. It should be remembered here, that this work was being done for Sgt POMEROY's squad with his materials. On finding that the work was completed, Cpl COUSINS picked up the detonators and fuses and held them in his hand in order to inspect whether the work had been satisfactorily carried out. It was at this moment that Spr BARTLETT noticed him.

When Spr BARTLETT turned his head to say something to his mate, Cpl COUSINS moved a few paces across the dugout with the intention of placing the handful of completed detonators and fuses with the heap of explosives then lying approximately in the north west corner of the dugout. On making contact with the heap of explosives, one of the detonators was accidently exploded and detonated the other detonators and the whole heap of explosives. There were many possible ways in which this detonator could have been exploded but the Court is of the opinion that the three most probable were:—

(a). Cpl COUSINS used undue pressure in placing the detonators on the heap causing one detonator protruding further than the others to explode.

(b). Cpl COUSINS when approaching the heap tripped and fell with the detonators between his body and the heap.

(c). Cpl COUSINS when placing the detonators on the heap, fainted and collapsed with the detonators between his body and the heap.

The Court desires to emphasize that the above is purely theory and, in its opinion, can never be proved.'

Seeking to determine some responsibility and blame, the Court arrived at the finding:

'... no responsibility or blame of any sort whatsoever can be attached to any person living at the time of the inquiry and there was insufficient evidence to show whether any person now dead was responsible or have blame directed at them for the accident.'

In addition, the Court's view was that had a coronial inquiry been held the only possible verdict in the case of all the deceased would have been 'accidental death'. It was a somewhat shallow finding. It lacked the necessary substance of apportioning blame that the families were perhaps hoping for.

In receipt of the *Final Report,* Major General Lloyd wrote to the Secretary of the Department of Army and forwarded a Minute with a summary of the Court of Inquiry outcomes. As a result of the accident, and the consequent Ministerial response to the report, the Directorate of the Engineer

In Chief released an immediate amendment to the safety precautions during instruction in the use of explosives and during demonstrations. The promulgated instruction resulting from the Court of Inquiry Report was an addendum to the extant précis. The addendum instructed a change be made to paragraph 12 of *Military Engineering Vol IV (Part I), Demolitions 1942 Appendix VI*, which now reads:

> '12. In a dugout or other confined space the following precautions will be taken:-
>
> (i). In a dugout or other confined space, there will be conducted only one type of explosive work at one time.
>
> (ii). In a dugout or other confined space, which contains a party of men exceeding three and a quantity of explosives exceeding 1 lb, there will not be any detonators.'

Reading this entire *Report of the Court*, albeit with significant hindsight 69 years later, I couldn't help but criticise the lack of what I'd term as the 'finer details'. I do recognise and appreciate that it was 1945, and perhaps such limited reporting was commonplace, but surely the largest loss of life in a single military accident warranted more substance than which was presented for the government and the Australian public? For me, the final report was found wanting in many areas; the least of which was the explanation, or the lack thereof, of adequate and appropriate medical evidence. Compounding my concern wasn't so much the scarcity of medical evidence and information concerning the men and cause of death, but the brevity with which it was concealed in the report.

Notwithstanding the amazing work by the ambulance drivers, medical orderlies and doctors who responded to the tragedy and, acknowledging that the dispensing of a New South Wales Coronial Inquiry meant no autopsies were performed, the report was so transitory with regards to the details of medical evidence that perhaps the medical response bordered on negligence?

It just appeared to me that signing off on the disaster was a matter of routine, courtesy of government pressures, rather than accurately investigating all possibilities and thereby accurately recording the history of the tragedy. Was there in fact a conspiracy of silence to erase from history the memorable aftermath and the senseless loss of young innocent life?

Apart from the omission of medical detail, after reading the investigation report I also arrived at the conclusion that some anomalies existed concerning the recording of statements from other key witnesses. Particularly, there was no evidence from Sheila Oehm, Frank Sim or any of his trainees. There was no statement from Sergeant McNabb or any of the experienced staff recovering torsos from the scene. Inexplicably, they were never interviewed. Why not? And if they were interviewed why don't their statements appear in the official report? Additionally, there was no evidence from the female ambulance drivers or the medical orderlies who ventured into the dugout to recover bodies; surely they had some evidence of interest? Even if it was just to support the overall picture of where the men were laying when they were recovered, what happened to their eyewitness accounts? Now, 69 years later, I'm trying to recapture the details of the worst training disaster to happen in the Australian Army. Even though swarms of people attended the scene that day, the investigative report leaves historians with just the bare details and enough loose ends to create an air of scepticism and speculation.

Another area of my concern, and perhaps my biggest criticism, was the lack of technical evidence surrounding the detonation. In saying that though, the two experimental explosions, designed to replicate the conditions of the tragedy, did allow the Court to make two careful and informed deductions. The first concerned experiment two. What they established was that a similar charge of quarry monobel and gelignite set off in the dugout replicated the same damage observed in the fatal dugout. And importantly, the second deduction, if only a smaller charge of ten pounds of gelignite detonated it would have still been sufficient enough to cause fatal injuries to most of the occupants of the dugout. The two experiments did successfully confirm the obvious presence of key facts in issue. Somehow, all of Jack Pomeroy's combined 110 pounds of gelignite and monobel detonated simultaneously. If, as some suggested, only the gelignite detonated and the monobel burned, the majority of men still would've perished—but perhaps, more than just one lucky man may have survived!

But the investigation and report was devoid of any form of technical evidence or experimental testing concerning or explaining the theory of 'sympathetic detonation'. Just how and why did the entire parcel of Jack's high explosives detonate? What was the technical process required? At what distance would this so-called sympathetic detonation occur?

Again, all things supposedly being fair and impartial, the Court relied on the testimony from engineers at the RAE Training Centre and, more bizarrely, only heard evidence from those engineers who were involved. They didn't seek any 'independent' evidence from non-involved parties or further experiments with the sympathetic detonation theory even if it was to come from another unit or someone with extensive knowledge of the instability of quarry monobel. So, with no impartial evidence from external engineers about sympathetic detonation, or explanation of how the detonation occurred, that meant that the Army's finding was in fact the sole finding.

That being the case, it does beg the question when reviewing the investigation and final report in its transparent totality; was the evidence gathered during the Brigadier's investigation fair and devoid of bias? Were the families of the men afforded the very basics of natural justice in getting to the truth of just what occurred? It was a case of 'too little too late' for Jack Pomeroy and his charges undertaking training that day. Whilst the safety precaution already existed in extant doctrine, Jack chose not to follow that doctrine and procedure. Like other instructors before him, Jack relied on the little discretion afforded to him as a SNCO to make a call on where the explosives were to be placed during the afternoon's lesson. The safety-conscious Sergeant instructor was inherently trainee-focused, but he appears to have been blindsided by an over-exuberant Corporal assistant. Or was he?

Closely examining the photographs supplied by family members in putting this story together, I noticed one piece of finite detail in a single photo. It was enough to make me wary of my cynical self and my own personal theories as to how the tragedy may have unfolded. It was an image of a youthful looking Jack Pomeroy with a cigarette in his hand. Many soldiers, who were initially non-smokers, eventually turned to smoking as a method of relaxing nerves. Even if Jack was a champion cyclist, did he also smoke? I began thinking that I was perhaps looking too far into it? What if Jack safeguarded all trainees by removing lighters and cigarettes, yet neglected his own matches and cigarettes in his pocket? Was it at all possible that standing at the front of the lesson with a four-ounce plug of high explosives in his hand, that somehow, he caused the initial flash? In the end my theory was highly unlikely, but certainly gave me food for thought. What if it was a case of complacency or lack of respect for the stores by person/s still unknown?

Apparently, and as reported, there were some blasé attitudes from some of the instructors. Attitudes which failed to instil and inspire trainee confidence. It was well-known throughout the engineering ranks that some Field Coy engineers were shell-shocked and 'bomb happy'. They were demolition cohorts who'd come back from overseas service a little 'worse for wear'. No longer frightened of explosives, they became the direct opposite; treating explosives with dangerous contempt. That being said, they were definitely not what the RAE Training Centre was looking for in an instructor. They didn't want 'cowboys'. One story I came across in the *Australian's at War Archive*

spoke of one such broken cowboy instructor teaching trainees of 2/9th Fd Coy the art of fishing … with demolitions:

> ' … We were up around the sand dunes, up around the Murray River, up at Echuca, and he was making these jam tin bombs and he was throwing them at us, with short fuses, and they were blowing bloody sand … I thought we would never get anywhere, I thought he would blow us all up. And it turned out; he was in the 2/8th Fd Coy.' ' … We used to make jam tin bombs and throw them in and get fish. But this silly old bugger … I said to my brother, ' … Did you have a bomb happy with you?' … He said, " … Yeah, I remember we sent him home." I said, ' … You sent him home? … You bloody well sent him to us.' I said, ' … We've got him.' ' … He was bloody mad. He nearly killed half of us throwing these bloody jam tin bombs at us—the silly old bastard ….'

Two further stories reinforced the danger faced by the Sappers everyday in dealing with mines and explosives during their training. Both stories come from former engineers who were interviewed for the *Australians At War Film Archive*. The first describes a training incident typical of some of the hijinks that occurred at Kapooka:

> ' … we were sitting around one day, a day like this, nice and warm sunny day. Walking around this Corporal's giving us a lecture on Mark II Anti-Tank mines. He's got one in his hand and he's primed it and someone over here says, " … Mind the snake behind you Corporal!" He threw the mine in the air, turned around, saw the snake and caught the mine about eighteen inches from the ground, well if it had hit the ground that would have been the end of that—and us!'

Unfortunately the second story has a sad finish. It's a reminder also of how complacency in dealing with the explosives almost always has a disastrous outcome. It's a story about engineers providing 'realistic' support for infantry training at Greta Camp in New South Wales:

> ' … they had the engineers, our boys, we made quarter plugs of jelly (gelignite) with a short fuse on them and you'd have a box of quarter plugs there and you'd pass them up to the bloke, you had a bloke

behind you with a smoke, and he'd just light the fuse, they picked them out by height, you know young strong blokes about my size or Macca's size.'

' ... Young Macca's throwing them but something happened, Blackie was passing up to Macca and whether Macca was looking out to where the last went or something but they dropped it and it fell in the box with all these bloody quarter plugs of gelignite. And Blackie was only a young bloke but he knew all about explosives and that but he started looking for it and of course up it went and blew his hands off. ... They were both stripped, just blew his hands off him ... Blackie couldn't see, hear or anything last time I had seen him.'

After learning this I was perhaps beginning to question the safety of monobel and its erratic explosive behavior, especially now as I understood that there was a history of blasé instructors. One particular story heightened my interest concerning the safety and unpredictably of monobel. The story goes that a man by the name of Andrew (Andrew Barb) LeBlanc was working in crop pits in Cape Breton, Nova Scotia in the 1940s. Mr LeBlanc placed a detonating cap into a stick of monobel (dynamite). Carrying the monobel in his hand, he walked near an operating car engine that was running the water pump for the pit. As he approached the engine, sufficient electric charge was generated to the attached detonator cap causing the monobel in his hand to explode. The subsequent explosion blew his hand clean off. Was monobel itself a contributing factor? Was it unstable and unpredictable as Mr. LeBlanc discovered?

Supported by the circumstantial evidence gathered via the Court of Inquiry report, including an analysis of the explosive properties of detonators, monobel and gelignite, the fatal high powered explosion *most* likely occurred due to the so-called 'sympathetic detonation'. Considering Jack Pomeroy wasn't present in the dugout when Colin Kendall accessed his explosive stores in preparation for the night activity, what did or didn't Jack notice?

Was he unaware that his explosives parcel had been opened and now exposed? Enter Bill Cousins' involvement. After Bill picked up the detonators, being prepared by Colin Kendall's three trainees adjacent the entrance of the dugout, the six-foot one inch Corporal moved across the dugout to place the detonators next to or near the explosives on Jack's right side. The rest is now tragic history.

For the last 69 years, families, friends and many others learning of the tragedy for the first time have all tried to understand just what happened in that split moment Bill Cousins moved across the dugout. Conspiracy theories and calculated guesses have been bandied about. One such theory was the 'horseshoe theory'. Steel horseshoes in the old style Army boots, possibly boots worn by Bill Cousins, reportedly created a spark. So when Bill walked towards Jack and the vicinity of the explosives his boots caused a spark and in turn caused the explosives to detonate. However, the theory was unfounded or never considered by the Court of Inquiry or any other person searching for the real reason 110 pounds of normally safe high explosives detonated. A lack of experimentation and replication of the theory consigned it to just that—one man's theory.

But the Court's notion was that Bill Cousins somehow tripped or stumbled whilst carrying the detonators with blue fuses. As he tripped he somehow managed to bring the fuses in direct contact with the opened explosives, which is ultimately necessary to cause detonation. The Court arrived at this theory as the only explainable method in which the explosives could've detonated. But what caused the strong athletic man to trip and fall in the vicinity of the explosives? Was he weary in the legs from the afternoon's physical training session causing him to stumble? Did he trip over the blackboard trying to squeeze behind his Sergeant as he gave his lesson? Was he weakened in the arms causing him to drop the detonators? Isn't it more likely that he dropped one or two of the detonators into the opened explosives as he prepared to place them carefully on the ground? Isn't it also more likely that the Corporal knew the dangers of having the detonators close to the explosives and was directed by his Sergeant to put them down immediately at a safe distance away from the explosives? Considering also Jack had a small plug of monobel in his hands and was perhaps distracted by his instruction, did he omit to give this instruction to his assistant? Did Bill Cousins do something wrong?

And what of the revelations by some of the Sappers who'd completed that course and complained that complacency and a lack of safety was creeping into instruction? They reported that some instructors, with an air of elitism and misguided bravado, openly 'thumbed' their noses at safety precautions. Crimping detonators with their teeth and smoking in the dugouts certainly doesn't paint a picture that safety was perfect at Kapooka camp during the war-years. What we do know is that Sergeant Pomeroy was professional and generally extremely careful of safety, what we don't know is why he put himself and his squad in that dangerous position.

This line of inquiry leads to one aspect of the tragedy that remains a grey area in the Court Findings and opinions and therefore leaves me a little confused. Why did experienced men like Jack Pomeroy, Roy Tafe, and Colin Kendall allow detonators to not only be stored in the dugout, but also allowed them to be prepared in the dugout by inexperienced trainees? They all knew it was a perfectly clear day outside so weather wasn't an issue. They knew Colin Kendall himself was working above ground and Roy Tafe in the nearby store, so security wasn't an issue. Why then couldn't Colin Kendall supervise them above ground? Were these three men, Stan Ross, Colin Boyd and Colin Hurley, distracted by the instructions being given to the squad by Jack that they incorrectly assembled one or two detonators? Is it possible they handed Bill Cousins a fragile detonator and with an innocuous bump, even the slightest footstep, that one or two of the detonators caused sympathetic detonation of the gelignite and in turn the monobel?

Whilst I think I have uncovered many more causal questions that will forever remain unanswered, my view is that the truth may never ever be known. The time and space continuum between Allan Bartlett's sighting of Bill Cousins carrying handfuls of detonators and the detonation of the explosives is historically crucial. Unfortunately we may never know what took place during that decisive one-second interval.

The wretched irony of this story and my concluding comments surrounding the multitude of pure theories, including the findings of the Court, their addendum action to re-write engineering doctrine and the one-second interval in time that will forever remain unanswered, should be directed at the underground dugouts themselves. Their design, function, and location on the demolitions area at Kapooka were to protect trainees and instructors from flying debris and noise, and also when appropriate, the painfully bitter Riverina weather conditions. Sadly, on this day this particular dugout failed to assist them, rather, it killed them.

Future records may never ever ascertain or acknowledge just exactly what happened at Kapooka all those years ago. All the speculation over the years, the allegations of blame, criticism and theories may, in time, eventually lead to a purposeless ending. But at the time of the disaster it was perhaps expected by a well-informed public, that the largest loss of life in a single military accident would be afforded some historical recognition. So, when the Final Report into the tragedy was released, and the public became aware of the more finite details, what type of memorial would be a fitting tribute to the men and where should it be erected? How was the Minister for the Army and the Engineering hierarchy going to recognise the sacrifice of 26 soldiers and one of the most tragic days in Australia's WWII account?

Chapter Nineteen
Something Strange Happened!

'... Let us remember those whose lives were given that we may enjoy this glorious moment and may look forward to a peace which they have won for us.'

Ben Chifley, Prime Minister of Australia,
15th of August 1945

In the days following the funeral, amongst the overwhelming outpouring of public grief, the Minister for Agriculture, the Honourable E.H. Graham forwarded the following message of sympathy to the nation via the *Wagga Advertiser*:

> '... I desire to extend my sincere sympathy to the parents and relatives of the young men who in the course of training to equip themselves for the carrying out of a dangerous and most important service to their country in its hour of need have fallen victims of a grievous misfortune. They have given their lives in the course of freedom just as assuredly as had they fallen on the battlefield. We will remember them with gratitude and by honouring them, we honour ourselves.'

Regardless of Mr. Graham's sincere message, and the work of the *Daily Advertiser* to report extensively on the funeral in its Friday edition, including double-paged reports and photos of the afternoon's solemn service, public interest in the tragedy soon subsided. A few short months following the mass-interment of Kapooka's 26 victims, allied victories against Japan and Germany were an almost certainty. And when two atomic bombs were dropped on Hiroshima and Nagasaki on 06 and 09 August 1945 respectively, not only did they deliver an incinerated end to 104,000 Japanese souls, it successfully accelerated the end of the war—ending it even quicker than it had started. Rejecting demands for surrender, Japan was finally brought to its knees. On the deck of the USS *Missouri* in Tokyo Bay on the 2nd of September 1945, Japan formally surrendered. Ben Chifley, the nation's prime minister at the time, announced to the nation, via radio on the 15th of August 1945, that the Japanese Government accepted the terms of surrender. Finally the war in the Pacific—Australia's backyard war—was at an end.

On the home front, by the end of 1945, all forms of military service had ceased. Thousands of servicemen and women began returning home; including surviving Australian prisoners of war who were released and repatriated. By the end of November 1946, almost all Australian military personnel had been released from service. Such was the darkness and consequences of war; the so-called war years became a defining era in Australia's sense of identity. It was especially poignant when the frequency of military deaths forced newspaper obituary columns to add a special column—'Killed on Service.' But being recognised by a respectful and appreciative nation, via emotionally-charged ceremonies, made grieving a little more bearable. Played out in the public domain, all too frequently moving memorial services displaying raw and powerful emotions to fallen local 'Diggers', meant that families were never alone in their grief. In rural country towns and in big cities alike, men were emotionally farewelled in the hundreds. Tree-lined streets were named in their honor. And, for the best part of seven years, it was a familiar scene. But, when the war ended, almost 40,000 families were left to lament in silence for what now seemed like a 'needless sacrifice' of their loved ones' lives.

When the 26 freshly-painted white wooden crosses, poignantly marking the resting place of the 26 unfortunate sons in a lonely Wagga Wagga War Cemetery, started fading against the Riverina weather—interest in the events of 21 May 1945 similarly faded. And, when all the emotive news reports of Kapooka eventually dried up, consequently resigning the tragedy to a by-product of the war, the nation's interest was almost non-existent. The lack of remembrance was swift, it was as if the deadly explosion had never happened. Disappearing from public recognition, the forgetting was a huge blow for families. For them it felt like the story of their lives had slipped into fast-forward. It was if no one cared anymore about them or the unimaginable loss of life. Just how such a significant event in Australia's military history achieved such an underwhelming recognition and remembrance, remains one of the most unexplained actions of a post-war government.

Amazingly, entire generations of Australians, those who live outside the boundaries of interest of the New South Wales Riverina district, have lived a life not aware of the details of the Army's worst training mishap in more than 100 years of its history. To add insult to the lack of recognition, for fifty years, Kapooka, the historic 'home of the soldier' and the military custodians responsible for maintaining its identity and history, primarily did nothing to commemorate the memory of these men. At the conclusion of the war, the violent accident site was rendered inaccessible, meaning families didn't even have a spiritual place to visit and reflect in the loss of their loved ones. Instead, the Wagga Wagga War Cemetery, their final resting place, provided the only monument. The lack of recognition of the tragedy is even more sadly pronounced when visitors to the Cemetery question why 26 headstones all reflect the same date of death. Sadly, no one could really explain just what happened on that day-hopefully now they can! Public interest in the tragedy is thankfully improving. With more than a passing interest into the history of Kapooka, new generations of Australians are learning of the importance of 21 May 1945 which brought a country to tears during the war years. Only now, after more than 50 years, the mystery surrounding just why so many men died on 21 May and now buried at Wagga Wagga can be addressed. Visitors to the Cemetery can now put faces to names and remember with a sense of patriotism those amazing men who perished that day. They can now reflect in the professionalism of Jack Pomeroy, the charm of Bill Cousins and the day 110 pounds of highly volatile military explosives ended their lives and 24 other proud 'Rising Sons' of Australia.

Part Three
The Story Of Their Lives

The Last Post

In military tradition, a number of bugle calls mark the phases of the soldier's day. The Last Post signifies the end of the day's activities. Incorporated into funeral and memorial services as a final farewell, it's a symbol the duty of the dead is over and, they can now 'rest in peace'.

Note to readers:

Part three of this book: *The Story of Their Lives* is written to fulfil a promise I made to surviving families to mark and recognise the 25 men who perished alongside Jack Pomeroy that day. It was about recognizing that they were more than just military statistics, but precious souls robbed from loved ones and each with their own life journey to tell. The pages contain dedications to the previously unknown names and identities surrounding this tragedy. I've included their life stories in the hopefulness that we, as a country, never 'again' forget them or the countless other unknown victims of military tragedies. It's also an opportunity to reflect in the miraculous life of the tragedies sole survivor, Allan Bartlett. His life journey was repressively blackened by the tragedy, but such was his amazing strength and courage of character, his fulfilled life included fatherhood and grandfatherhood. Having been spared, Allan was gracious for his survival and a second chance at life, but for sixty years, he never ever forgot those he left behind.

Part three also contains more than just a collection of short stories of the lives of the unfortunate men and their individual journeys from far reaches of Australia to become soldiers at Kapooka. Their lives were much more than that. They're symbolic. They represent the type of immeasurable excitement, anxiety, fear and tragic loss experienced by *all* Australian WWII soldiers who courageously responded to a war in Australia's backyard and in doing so changed their lives forever.

Chapter Twenty
Our 'Jack'

1—VX57880 Herbert John 'Jack' Pomeroy (aged 31)
Plot A, Row C, Grave 8

Born on the 21st May 1914, in Reading England, Jack was the second-born son to William and Susan Pomeroy. In the early years of his extraordinary life, William Pomeroy was lured to Australia by a promise made by *his* parents that they would be supplied with a life of 'milk and honey'. Departing the family home in Oxfordshire, England in November of 1920, William and Susan were seeking a quality of life far removed from the one they were experiencing in post-war London. With his parents' guarantee of a new existence, the 31 year old former cattle manager, and his somewhat reluctant 3-month pregnant 30-year-old wife Susan, boarded the Peninsular and Oriental (P&O) Passenger Liner *SS Benalla* in London on their 'via the Cape Service'. Accompanying them on their long sea journey were their five children, all aged under 10; the eldest being William or 'Bill' aged seven, then Herbert or 'Jack' aged six, followed by 5-year-old Edward or 'Ted', their first and only daughter Anne aged three and their baby, George a mere eighteen-months-old.

William's parents had previously made the same life-changing journey. Immigrating to Australia from Britain at a time when the government sought new settlers from the Motherland as a means of 'cementing' their friendship after the Great War, the senior Pomeroy family settled in Victoria. Enjoying the riches and spoils of life in this most welcoming country, they were determined their son and his family should also experience the utopia they'd discovered in this rich far-away land. Convincing William and Susan that this was the ideal place to raise their five children, both governments contributed to ease the financial burden, subsidising the passage, even offering loans to cover costs. Enthused by sponsorships and financial assistance, the young Pomeroy family was hoping to sail into a new and bright future. It was, after all, a far cry from a bruised and battered post-war Europe.

Rounding the Cape of Good Hope in early December 1920, reaching Adelaide by the 22nd of December, Christmas of 1920 was spent on board somewhere on the high seas between Adelaide and their disembarkation port of Melbourne. Considering their nautical surrounds, the family celebrated the festive season as best they could. Finally, after more than six weeks at sea, suffocated by the cramped and cold living conditions in the endless pitch and sway of an oceanic voyage, they arrived at Port Melbourne. William and his young family were finally on solid ground. More importantly, the weary young family was cautiously preparing to rewrite their future. Greeting his parents at the busy port on the 29th of December 1920, William registered his young family's immigration intentions.

Settling amongst the dairy and cattle areas of southwest Victoria, the family lived in a small cottage in Grey Street, Terang. It was an attractive, medium-sized village set in a dairying and farming district about 130 miles from Melbourne. The Terang area was opened up for soldier settlement under the government's generous scheme assisting repatriated warriors following the Great War. For the Pomeroy clan, the rural Victorian setting reminded the family of the British countryside. There was a little bit of the *old country* in their *new* country.

William Pomeroy was a hard worker. He was determined to make the most of his opportunity to forge a new and comfortable life for his young brood. Back in England, holding an important job managing cattle on large wealthy properties, he was a well-respected man. Standing six foot tall, he was normally strong and sturdy and would've made a great soldier. But accidently gored by a bull on the job back in England, his serious wounds rendered him unable to join the British Forces. His injuries disadvantaged his employment prospects—both in Britain and Australia. Seeking the type of work he was comfortable with—and good at—William struggled. Eventually resigning his desires to a non-preferred job, he settled on working as a labourer for the local Terang Council. Although her father-in-law and husband appreciated the new start, Susan wasn't entirely convinced she'd made the right choice. She missed England. She missed her parents, her friends and most of all she missed her old life terribly. ' … One day', she thought to herself, ' … she'd return and once again walk on home soil!' In the meantime, she had five hungry kids to feed.

Life steadily improved for the large struggling Catholic family living in rural Victoria, but two more children adding to the hungry brood of mouths to feed made the Pomeroy family resourceful beyond their own expectations. Charles, after having been carried from London on the six-week journey, was finally born in May of 1921 and Leonard or 'Len' as he was known was born in May of 1923. Len's arrival added yet another boy to the family. The Pomeroy's large sibling composition finally consisted of six boys and a single girl. Little did they know they'd be rearing six strong boys who'd grow into six men of fighting age when the Second War arrived sixteen years later. Ironically, with the arrival of Len in May of 1923 this made four of the Pomeroy boys born in May; Jack on the 21st, Len on the 8th, George on the 12th and Charles on the 25th. Regrettably, in the years to come, the month of May would no longer bring happiness to the Pomeroy family—it became a month marked by tragedy.

Susan Pomeroy was an attractive and statuesque woman. Not only tall and eye-catching, she was also an accomplished musician possessing the loveliest of singing voices. Like most talented musicians, from mouth organ to piano accordion, Susan could play almost any instrument she picked up. Her talents had been previously recognised. As a young woman, long before she'd left England for Australia, a request had been made for her to perform on the London stage. It was a request she unfortunately turned down. But her passion for music continued in Australia. She ensured a Pianola adorned her home at all times and family sing-a-longs was commonplace in the Pomeroy household. Susan's voice and love for music inspired her children to sing together at church on Sundays in the choir. Even more amazing was the fact she passed her musical talents on to her children. The family was so good that often they were requested as a singing group vocalising together at local weddings. Susan's musical pedigree would also later surface in her third youngest son, George, especially during his service in the Army during World War II. George Pomeroy would entertain troops as part of the 2/24th Battalion Concert Party. It was the type of artistic performing life Susan had once hoped for herself.

Being a mother of seven, Susan was intelligent and resourceful but equally loving and caring. When required she was also a shrewd authoritarian. In order to manage her large brood, discipline was required to be effective and it needed to be swift. Susan was a woman who could deliver. She earned her children's love and respect through discipline, and ruled the family home with 'thunderous velvet gloves'. Straw brooms would mysteriously appear from nearby cupboards when the unruly Pomeroy boys became uncontrollable. The mere sight of the upturned broom, in the willing and capable hands of their mother, signaled that mum had literally reached her final straw. At the sight of the upturned broom gripped in her hands like she was batting alongside Bradman, knowing full well the powerful wrath of their mother, normally brave Pomeroy boys would scatter. The long handled sweeper was a menace in their mum's capable hands. Every obstacle was negotiated until such time as the house resumed harmony. In the small but modest home, rambunctious Pomeroys were soon tripping over each other, hastily making their retreat, even reaching as far as the backyard. Susan's methods were both efficient and prompt. Yet it was the type of discipline her boys needed, and, as the future would dictate, the type of discipline they needed as soldiers.

Immigrants like William and Susan were attracted by Australia's promise of guaranteed employment, good wages and plenty of opportunities for their children. What they found was demonstrably different. The economic distress, unemployment and poverty of the 1920s and 1930s during Australia's Great Depression was not the promised land of milk and honey that William's parents had portrayed. Compounding the enduring hardships was the toughening community attitudes against immigrants—especially English ones. Communities didn't originally embrace families like the Pomeroy's. It was hard work to build meaningful relationships and meet friends. But one of life's greatest levellers; tragedy, soon found its way to the Pomeroy household. It briefly changed Susan's attitude towards what she thought of Australia's so-called hospitality.

In 1923, William, the strong husband and father, the ambitious immigrant with an adventurous spirit had fallen ill—pneumonia. He was a victim of the notorious and deadly flu epidemic sweeping the country. Sadly, he never recovered. The young father of seven passed away at just 34 years of age. Travelling to Australia in the pursuit of an exciting life away from the misery of war-torn Europe, within three years, Susan Pomeroy was left alone with a large young family. With her youngest less than a year old, Susan had to make a life for her children without their father in a less than 'hospitable' new country. Unfortunately, despite the fact that they were living just down the road, William's parents didn't provide a great deal of support for their daughter-in-law or grandchildren in their time of need. On the brink of financial and physical despair, Susan found herself with an insurmountable mountain to climb. She missed England and her supporting family even more now. How was she going to cope now that her husband was gone?

With no father, the Pomeroy family struggled like many of the other families who'd lost the key breadwinner to the flu or killed during the Great War. Short of funds to buy clothing, it was always a hand-me-down from sibling to sibling which kept them clothed. But in a large struggling family that's the way it was. They battled along and went to school in uniformed clothing that had itself become quite 'educated' over the years.

Amid William's death, with her children running around in patched hand-me-own clothing, Susan witnessed a disturbing side of community attitude towards woman like her and their plight. It was during this period in time that young British boys were being encouraged to migrate to Australia to work as 'farm boys'. Destitute and deemed unsalvageable, including war-caused orphans from poor and neglected backgrounds in the UK, many boys were forced to re-settle in Australia. With more than 3000 orphans arriving in Australia during the 1920s and 30s, a number of children's and boy's homes, orphanages and work farms pressured Susan to give her boys up into the care of the State.

Suddenly Susan experienced a profound sense of despair. Churches and local welfare societies, believing she was a neglectful mother, wanted to take her children away from her. They assumed the children's wellbeing was at stake, and wanted to place them into the very 'un-family like' homes Susan detested. The family matriarch remained resolute. She wouldn't allow it to happen. She wasn't going to watch her whole family become estranged. Out-and-out stubborn English conviction gave Susan the strength to keep her young family together. Deciding they would 'always' look out for each other, regardless, the family unit remained complete. Washing and housework jobs kept the family together and somehow they survived. To keep the house in order all the family had chores; William, Jack, and Ted guided the younger ones in feeding the chooks, and chopping the wood for the stove, keeping them in line and, with the added responsibility of being the oldest, they assisted their mum.

Growing their own vegetables, chickens provided eggs and an annual Christmas fare. Not able to enjoy some of life's luxuries, life was simple, and hard to put nutritional meals on the table. Susan struggled to pay the bills and keep a roof above their heads; she got a small amount of money from State Assistance for the boys. It wasn't a generous sum, but at least her boys were with her. His father's death meant that the eldest son, Bill, had to leave school early. Nicknamed 'Cyclone' because he used to whiz around everywhere always in a hurry, Bill was a strong and tall man much like his father. Strong of character, he was also supremely clever and would grow to be an accomplished and well-respected engineer. As the family grew up, in turn, one by one the Pomeroy boys left home. After one boy left the next in line became the eldest son and assumed the responsibility as the male head of the family. As the head of the siblings he'd be responsible for watching over the younger ones making sure things went along all right. It was the type of support from her adoring sons that Susan proudly accepted. Although as each son grew up and left the home she felt empty, her next son in line simply took over. It made her immensely proud that she'd done a good job with her sons.

When Australia's crippling Depression years arrived, Pomeroy boys were struggling to find jobs and get work to help the family survive. Jack started working at a firm called 'True Food'. It was a local milk-producing factory where the majority of the younger people in the town also found employment. Throwing six-foot logs into boilers in order to make steam to run the plant, it was here that William and Jack both became interested in engineering. 'Cyclone' later became an electric engineer and managed his own plant in Adelaide for almost the next thirty years.

In 1939 the war arrived. For the Pomeroys, life would never be the same again.

Jack Pomeroy is buried in Plot A, Row C, Grave 8 of the Wagga Wagga War Cemetery.

Chapter Twenty-One
Jack's Squad:
A Duty Nobly Done

'… men will come out of this war as gloriously unequal in many things as when they entered it.'

The Forgotten People—a speech by Robert Menzies, 22 May 1942.

As Jack Pomeroy's remains were ceremoniously interred into the Riverina soil at the Wagga Wagga War Cemetery by fellow members of the RAETC, on the afternoon of 23 May 1945, he was accompanied by 25 other lives cut horribly short in that fatal fifth period of instruction. The gathered mourners remembered them all as soldiers and Australia's sons. But more importantly the men were remembered by grieving families. Taken from them without a chance to say goodbye, it was a painful and solemn farewell for their lost loved ones.

Signifying the end of their brief but inspirational military journeys, the bugler sounded the 'Last Post'. Emotionally reverberating in the still air of the afternoon, its message was quite clear. In the hearts of the thousands of attendees they knew it was the final dignified act—a signal to all. For the dead now lying in marked graves, their 'duty was well and truly nobly done'. After experiencing the nervousness and excitement of enlistment into the AIF, their 26 individual journeys of transformation from civilian to soldier, including their fleeting time at Kapooka, ended prematurely. Now, with a life cut short, they left families behind only to be reunited with the thousands of fallen comrades sacrificed in all of Australia's past military actions. Alongside Australia's growing casualty rate of WWII, thankfully their final eternal resting place was amongst friends.

Gone, but not forgotten, their short lives embodied and represented the type of men whom Robert Menzies referred to as 'gloriously unequal' when describing the dynamic middle-class Australians who would go on to serve in the war. Regardless of class, religion, marital status or station in their life, when Australia's sons volunteered to fight they faced an uncertain future. Kapooka's 26 victims were no different.

They were the strivers, the planners and the ambitious ones; *'The Forgotten People'*. Coming from as far away as Western Australia, Victoria and Queensland, including local men from New South Wales, they were middle-class stories and lives about wives, marriages, girlfriends, fiancée's and the children they left behind during their individual acts of patriotism. Their individual life stories represent the diversity, logistics, and problems of WWII enlistment and the honour of serving. However, as they were training to fight a war for their struggling homeland, their untimely deaths added to family heartbreak. If they hadn't been killed alongside Australia's 40,000 WWII souls, you can't help but wonder what sort of lives they might have had?

Buried in Plots A & B, Rows B, C & D of the Commonwealth War Cemetery of Wagga Wagga, some lay side-by-side, others are separated by the spectacular lawns of the garden setting. In uniformed rows of white headstones, they deservedly rest in peace. Like their fallen comrades in Commonwealth War Graves around the world, individual memorials are beautifully and proudly embossed with the 'Rising Sun' badge of the Australian Imperial Force. It's a universal identifier that stirs emotions in all who gaze upon the countless rows of Australia's war dead. Underneath their symbolic and unique identity of rank, name, unit, date of death, their age is inscribed. The most noteworthy of the details is their age. The explosion on the 21st of May 1945 instantly deprived the world of young amazing men in the prime of their lives; and their senseless deaths cause all who gaze upon their headstones to consider the human cost of war. The sheer waste of young lives is both overwhelming and emotionally confronting.

On their headstones, revealing their identities and details displaying the shortness of their lives, an appropriate religious symbol and personal dedication by relatives narrates powerful messages of painful loss and messages of goodbye.

Lest We Forget.

Victorians

2—VX96197 Kevin Francis Pierce (aged 18)
Plot B, Row D, Grave 4

Aside from Jack Pomeroy, 18-year-old Kevin Francis Pierce was the only other Victorian killed in the tragedy. Before arriving at Kapooka, influenced somewhat by his troubled father, Kevin experienced a tumultuous life as a young adult.

Like many other men of his era, Kevin's father experienced first-hand the horrors of the Great War. He was a Gallipoli veteran, a true-blooded Anzac. Although blessed with extraordinary luck in combat, his military career, and post-war life, reflected the lives of many Anzac Diggers who came home a little worse for wear. His is a story worth telling. Not to stagger at the horrors of war, but to understand the vicarious damage it causes families. In this case, the life of Kapooka victim, number two, Kevin Pierce.

As Frank Pierce (instead of Francis), a name he went by in his early life, Kevin's father was one of the original Anzacs. He enlisted in Ballarat, Victoria, as a 20-year-old lorry driver in November of 1914 and joined the 8th Infantry Battalion as it prepared to go to war. Arriving in Egypt before travelling to Gallipoli, Frank Pierce and his 8th Battalion mates landed on the volatile peninsula on the 25th of April 1915 as part of the second wave of troops. Enduring his initial combat baptism, within a month Frank had been shot in the face and evacuated rearwards. Surviving the gunshot wound, he returned to the unit, again ready to fight. Later serving with the 46th Battalion during the fighting in France, Frank was wounded a second time in August of 1916. Immediately admitted to hospital, he recovered within two months and was sent back to the battalion. However, the cold and wet conditions experienced in the French trenches saw Frank evacuated from the front lines in November of 1916 to treat a multitude of ailments. He rejoined his unit five days later. Sometime after February 1917, he was seriously wounded for the third time. Shot in the right leg caused serious fractures. By now one would expect Frank's luck to be running out. But he recovered and returned to the fighting. In April of 1918, he was wounded on a fourth occasion. Treated at a Field Ambulance Station for a minor wound, it was this incident that made Frank think. After three years of fighting, and being wounded four times, '… maybe it's time to go home'.

By August of that year, following the longest sea journey of his life, after nearly three and a half years away, Frank was able to touch his home soil of Melbourne once more. Despite being wounded four times and surviving, Frank Pierce, the Anzac veteran, the distinguished combat soldier and witness to the horrors of the Great War, *tried* to return to some semblance of normality in his life. He was a war hero, yet walked the streets of Melbourne in anonymity. No doubt contributing to his struggles with the very basics of everyday life, his experiences of fighting and surviving multiple injuries became an unwanted legacy. Suddenly, the simple things we take for granted, such as raising children, became a grim prospect for the battle-scarred veteran. It was the invisible wounds of his service that affected him greatly and, in turn, the life of his only son.

He got on with life—considering the circumstances. But inside he was a disturbed man. Later reverting to his birth name, Francis Pierce stepped back into his old life driving taxis. He soon met Ruby Wilson. The pair of 26-year-olds married a short time later and the couple had two children; Gladys born in 1924 and Kevin in 1927. Entering his short time on earth on the brink of Australia's economic depression of the 1930s, Kevin arrived on the 5th January 1927 in Melbourne, Victoria.

It was during the time 4-year-old Kevin was living with his parents in Somerset Street in the Melbourne inner suburb of North Richmond, that his life changed forever. His father's reckless actions led to the death of an innocent man. Shattered by his actions, Francis attested it was nothing more than a horrible accident. Luckily the Deputy Coroner agreed. It was a case of death by misadventure; Francis was relieved. The tragic event, however, changed the Pierce family forever.

No less than four short years later, 1935, when he was but a small boy of eight, Kevin's beloved mum Ruby passed away aged 41. Following her death Francis' life unraveled. Shifting between jobs, as a driver and labourer, even enduring a short five-month stint in the part-time Army in 1936, looking for purpose he rejoined his trusty 46th Battalion. But through the tragedy of his father's struggle, Kevin was thrust into the circle of juggled care between local family members. His guardians included his Aunty Mary, his father's sister who was living in Hoddle Street, Collingwood and her husband, Bill. It's unsure how, when and why Francis lost control, but eventual custody of Gladys and Kevin, and no doubt the loss of his wife, placed an insurmountable strain on him. Suddenly, the proud Anzac veteran and war hero was not coping with his own battle for life.

Young Kevin was forced into a foreign world as a ward of the State. Among the throng of downtrodden and unwanted Melbourne children, Kevin couldn't believe he was a discarded and unloved child. Unlike the steely resolve of Susan Pomeroy and her conviction, ' ... that no man would take her boys from her', by December of 1936 Kevin had become one of 187 kids living in the type of home that Susan Pomeroy detested.

Commonly known at the time as a refuge for 'destitute' children, Kildonan Home for Children in North Melbourne was established in 1890 and run by the Presbyterian Church. The home provided congregate, institutional care primarily due to concerns about the care of children living in the central area of Melbourne. It became a 'Placement Centre' for neglected boys and girls aged between two and fifteen. They would live at the home until such time they were 'boarded out', the term given to boys who'd found a temporary home with a family. Usually the placements were in country Victoria, where the boys could be put to work on farms. But it wasn't uncommon for some children to live at Kildonan for longer periods of time beyond the normal time of 'processing'. The unruly and the unwanted had nowhere else to go. It's unsure how or even why Kevin ended up at Kildonan and not in the care of surviving family, but the tragedy of his young life doesn't end here. By December 1937, aged just ten years old, young innocent and unfortunate, Kevin was introduced to a world of corporal punishment, discipline, community-funded holidays and group activities with many boys just like him. He was now sharing his life with strange kids, all of which had a similarly sad story to tell.

But boys who were aged eleven and over, who hadn't been 'processed and placed' into homes, were sent away to Kilmany Park Farm Home for boys in Sale; a small Victorian country town about 120 miles east of Melbourne. Kilmany Park was also owned by the Presbyterian Church and used as a training farm setting for boys aged ten to 16. With the rare highlights being an egg once a year on your birthday and perhaps butter on Sundays, Kilmany Park was not atypical of the type of out-of-home care system ridiculed for its mistreatment of the thousands of young boys in their care over the years. Mistreatment, which also included a disdain for the boys and a 'shocking' lack of interest, affection and genuine care for their welfare. On the 5th of May

1938, at aged 11, the 'forgotten' Kevin was admitted to Kilmany Park. For Kevin, it became a horrible existence. He missed his family and seemingly, he'd lost contact with his adored sister.

Often referred to today as the 'forgotten Australians', the goal of Kevin's new home was to transform young delinquents from corrupted atmospheres where they were being exposed to criminal behavior, and through discipline and a certain 'brushing up' of farming skills, make them useful and trustworthy. The idea was not just to provide labour for farms, but in the words of the Presbyterian Church ' ... to make out of what may become waste human material valuable citizens of the state'. Sadly, in the eyes of the church, Kevin Pierce, the youthful son of an Anzac hero was referred to as 'waste human material'. As a product of the 'system', he was a young, impressionable and blameless youth who somehow earned a most undignified mantle.

For very good reasons, the personal files of young boys sent to homes in Victoria as 'wards of the State' in the 1940s and beyond have been closed by the Victorian Government to outsiders. Unfortunately or fortunately for that matter, Kevin's time at Klimany Park over the next five years will perhaps forever remain an unknown period in his life. Perhaps the less we know the better.

Fortunately we're able to re-enter Kevin's life in 1943. He'd survived the torturous and somewhat enigmatically mysterious life as a 'wasted material project' and was now forced to leave Kilmany Park. Acquiring sufficient farming skills and now finally no longer a minor, Kevin was able to break the grip Kilmany Park had had on him for the last five years. He was now permitted to seek out independent work and make a living for himself. He became human again; it was the break he'd been waiting for. He managed to secure a job working on a larger farm in the Traralgon district about thirty miles away from his former 'temporary' home. But during his years at Kildonan and Kilmany Park, Australia had committed itself to war. Kevin had entered Kilmany as an 11-year-old in the year before the war started, but in 1943, he'd turned 16. The potential glory found in the sacrifice for King and Commonwealth, by volunteering for the 2nd AIF, played on every teenage boy's mind—Kevin was, of course, no exception. As an impressionable teen he was fast arriving at a decision to become a soldier. It was especially important for those teenage boys whose dignity had been stripped away as a 'ward of the State'. Joining the Army would mean becoming young men and a chance to at last feel worthwhile contributors to society.

In January 1945, shortly after he turned 18, for one final time Kevin returned to Kilmany Park. Seeking his greatest adventure, he returned to say a final goodbye. On this final occasion, he decided to call in because he was actually on his way to the local Drill Hall in Sale to put his name down as a volunteer in the 2nd AIF. Upon his return to Kilmany he bore no malice towards his family who put him there, or towards the so-called 'Kilmany family' that was forced upon him. He was now a young man who knew he'd had a rough start in life, but the war was giving him a great opportunity to start afresh. He returned to say goodbye and good luck to those he formed friendships with and some of the genuine people who cared for him over the years. Perhaps his return was to attempt to rid the demons from his past?

After his emotional farewell, Kevin caught the train to Spencer Street in Melbourne, and then found the electric train that was heading out to the bustling Royal Park camp. Upon his arrival at camp, he was thrust into a foreign world far removed from anything he experienced on the farm or at Kilmany. At least Royal Park brought new beginnings and friendships. Kevin knew his eventual life, as a soldier, would deliver a much brighter future than his darkened past.

Undergoing the rigourous selection processes in order to be to accepted into the Army, Kevin's testing included mandatory medical examinations. Looking for the ones with flat feet, the medical board tested every soldier. In excellent health, Kevin got through enlistment unscathed. His service record was annotated with his official number: *VX96197*. Enlistees like Kevin were accommodated in tents that lined the side of the road. In the tent of course was the dreaded straw paillasse. The paillasse was simple hessian bag that young men like Kevin filled with straw and spent many nights on it on hard-boarded floors in Royal Park tents. For Kevin the less than inviting bed was already there and each morning, as part of camp routine, recruits had to make it up. It was simply a case of resting the supplied blanket on the straw filled sack and then folding it all up neatly during a daily morning parade. On the parade, the camp NCO instructors inspected the tents. For Kevin, who'd spent his teen life making his bed every day in the home to inspection standard, his self-discipline meant that his morning routine in the bustling camp was a breeze. Thanks to this instilled self-discipline, Kevin loved being a soldier.

After four-days of introductory training Kevin was supplied his uniform. Surrounded by thousands of strange faces, dressed resplendently in his uniform he found himself standing in the middle of the large encampment. Dwarfed not only by the physical enormity of the camp, he felt small in the very moment he was standing in. At five and a half foot, skinny-framed farm labourer Kevin was easily overburdened by the weight and uncomfortable nature of his new 'kit'. He didn't mind it though. When he looked around at where he was, and importantly where he'd come from, his demeanour couldn't hide the fact that he was excited to finally be a soldier.

It wasn't just his tractoring and engineering skills that had prepared him well for life as a soldier. Kevin's life as a 'ward of the State' not only disciplined him but it thickened his skin, made him resilient to change and strengthened his acceptance of authority. As expected, Kevin was due to be posted to the

3rd Army Recruit Training Battalion at Cowra for initial recruit training. Accepting his orders just before departure, Royal Park authorities sent him home to gather his personal belongings, get his affairs in order and to say his final goodbyes to family and friends. After many absent years, Kevin finally caught up with his big sister. In an emotional Pierce family greeting it was an equally painful farewell—again!

On the 28th of February 1945, Kevin marched into Cowra. For the next six to eight weeks of intensive training Kevin experienced yet another different kind of 'home'. But discipline and organisation were Kevin's strong points. No doubt courtesy of his treacherous days at Kilmany Park, Kevin developed a steely resolve for doing what he was told, when he was told, and both to an impeccable standard. At Cowra, Kevin excelled—he was going to be a great soldier. Completing his training in April of 1945, aptitude-testing identified that he was suitable for the Royal Australian Engineers. The journeyman was soon bound for the Engineer Training Centre at Wagga Wagga, aboard a crowded troop train. Once again he had time to reflect on his achievements thus far. Excited to be finally free of his most unwelcome tag as 'waste human material', he welcomed the opportunity, with a wry smile gazing out the window of a rowdy train, to become an Australian Army Sapper.

Having been forgotten for the best part of a half a century, following his tragic death in the Kapooka explosion, and, coupled with his early life as a 'ward of the State', you could say Kevin Pierce was a 'twice-forgotten' Australian. To the best of my research, no family members attended the mass funeral to say goodbye to the former Kilmany Park boy. Sadly, no inscription or message from relatives adorns his headstone. But for Kevin I'm sure it mattered not. He was laid to rest in the Wagga Wagga Cemetery surrounded by his 'new' family—his fellow soldiers and Sappers.

West Australians

3—WX25792 Ronald Irwin Linthorne (aged 25)
Plot B, Row C, Grave 3

Born on the 5th of May 1920 in Claremont, Western Australia, to Irwin and Elizabeth Linthorne, Ronald Irwin was the second child of the civil servant and his wife who were living in Reserve Street, Claremont. His arrival delivered the little brother who was so desperately wanted by his big sister Eileen. Just three years of age, she finally had a young playmate. But when Ron was just three months old, Eileen became ill. Admitted to the Children's Hospital in Subiaco, the young Linthorne family experienced their first immeasurable tragedy in the early morning of August 17 of 1920. Irwin and Elizabeth's little angel never woke up. She died in the warmth of her hospital bed that morning. She was aged just three years and nine months old. Irwin and Elizabeth Linthorne lost a little of their heart and soul that tragic day. Their baby daughter's death rocked them to the bone. They became more protective of their young son. For Ron's father, it would be a heartbreaking tragedy from which he struggled to recover. Sadly, one of the most loved and respected families in Fremantle had been dealt the cruellest of shocks. Little were they to know at that time, but in years to come further agonising tragedies would befall this luckless family.

Growing up under the watchful eyes of his parents, Ron lived a full and loving life in Claremont. As a child, he wasn't a spoilt brat. In fact, his working class occupation as a carpenter's improver showed just what type of character he was. It was a noble trade that allowed him to be creative and, unbeknownst to him at the time, it would also facilitate his eventual enlistment into the full-time military in the years to come.

On the 25th of July 1940, Ron attended the Drill Hall in Bazar Terrace, Perth, Western Australia to undergo a medical examination for service in the Militia. Passing fit, he was called up more than five months later for compulsory service. On the 2nd of December 1940, whilst living with his parents at Princess Road, Claremont, Ron Linthorne, a single 20-year-old with brown hair and hazel eyes, was conscripted into the Militia. Listing his father Irwin as his next of kin, he was given Army number, '*W5283*' and posted to the 55 Anti-Aircraft Company. After taking a couple of days off after training, by June 1941 he was posted to Rottnest Island with the Search Light Batteries. As the guns of the Batteries were manned 24 hours a day, it was somewhere for a young Ron Linthorne to cut his military teeth

Spending a month detached to RAAF Base Pearce in 1941, Ron returned to Rottnest Island and transferred the following year in February to 66 Anti-Aircraft Coy. Ron was only a baby when his 3-year-old sister Eileen died suddenly, so when he had to spend a month in the 110 General Hospital at Karrakatta in April 1942, it was adjacent to the cemetery in which his big sister had been laid to rest. Not missing the opportunity, Ron regularly paid her a visit. Reciting a prayer or two at her graveside, he regularly brought flowers and carefully placed them at her headstone. Dressed in his uniform, Eileen would no doubt have been proud of her little brother.

In July of 1942, aged 22, Ron attended the Mosman Park Drill Hall and transferred his Militia service into the 2nd AIF. Completing his Attestation

Form he was given his new identity *WX25792*. Enlisting in the 2nd AIF complemented his desire to continue the excitement and challenges he found in life as a soldier. Alongside him also at that time was another 21 year old, Harry Ballingall from Mosman Park. The two men had joined up together and became good mates. At the unit, Ron befriended a man who was both charismatic and unique—his name was Alfred (or Alf as he was known) Woods. With a life paradoxically opposed to that of Ron's, the two men found comfort in their differing upbringings and eventually became great mates. Ron wasn't to know of the pain and burden that Alf carried on his shoulders in those early years, but in the years to come, the men would talk about everything and everyone and become inseparable mates—almost brothers.

When Ron volunteered to join the 2nd AIF, suddenly, living remotely in the nation's most expansive State meant two things. First, it was inevitable that training would be undertaken in the eastern States and secondly, to get to the eastern States, it was a five-day train ride—on five different trains! To get from Perth to Cowra for recruit training the first train is from Perth to Kalgoorlie; then, change trains and do Kalgoorlie to Port Augusta in South Australia; once again change trains in Port Augusta to get to Adelaide; change trains again in Adelaide to get to Melbourne. In Melbourne, change trains again to get to Albury/Wodonga in Victoria; then on to Sydney and eventually, Sydney to Cowra.

Crossing the Nullarbor Plain was the most boring leg of the journey—crossing the barren plain, the train would slow down to an almost full stop at times. This allowed the men to jump off the train, run alongside, whilst grabbing large stones off the track and when hands and pockets were full jump back on the train. Then, as the train accelerated, empty bottles littering the side of the track would become opportune targets for their pocketful of rocks. Sleeping head to toe in row upon row in each car, there was ample opportunity to get agitated and frustrated on the long journey. However, the men kept their cool playing cards and testing their accuracy with a stray rock or two. At least the boredom was passed and no doubt a few wagers made. Whilst some men made a small fortune, courtesy of their accuracy with a stone, many were fleeced of their tiny wages courtesy of a game or two of crown and anchor.

Ron would become an excellent 'full-time soldier'. He was well-respected by the unit's officers and men who'd had the opportunity to work alongside him, including his good mate Alf. Ron especially impressed his supervising officer, Major Gordon Benporath and Ed Bailey. All the men, Ron, Alf, Harry and their supervising officers enjoyed great times at the unit, but the country was at war and there was a more serious side to their training.

Influencing his chain of command by his maturity, Ron was promoted to Lance Sergeant by August of 1942 and was acting Sergeant by April of 1943. 23-year-old Ron, who was still living with his parents at 22 Thompson Road, Claremont, met and fell in love with a 21-year-old English girl, Pauline Graham in June of 1943. Having already enlisted in the Women's Auxiliary Australian Air Force (WAAAF) in May of 1942, Pauline was a local girl from neighbouring suburb Palmyra. The young couple fell madly in love. Wedding bells were soon being spoken of. For an engagement present, Pauline gave Ron a beautiful wristlet watch with his initial 'R.I.L' engraved on its rear. Ron wore his watch with pride serving as a constant reminder of his beautiful bride to be. Often seen staring at its significance, it was a momentary welcome distraction for Ron from the perils of his sacrifice as a soldier in war-time Australia.

Tragedy for the Linthorne family, especially Ron and his mum, never was far away. On the 22nd of September 1943, Ron's dad Irwin passed away suddenly in his sleep at home. He was fifty years old. Suddenly the wedding joy and excitement in the Linthorne home had been shattered by the most horrible news. Just three short months after Ron and Pauline announced their engagement to the world, placing a notice in the West Australian newspaper on the 28th of June, they were now writing obituaries for his father. With the kind help of neighbours, the letters, cards, floral tributes and telegrams helped the much-loved Linthorne family, especially Elizabeth, ease through the pain of another family loss. Laid to rest with his young daughter, Irwin was reunited with Eileen at the Anglican Cemetery, Karrakatta. With half of his family gone, Ron now visited the cemetery with two bunches of flowers.

By October of 1943, with his father's death still fresh in his thoughts, Ron transferred to the 69th Mobile Search Light Battery, which at the time was operating out of Fremantle. In March of 1944 Ron was detached to the 3rd Aust Corps of Signals in South Australia and, with his Sergeant rank confirmed, by June he was on his way to the Northern Territory. He was travelling north to attend the No 41 Field Works Course conducted at the NT Training Centre from the 9th to the 25th of October 1944 in the hope of becoming an engineer. But it didn't *all* go according to plan. Failing the written component, but passing the oral and practical instruction of the course, his confidential report from his instructing officer noted Ron was *'a keen and willing worker who was confident and capable in practical work but found difficulty in manipulating figures'*. It was sufficient enough; Ron was qualified in basic field engineer work. Marching out of the Northern Territory, he was back in South Australia about a week later before returning home to Pauline and his mum in the West.

January 1945 brought some much-needed happiness to the Linthorne and Graham families. On the 30th of January, at 6pm, surrounded by a large congregation, Ron and Pauline were married at the Scots Church in Fremantle. During the ceremony, as customary, the couple exchanged rings. Pauline presented Ron, by then a Sergeant in the Army, with a beautiful gold ring. Pauline had taken great pride in buying the ring for her husband and having Ron's initials 'R.L' inscribed. His inscribed gold wedding ring was his constant reminder of his beautiful wife and the long and happy life ahead with the woman he adored. Enjoying some well-deserved time off to honeymoon with his bride, it wasn't long before duty would once again call for them both.

With the end of the war a near certainty, Ron and his new bride were looking forward to a future in the West. But somehow, the Army always got in their way. For some reason the inseparable Ron and his good mate Alf decided to become combat engineers. They knew they'd be leaving wives behind and were required to travel to Wagga Wagga to undergo the next phase of their conversion to become Sappers.

Before departing for Wagga Wagga, Ron altered his Next of Kin details in March of 1945 to reflect his marriage to Pauline. By this time Pauline had returned to her job in Melbourne with the Air Force. Experiencing a painful separation due to the war, Pauline was living at Nerissa Street Ashburton, a north-western suburb of Melbourne, whilst Ron was more than 2000 miles away on his way to the Riverina district in New South Wales. No doubt twisting and twirling his golden ring as he pensively stared out the train window during his exhaustive five-day trip to the east, Ron was regularly awoken from his daydream by the raucous noise of trackside bottles being smashed. But his thoughts almost always returned to his bride and the lonely mum he'd just left behind in Perth. Thankfully, on this journey good mate Alf Woods was beside him all the way.

When I first observed the studio portrait of Ron Linthorne, I couldn't help but be drawn into thinking just what type of man was the baby-faced 25-year-old staring back at me? With a vibrant bubbly welcoming smile, he had that special hint of something in his eyes. It was that kind of look that leaves you thinking, '... here's a man I can trust.' I pictured Ron as a gregarious man with an infectious personality to match. Above all else, he was a courageously honest and honourable man; evidenced by his care, support, and respect for good friend Alf Woods. He was no doubt well-respected, a loving husband,

and one hell of an excellent soldier. He impressed upon me as the kind of man who others gravitated towards and wanted to be near when things got tough. Like Jack Pomeroy, he'd be the perfect son, and a man who'd be an ideal dad, a fun uncle and a charismatic cousin. From that one photo alone I had the impression that Ron Linthorne was probably one of the world's true gentlemen.

Ron's infectious character was best summed up on his headstone inscription, ' ... I Shall Always Remember You Smiling Mr. Darling." Following his death, such was the esteem in which all held this charismatic man; Western Australian newspapers were inundated with obituary notices for Ron and his mate Alf Woods. For Pauline, after just three short months of married life, the tragedy robbed her of the most wonderful of husbands. In June of 1945 Pauline posted the following message to her late husband in the *West Australian* newspaper:

> ' ... MRS RON LINTHORNE WAAF Melbourne wishes to THANK all relatives and friends, especially Mrs. A.Smith (Wagga) for their love and expressions of sympathy in her recent sad and sudden loss of her dearly-loved husband (Sgt Ron Linthorne, AIF) with all kinds of friends please accept this as a personal expression of her sincere gratitude.'

Ron's mum also left this most fitting tribute:

> ' ... In proud and loving memory of my dear son, Happy and smiling always content, Loved and respected wherever he went, His life was great his heart was kind, A better man no-one could find, What beautiful memories left behind, He lies now in Wagga in a soldier's grave, Honoured with Australia brave. Our Ron.'

And from Pauline on the second anniversary of his death:

> ' ... Two years ago you left me, Left me without a word, But my memories of you and your ways, Will linger with me all my days.'

In 1972, Ron's mum, Elizabeth, died in Western Australia.

4—WX27166 Alfred 'Alf' Edward Woods (aged 32)
Plot B, Row B, Grave 5

Of all the men in this story the early life of Ron's fellow Western Australian, Alfred 'Alf' Woods', provides that one necessary charismatic character who delivers what I describe as an 'enigma under a slouch hat'! With an air of mystery and intrigue enveloped in a cloak of international destinations and debonair dispositions, not only was Alf Woods the son of clergy journeyman, he found himself in an occupation not normally reserved for men in a true-blue pioneering Australia. This unique fact alone shrouded greater mystery to complement his already mysterious existence—just who was Alf Woods?

Alf's mother, Magdaline, was born on Christmas Day 1890 in the small village of Rewa; a nondescript village just north of the Fijian capital city of Suva. In contrast, his father Edward was born in 1885 in Sale, a small country town in Victoria, Australia. It's not completely known how a Victorian boy from the Gippsland bush could meet and marry a Fijian girl, but it's believed that divine intervention played a small part. At a young age, Alf's father Ed became involved in the congregation of the Church of England in the Gippsland area. By 1909, as a 24-year-old, his father was employed as a lay reader in the Church. Called upon to preach and lead services, his father was not yet a full-time minister, but he clearly had the dreams, desire and passion to be one.

It was somewhere during the period 1909 to 1911, Ed was doing missionary work with the Church of England and somehow found himself in Fiji—the rest was left up to Ed. It was here he met Alf's mother, the beautiful local girl Magdaline. The couple courted and were married on December the 14th 1911, in Suva. Within two years, the couple would have their first-born son, Alfred (Alf), born on the 9th of April 1913. By the 16th of June the following year, Alf would have a baby sister Doris, also born in Fiji.

The outbreak of the Great War in 1915 saw the young Woods' family back in Australia. After making the journey from Suva to Melbourne aboard the passenger cargo vessel *SS Levuka*, a ship from the Australasian United Steam Navigation Company in Sydney—the family settled back in Alf's father's hometown of Sale. Before too much longer family joy and anticipation was at an all-time high as they prepared for the arrival of another baby. Born on the 19th of January 1916, David Woods—a brother for Alf and Doris—was born at the St Helen's Private Hospital. But something was wrong with David! Surviving for only two days, he never saw the real world and passed-on in the very hospital he was born in. It may have been a short life, but it devastated Magdaline and Ed. For Alf and Doris they'd lost the most important anticipated arrival in their life—their little playmate.

Still distraught by the death of the infant, the family returned to Fiji to seek the solace and comfort of Alf's mother's family to help with the grief of losing a child. Nevertheless, they also sought the comfort and homeliness showing off their beautiful babies, Alf and Doris, to the extended Fijian family. However, for Ed, church commitments were looming. Ed returned to Melbourne in December of 1917, thereby leaving his wife and two children in the safety and sanctity of Fiji and his wife's loving and compassionate family.

The following April, shortly after celebrating his fifth birthday, Alf and his mother and sister boarded *SS Levuka* once again and made the journey back to Australia. Arriving in Sydney with young Alf and Doris, the children had become good travellers. Eventually reuniting with their husband and dad, by 1919 the family moved 216 miles west of Sale and established themselves in the Victorian town of Cobden. The following July, the family welcomed another arrival to the Woods' brood. It was to be their third and final child, this time another son—Clifford. Shortly following Clifford's birth, the transient clergy family were again on the move. This time their destination was Bealiba, a small Victorian gold-rush town 113 miles north of Cobden and about 46 miles from Bendigo. The move was important for Ed's career. He was appointed the Deacon of the Church of England and was now a registered clergyman in the State of Victoria. The history of Bealiba suggests that the town wasn't very well-populated in the 1920s and at eight years of age, it provided Alf with limited schooling opportunities. But the travelling clergyman and his resilient family wouldn't be calling Bealiba home for too long. Before the family really had time to settle in the sleepy hollow, they were off again. Enduring a painfully long journey, the Woods family landed in Western Australia.

By 1925, Alf was a 12-year-old boy living in the Rectory (the residence associated with the church of the clergyman and his family) at Northampton. The town was a sparsely-populated wheat belt township about 32 miles north of Geraldton in the mid-west region of Western Australia. Then, in 1931, when Alf was 18, the family finally found itself in the Perth suburbs and was now living in the Rectory in Middle Swan. For Alf he was thankful to be in the larger cities, especially as a young man looking for a girlfriend. The isolation and arduous life in the remote towns couldn't compare with the opportunities he would face in the large city. All he had to do now was meet a girl.

Over the coming years, Alf didn't follow his father's calling as a clergyman. Instead, the Fijian-born traveler became of all things—a hairdresser. By 1936 the stylist had met and fell in love with Lynda Jean Oliphant. Due to her mother's first name also being Linda, she preferred to be more commonly known as Jean. Thankfully the pair met before Alf's mother and father took off again on yet another church 'mission'. On this occasion they were bound for Currie, King Island. It was a small island anchored in Bass Strait between Victoria and Tasmania's North West coast. It was not somewhere Alf wanted to go; besides, the love of his life was in Perth. Before departing for the isolated island, the family celebrated Alf's marriage to Jean.

After his parents departed Alf and his new bride moved into a residence at 200 Lord Street in East Perth. The house was just a short stroll from the banks of the beautiful Swan River. It was an idyllic life for the newlyweds. In 1937, the recently-married Alf and Jean insisted that Linda, Jean's mother, move in with them in a bigger house. Both Jean's mother and father had been born at the Wallaroo Mines in the late 1880s and were married in the mining town of Boulder, near Kalgoorlie, in 1908. Being a miner at the turn of the century, Jack Oliphant had a hard, but disastrously short life. He died working in the mines in 1917, aged just 36. His widowed wife, Linda, remarried Clarence Naulty in 1929, and moved into a residence in Pier Street, Perth. Regrettably, Clarence was also to die early, passing away in 1932 aged just 40. Linda, a widower once more, was forced to live alone during Australia's immediate post-Depression years. Later moving in with her daughter and new husband, it was just the family connection she needed.

The new family situation resulted in Alf and Jean finding an even bigger home to accommodate Linda; besides they were also planning for a baby. They found a more suitable home less than a mile and a half away. This time the house was in an adjoining suburb close by at 18 Elizabeth Street, Maylands. Although this house was even closer to the Swan River than the last, it had a dark past. Unbeknownst to the new occupants, the previous occupants, the Gordon family, recently faced their own personal heartbreak causing them to vacate the home.

With five daughters and one ten-month-old infant son, the Gordon house was full of life. But without warning after being admitted to hospital, the Gordon's lost their only son in January of the previous year. It appeared as if this particular house in Maylands was troubled or cursed by tragedy. Believing the dwelling cursed, they packed up their belongings leaving the bad memories of the house behind. Their departure made room for the extended Woods family to move straight in. Unaware of the tragedy in their latest home, Alf Woods and his family found life in pre-war Western Australia to be comfortable.

Alf was cutting hair, whilst Linda and Jean tended their house. Although it was Australia's gloomiest years, the family remained tight-knit and survived together, guided by the Church. Sadly, Linda didn't get much time to enjoy living with her daughter or getting to know her son-in-law Alf. She passed away the following year. Laid to rest at Karrakatta Cemetery, aged 52, it appeared as if the ill-fated house had claimed another victim. But further distressing tragedy was still to visit the occupants at 18 Elizabeth Street, Maylands.

An avid horse lover, rider and owner of a trotting cart, Jean decided it was time to sell her equipment, including road wheels, boots, rugs and whips. Falling pregnant in late-1940 and expecting their first child in mid-1941, she was selling her equipment for thirty pounds, so she and Alf could afford to have the baby. Placing an ad in the West Australian newspaper in April 1941 it ran alongside a spread advertising the latest-released movie from MGM studios playing at the local cinema, 'Marx Brothers Go West'! On July 4th 1941, Alf accompanied Jean to the King Edward Hospital for Women in Middle Swan. The excited, nervous and anxious couple was expecting to witness the amazing miracle of the birth of their first child.

1941 brought not only the war to Alf Woods' life, but inconsolable heartbreak and pain. Jean and Alf knew something was wrong with their baby when the medical staff in the delivery room wasn't smiling. The normal happiness and emotion of a new life was replaced with hushed tones and whispers. On July 4th 1941, Alf and Jean lost their baby; it was a stillborn son. Heartbroken and inconsolable, Alf and Jean returned home and remained in the Elizabeth Street address. But the house was always a little empty. Alf and Jean were left traumatised following the death of their only son. Scared of the consequences of a recurrence, they never experienced the joy of having their own children. The cursed house never again heard the patter of tiny feet, or the innocent laughter of a child.

Consumed by grief Alf tried to block out his pain by becoming a more-involved member of the local Militia. The war in Europe raged on and West Australian men were being summoned to support. It was the type of distraction Alf needed. He tried to put the pain behind him and it was made a little easier when on the 18th of February 1942 he was called up for full-time duty with the 55 Anti-Aircraft Searchlight Battery. Throwing himself into his work, by the end of the year he was promoted to Acting Corporal. Within five months, Alf had signed on with the 2nd AIF and allocated his new Army identity. It was there, working with the Anti-Aircraft Searchlight Batteries, Alf met Ron Linthorne and another good mate Harry Ballingall. They'd all became great mates, but the relationship between Alf and Ron was special. Like Ron and Harry, Alf was a good soldier. Between the three of them, they'd all become well-respected men of the unit. Alf and Harry would eventually be promoted to Corporal, whilst Ron was a little more ambitious reaching the rank of Sergeant.

In April of 1943 Alf was detached to the Regimental Instructors' School at Karrakatta. Here he learnt valuable leadership skills before re-joining his unit in May that same year. In September, the 3 Australian Corps Mobile (Chemical Warfare) Training Unit arrived at 55 Anti-Aircraft Searchlight Battery where Alf participated in the Regimental Gas Instructors' Course. And with a total of 63%, he qualified as a Unit Gas NCO. For the next six months he was backwards and forwards between Searchlight Batteries, eventually ending up at 69 Mobile Searchlight Battery alongside Ron where the two men hatched a plan to become engineers. In February 1944, Alf completed infantry training and by May of that year he had been promoted to Bombardier.

In June of 1944, he travelled to South Australia for onward movement to the Northern Territory. He wouldn't be returning home to Jean for a very long time because the plan he hatched with Ron meant that he was also due to go to Wagga Wagga after his time in the Territory. Before he knew it Alf was on the train bound for the East. Thankfully beside him was his great mate Ron, still twirling the ring on his wedding finger whilst staring out into the boredom and emptiness of the Nullabor. Shortly after, Alf and Ron arrived at Wagga Wagga and after a couple of weeks waiting joined in with G Coy during demolitions week.

Much like his good mate Ron, Alf's family received remarkable outpourings of notices and well wishes in the Western Australian press following his death. His inscription simply reads;

'... *His Duty Nobly Done*.'

His mother died in 1949 and his father passed away in 1966. His little sister Doris passed away in 1986 after a full life aged 82. As for his wife Linda, (Jean) she remarried in 1948 and died in Perth in 1995, aged 79.

5—WX23101 *Alfred George Witt (aged 20)*
Plot B, Row D, Grave 8

According to the dictionary, 'destiny' means '*a predetermined course of events considered as something beyond human power or control*'; for Alfred George Witt, his destiny was sealed when he was born on 25th of April 1925 in Western Australia. Exactly ten years before the day of his birth, long before sunrise, nervous and well-trained Australian and New Zealand soldiers—ANZACs—sailed from Egypt, and, in the darkness of the morning, quietly climbed down rope ladders and stepped into small rowboats which were towed to the shores of Gallipoli. The men rowed the last part to the shore. It was April 25th 1915, and the men were rowing into a seminal moment in forging Australia's military ethos.

10 years later, and 53 miles south of Perth in Pinjarra, Frederick (Fritz) and Ella Witt gave birth to their son Alfred. But like most children born Alfred, he was known by all, as simply, Alf. When he arrived he already had

two older brothers; Percy aged seven and little Leslie aged four. The family had moved from Fremantle to join his father's parents in Pinjarra sometime after they were married in April of 1915. With no more than five families living in town, the area was an isolated farming community. And, much like other pioneering families, the extended Witt families were scattered all around the region. Occasionally, normally during big events like dances, outlying families would gather in town. Arriving from an assortment of wheat farms, some would travel in from up to ten miles away to enjoy community events in town. Ella, Fritz and the three Witt boys were no exception.

Alf came into the world when his parents were living in West Coolup, a small farming community just over ten miles southwest of the larger community of Pinjarra. His father Fritz found work as a farm labourer, whilst his mother kept the home running. Young Alf and his brothers attended the nearby Pinjarra State School and following his formal education, Alf moved to Waroona, a larger town with more work opportunities in the local dairy industry where Alf found work at the local Nestle Milk Factory. He was now living and working sixteen miles south of his parents who were still in Pinjarra. For Alf, living in a room at the nearby Whitakers Mill and working at the factory became his existence—he enjoyed the freedom away from mum and dad. Coincidently, and crucial to the town in the 1930s and 40s, the factory where Alf worked was located in the famous and renowned McLarty Street. Having attended the same Pinjarra State School as Alf Witt, Duncan McLarty was a decorated soldier in the Great War and a great role model for the impressionable Alf, particularly when the war rolled into Western Australia in 1939. Later, McLarty became Sir Duncan as the 17th Premier of Western Australia.

By August 1940, his eldest brother Percy, at age 21, had been conscripted and became a craftsman in a mechanical workshop. In July of 1942, Leslie now also 21, joined the 2nd AIF and was sent to a field ambulance unit.

In January of 1945, with visions and aspirations of heroic deeds, similar to his former school hero, Sir Duncan McLarty, and now his two older brothers, Alf Witt decided to join the 2nd AIF. With his parents' consent, he made his way to the Drill Hall at Karrakatta in Western Australia and enlisted on the 23rd January. On his Attestation Form, Alf listed himself as a 19-year-old single truck driver with no previous military service. He listed his father as his next of kin, and by the 27th of January, Alf marched into the recruit reception Centre at Karrakatta. For the next two weeks he remained at Karrakatta as WX23101 Private Alfred George Witt waiting for more West Australian volunteers to fill up the troop train heading east.

Bidding his parents farewell, the proud Western Australian joined the 2nd AIF ranks hoping to meet up and serve alongside his brothers. Undoubtedly, the unassuming truck driver was just another uniformed passenger on that infamously boring train ride to the eastern states. At least there was the sport of bottles and rocks, which kept Ron Linthorne and Alf Woods busy a few years back. For Alf it was all an adventure. Five days of hell on five different trains saw him finally arrive in the east. On the 19th of February 1945, Alf marched into a foreign world of discipline, dust and danger at the 1st Army Recruit Training Battalion at Cowra.

Surviving recruit training, doing reasonably well in aptitude testing and interviews, Alf was allocated to the Royal Australian Engineers. Just over two months after arriving at Cowra, Alf marched proudly out of recruit training on the 30th of April.

Posting into Kapooka camp, he was allocated to the 1st RAE Training Battalion. After settling into camp routine, he was well on his way to emulating the feats of Sir Duncan and that of his two elder brothers, who he hadn't seen in quite some time. Alf was keen to catch up with them after he finished training at Kapooka. Sadly, after the events of week four down on the demolitions range, the three proud Witt brothers from Western Australia never met up in the east. And for Alf, they never saw him again.

Alf's parents and his two serving brothers, Leslie and Percy, never made it to the funeral. The two surviving Witt soldiers both discharged from the 2nd AIF in early 1946. Sadly their mother Ella passed away on 12 September 1959 aged 66. Their father, Frederick 'Fritz' lived until he was 85 and passed away in September 1971.

In the Wagga Wagga Cemetery Alf's headstone inscription simply reads;

> *'... His Life a Beautiful Memory His Absence A Silent Grief'.*

South Australians

6—S115574 William 'Bill' Reid (aged 36)
Plot B, Row C, Grave 16

Brown-haired blue-eyed William or 'Bill', Reid was born to his single mum—Ruby Irene Reid in Broken Hill, New South Wales on the 9th of January 1909. Arriving just before her 17th birthday, Ruby gave birth to her first-born son William whom she later referred to simply as Bill. Moving back to South Australia as a single mum with her small bundle of joy, Ruby soon fell in love with 22-year-old local man Joseph Height. The couple married in Mt Gambier at a small ceremony in the Church of Christ located on the corner of Bay Road and Railway Terrace. They settled down into married life and, although Ruby was a young bride with a baby, presumably Joseph unconditionally accepted baby Bill as his stepson.

An air of mystery surrounds Bill's infant life. Perhaps no longer in her care, mainly because Ruby and Joseph were now starting their own family, Bill's upbringing was far from ideal. By 1910 Ruby had given birth to a young daughter, Irene, and by 1912 her second-born son, strangely also named William, was born. 1914 brought more baby news for Ruby and Joseph when the more mature mum found out she was expecting twin sons. 5-year-old first-born 'Bill', not to be confused with baby William, was about to have two more stepbrothers. But tragedy was to cruelly strike Ruby. Her two long-awaited infant boys didn't survive childbirth. Little Jack and Edward Height died shortly after their arrival on 14 July 1914. Anxiously, Ruby escaped the pain of her loss by replacing her lost boys with another bundle of joy. Along with Joseph they welcomed another daughter, Margery, in 1916. By this time Bill, now aged 7, was being educated through the primary school system, but as the years wore on Bill realised he wasn't the academic type. He struggled to stay at school and left without achieving a secondary school Leaving Certificate. Throughout his teen and early adult years, Bill's life and education was forged on the streets of Mount Gambier and Port Augusta. He became street-wise and a troublemaker. By 1934, aged 23, Bill was busy forging a life beyond Australia's gloomy Depression in South Australia when his grandfather, Ruby's father Amos, passed away. It was a sad passing in the Reid family of a much-loved father and grandfather. Four short years later, in 1937 now a strapping 28-year-old man, and hard at work as a fettler (a railways maintenance worker) with the Commonwealth Railways, Bill watched as Joseph and Ruby's relationship deteriorated. The demise of the relationship forced his younger step-siblings to suffer at the hands of a broken marriage. By the end of the year Joseph, a ganger also with the railways, sought a divorce from Ruby.

The war dominated everything in Australian in 1939. By then Bill knew his time to serve as a soldier was fast approaching. He was however reluctant to sacrifice everything. He chose not to volunteer. Instead, like many other South Australians, he chose to work and wait it out—hoping that he also missed the feared conscription notice. But the Army's need for reinforcements meant Bill's plight and journey faced an inevitable outcome. At nearly 34 years of age, living in Adelaide with his wife (coincidently also named Ruby), he was conscripted in December of 1942. On his Attestation Form he declared his court conviction for drunkenness in Broken Hill in a bold attempt to be over-looked, but it wasn't enough. Completing his Militia training with the Australian Citizen Military Force, Bill returned to his work with the Commonwealth Railways. But *S115574* Private William Reid wasn't employed in a reserved occupation. This made him susceptible to being called up for his full-time service obligation in the 2nd AIF. What he didn't know was just how long he would have to wait.

On the 2nd of November 1944, almost two full years later, the mature-age South Australian, now aged 36 and living and working the rails in Port Augusta, actually fronted-up for full-time duty. Conscious of the fact it was time to do his bit, Bill didn't necessarily accept his obligation with enthusiasm. At his age, with many hard years of toil etched into his hardened brow, Bill found himself at the Wayville showground surrounded by much younger-looking men. Not exactly eager to blend in with other recruits half his age, Bill suddenly found himself surrounded by something he'd never experienced before. Enthusiastic young men, who hadn't yet experienced life, appeared strangely eager to sacrifice their young lives to serve their country proudly. Thankfully, by the time Bill got to don his uniform, in late 1944, the war in Europe was coming to an end but that was only one of Australia's commitments. Unrelenting Japanese soldiers were still posing a threat in New Guinea allowing Bill to realise that the role of the 2nd AIF was far from over. The following day he reported for duty and was granted a couple of day's leave without pay to organize his affairs in preparation for his life as a soldier and recruit training at Cowra.

On the 7th of November he departed Adelaide bound for Cowra. Arriving on the 11th of November, Bill realised his tumultuous life and upbringing prior to his Army service, that authority, and discipline would prove difficult. Amongst his much younger and somewhat overzealous military cohorts, the hard-working railway man struggled with life as a full-time soldier in the 3rd Australian Recruit Training Battalion. His internal struggles came to a head when he failed to parade at 0600 hours on the 5th of January—instead he'd decided AWOL was a better option. Declared an illegal absentee (I.A.) on New Year's Day of 1945, the 36-year-old reluctant 2nd AIF recruit was now on the run. But freedom was short-lived. On the run for more than three weeks he was finally arrested by the Provosts (Military Police) at 0645 hours on 23rd of January 1945, somewhere in South Australia; possibly hiding at home. Held in custody by the 'unsociable and uninviting' Provosts, he was soon delivered back into the custody of Cowra's 2nd Recruit Training Battalion. But Bill's life was about to become even more complicated. Not only did he go AWOL again, but he'd also allegedly 'lost' his issued military clothing and equipment. Within days he was again charged with AWOL. In addition, a further charge of negligently losing his military clothing and equipment was added to the charge sheet. The Commanding Officer of the 3rd Battalion handed down a hefty punishment. Fined five pounds, Bill's pay was stopped for a week and he forfeited 23 days of pay—a fine for each day he was on the run as a temporarily free man.

Following his infringement, he was sent back to the 2nd Battalion to continue recruit training. Within two weeks he was suffering from an undiagnosed illness (possibly dyspepsia) and removed from training. Initially evacuated to the 104 Australian General Hospital at Bathurst, he was later moved to the 103 Convalescence Depot at Baulkham Hills that March. By the 28th he was discharged and eventually marched back into the 2nd Recruit Training Battalion to resume training on the 7th of April 1945.

Completing his training in late-April of 1945, convincing Corps allocation officers that he would be well-suited to a life as an engineer, Bill marched out of Cowra and found himself on the troop train bound for Wagga Wagga. This time the young faces he scorned at when he first got to Wayville were now mates alongside him from the 2nd Recruit Training Battalion.

To Bill they were still young men, but at Cowra, many of them like Norm Dilley, Teddy Robson, 'Brickie Hurst, Allan Flood and Joe Collins, taught him that life was too short to be bitter. That same day, along with his new group of mates, the mature but somewhat 'reluctant' soldier marched into the RAE Training Centre.

After Bill was killed in the explosion, life for his mother Ruby was further tainted by tragedy and heartbreak. She not only lost her twin sons, she later divorced from her husband and with Bill's death, her first-born son was also taken from her. In February of 1946 Ruby was taken to court for failing to leave a rented property. She had been living in the home for four years when the owners decided they wanted to move back in; Ruby refused to go. After a life of tragedy, Ruby died and was buried under her married name of Height. Reunited with her three sons, she rests in peace in Carinya Gardens, Mount Gambier, South Australia.

Just like Kevin Pierce, Bill's burial was devoid of a family presence. His military headstone at the Wagga Wagga War Cemetery bears no inscription from a relative. He rests now among those many young enthusiastic men he now calls his mates.

7—SX34059 Denby Eric Grasby (aged 18)
Plot B, Row C, Grave 3

Since the turn of the 20th century, the Grasby family, of the Balhannah Hills area southeast of Adelaide in South Australia, have been well-known as generous pioneers. It's documented that in 1913, the family purchased and worked the land that became known today in Adelaide as the 'Grasby Memorial Park'. Apart from their pioneering legacy, the Grasby family also displayed an extraordinary spirit of community that also made the family well-respected. One particular story, which describes the family's selfless generosity, occurred during the Great Depression. It was a time in the nation's history where fiscal uncertainty crippled the livelihood of almost every Australian family. For Joseph and Beatrice Grasby it mattered not. They answered the call of support by providing Christmas trees from their property for every child in the area. Not only just a tree, but also a small gift to go with it. It was a selfless act and one that cemented their status as community doyens. Understandably then, when WWII arrived in 1939, the generous family organized for local ladies to pick daffodils from the Grasby land and sell them to make money for the Fighting Forces Fund.

But the Grasby's contribution to war was more than just daffodils in WWII. Just like Kevin Pierce's father, Denby's father Eric was also a soldier in the Great War. The impressionable 18-year-old was working as a gardener when he joined the Australian Imperial Force in June 1917. He served with the 10/50th Infantry Battalion in France and, after witnessing the horrors of combat in his two years of soldiering, was eventually discharged from Service in 1919. Unlike Kevin's father, Eric came home physically unscathed. Following the war, Edna and Eric Grasby were married. On December 2, 1926, Edna gave birth to their son, Denby Eric Grasby in Ambleside, South Australia.

As we fast-forward through what was by all accounts a normal life for Denby growing up in South Australia, he finished his school years without drama. In the early 1940s, he was working as a well-paid carpenter and still living with the family in Kitchener Avenue, Dulwich—a small suburb adjacent to Victoria Park and less than two miles from the city. But festering away in the background noise of the Grasby house was the constant newspaper and wireless radio announcements about the war. Courageously accepting his responsibility, Denby presented himself at the Wayville showground and volunteered. Considering his father's military history, Denby thought joining the 2nd AIF was a compulsory duty—it was just a matter of time and age.

Much like the many other Great War veterans, Eric also responded to the call to take up arms and sacrifice his life for a *second* time. This time it was January of 1944, almost 27 years after first putting up his hand. At 46 year of age, Eric joined the 2nd AIF and was posted as a member of a local Ship's Company. The Grasbys, by engaging all members of the family to support the war effort, were a good example of what Australia was trying to achieve.

Denby had his sights firmly planted on serving alongside his father, but it was another family member who beat him to the chase. His 19 year-old sister, Patricia (or Pat as she was known) was also patriotically doing her bit. Providing inspiration for Denby, Pat had enlisted into the AWAS approximately one month after her father re-enlisted in February of 1944. Posted as a private to the staff at the Recruit Reception and General Details Depot (RR & GDD) at Wayville, Pat was working in the recruiting offices when Denby and his mates sought out enlistment.

It was fast becoming inevitable, and with some irony, that Denby was going to serve alongside his father. Ably assisted by his older sister, Denby's enlistment into the Army was soon complete. Although 18 in the previous December, even though Denby was working as a well-paid carpenter, he was still under 21 and authorities considered that he still needed his parent's permission to enlist. It was a hard decision for Edna. Eric started it all when he re-enlisted into the 2nd AIF, but now her husband AND both of her children wanted to be in harm's way. With husband Eric and daughter Pat already in uniform, Edna Grasby really had no choice. Assisting her dashing son to serve alongside his family and friends, Edna signed the appropriate paperwork and gave her excited 18-year-old son the opportunity to have the adventure of his life.

Like Bill Reid, Denby signed on the dotted line volunteering to join the 2nd AIF on 15th of February 1945. Coincidently, it was on the same day two other South Australians, Allan Bartlett and Ivan Merritt signed on. Signing their declarations and receiving their 'kit', the four men were soon on the biggest adventure of their lives. Showing no signs of nervousness, the Wayville showground uniformed staff put Denby and his new colleagues through the rigorous enlistment procedures where he was finally allocated his official Army number: *SX34059*. He was ordered to report back in two days when he would commence his life as a soldier. Denby reported back as ordered. Spending a couple of days being inducted into military life, he was granted leave without pay from the 16th of February until 0800 hours on the 20th to go home and sort out his affairs before departing for recruit training at Cowra. Denby returned home and got organized. The well-disciplined son of a Great War veteran returned to the Wayville showground right on time. That morning he again ran into Allan Bartlett and Ivan Merritt who were both also returning from leave. Spending just one night in the rough Wayville camp, the following day, accompanied by his new mates, Denby caught the troop train and marched out of Wayville bound for Cowra. Three short and mainly sleepless days later, he arrived at Cowra and marched into the 2nd Australian Recruit Training Battalion. Beside him, as always, were Allan and Ivan.

Understandably, Denby was an enthusiastic soldier and breezed through recruit training. By the 7th of May his future as an engineer had been confirmed by allocation officers. That same day he marched out of recruit training and transported to Kapooka where he marched into the RAE Training Centre. Unbeknownst to Denby of the path he was now on, he was allocated to the ill-fated 1st RAE Training Battalion.

The events of 21 May 1945 involving his young son would leave Eric with a sense of anger towards the Army. Sadly the pioneering Grasby family of Adelaide never fully really recovered from the loss of Denby.

Eric discharged from the 2nd AIF as a Warrant Officer Class Two in the November following his son's death, whilst Patricia stayed in the Army for a little longer. She eventually discharged in August of 1946. Life after the tragedy was especially hard on Eric who'd buried his father just five short years before he had the heartbreaking task of burying his son. Eric and Edna farewelled their young son with a fitting inscription on his headstone identifying his life as a soldier,;

'... *Rest in Peace Beloved Son Until the Day Dawns.*'

8—SX34069 Ivan Walter Thomas Merritt (aged 26)
Plot B, Row D, Grave 10

One man Denby Grasby took a shine to at the Wayville showground recruitment depot, and later at Cowra and Kapooka, was 26-year-old Ivan Walter Thomas Merritt. Ivan was the second-born son to Walter and Olive Merritt who welcomed their brown-haired blue-eyed son to the small Victorian town of Kaniva, Victoria on the 13th of August 1918. The small isolated country town was less than fifteen miles east of the South Australian border. Over the years the family would grow to a total of five sons and three daughters. As Kaniva was closer to Adelaide than Melbourne, when the Great Depression approached in the early 1930s, Ivan's family was forced to search for work in nearby Adelaide rather than Melbourne. In his father's quest for stable employment, the family moved to Mitcham, a suburb about four miles south of Adelaide. Their transient life was a familiar episode for similar families struggling through the uncertain post-war period and into Australia's fiscally gloomy years of the Great Depression. It was in Mitcham where Ivan grew up and became a fully-fledged South Australian rather than a Victorian. Completing his primary schooling and gaining his Secondary School Entry Certificate, Ivan and his siblings had a hard, but joyful upbringing.

However, it was also in the small Mitcham township where the family, especially Ivan, experienced the despair of a family tragedy in 1936.

That year, Ivan was just 18 when his eldest brother, and best mate James, died in a car accident. His much idolized and loved big brother was just 25. Ivan was heartbroken. In that one tragic instance he became the eldest sibling in the large Merritt family. Suddenly, Ivan accepted an unwavering obligation to assist his ageing parents raise their four boys and three girls during the hardships of the Depression. Whilst, Ivan's early adult life was less than idyllic, happy and memorable times were soon to follow.

After falling in love and getting engaged to local girl Dorothy Joyce Haywood in June of 1941, 24-year-old Ivan and Joyce, as she preferred to be known, tied the knot at the Mitcham Baptist Church on May the 30th 1942. Coincidently, in the 1940s Ivan Merritt's elderly parents were living in Welbourne Street, Mitcham, less than four miles apart from the Grasby family home in Kitchener Avenue, Dulwich. But in the early 1940s Ivan and Joyce weren't living in Mitcham with Ivan's parents or anywhere near the Grasbys. Driving tractors for a living, Ivan was working more than 125 miles north at the Engineering and Water Supply Department, at Crystal Brook. Known as the southern gateway to the Flinders Ranges, Crystal Brook was a pleasant location in the heart of some of South Australia's most productive sheep and wheat country. With shady peppercorn trees lining the main street, Crystal Brook was a sleepy country town. It was so sleepy that visitors often reported time actually stood still when walking through town. For Ivan and Joyce this was their slice of the quiet life. It was, thankfully, miles away from the horrors of war which had gripped the country.

But war was still raging overseas and Australia was well and truly in the thick of it. It would also soon have an effect on the entire Merritt family. A far cry from the serene scenes of downtown Crystal Brook, Ivan's younger brothers, Frank and Richie (or Dick as his Army mates called him) were already serving in the 2nd AIF. At 21, Frank, a butcher, joined the Army in April of 1941 and was posted to the 2/1 Field Butchery Unit. A couple of months after Ivan married Joyce, his 21-year-old younger brother Dick joined the Army in August of 1942. Dick was an engineer, a Sapper, who was eventually posted to 6th Australian Army Transport Company and sent to Darwin.

It was around Christmas of 1944 when the news that Joyce was pregnant lured Ivan back to the big smoke of Adelaide and needing family support. News of the pregnancy forced the couple to return to Mitcham and they moved into a house about 150 metres around the corner from his parent's home. Shortly after returning, like his hero younger brothers, Ivan was enticed to enlist into the 2nd AIF. Whilst providing some form of stability and security for his new family, his life as a soldier would test his pending fatherly resolve. But hearing the stories of adventure from his heroic brothers, his endeavor to join their ranks almost outweighed and overshadowed his first-time fatherhood.

At 26 years of age, with two younger siblings already serving and his three-month pregnant wife at home, Ivan sensed it was a time for him to show a greater responsibility to his family. With pending fatherhood he set off for the great adventure. The date was the 15th of February 1945. With his farming and engineering experience Ivan was hoping to land a good military job. Before he knew it he was processed through the Recruiting and Reception Centre at Wayville and quickly became *SX34069 Ivan Walter Thomas Merritt,* the soldier. Ironically it was the very same day Allan Bartlett and Denby Grasby stood together in the volunteer line at Wayville. At 26 and married, setting off on the journey of his lifetime, the mature Ivan certainly stood out from the majority of baby-faced recruits that day.

The following day Ivan reported for duty at the Wayville showground. Issued with his clothing and equipment he bedded down in the primitive facility, a far cry from the comforts of home. As he settled in, the more mature and well-weathered face of a working man couldn't help but acknowledge that he'd found himself amongst much younger-looking men; men like Allan and Denby who were nearly ten years his junior. Soon, after three days, at midnight on the 16th of February, Ivan was released from training on leave without pay to go home and get his affairs in order before setting off to the east for recruit training. The expecting father didn't hesitate. Finding the quickest way home Ivan quickly raced to Mitcham to check on the welfare of the pregnant Joyce. He was however due back at Wayville by 0800 hours on the 20th of February. Ivan had three days with Joyce and his family before making the journey back to Wayville and then onto Cowra. After a heartbreaking farewell with his pregnant and emotional wife, Ivan tearfully left the embrace of Joyce and returned to Wayville. Spending a restless night in the showground, Ivan's friendship with Allan Bartlett and Denby Grasby helped him through the pain of missing his wife. The following morning the three previous strangers were soon on a train bound for Cowra and recruit training for the 2nd AIF.

On the 24th of February, Ivan marched in to the 2nd Australian Recruit Training Battalion to commence a training regime that was perhaps designed for much younger men than he. In the following hard, hot dry months Ivan and the other South Australian boys, including Bill Reid, would form a special bond during the highs and lows of the difficult recruit training. For Ivan, as an older recruit amongst the young men, he would find a friend and confidant in the equally mature Bill Reid. Both the men would become a mentor for Allan Bartlett.

Ivan's training was interrupted after he was admitted to the 11th Camp Hospital at Cowra on the 17th of April suffering from an acute case of tonsillitis. After a week in hospital enjoying the break, Ivan recovered and returned to full training. He still had about two weeks of the program to complete. By the 7th of May, he had finally completed his training and convincing allocation officers that he'd be well-suited to engineers, along with Allan, Denby and Bill, the four South Australians were soon inseparable. On the short train ride to Wagga Wagga the men exchanged the highs and lows that they'd just experienced at Cowra, all the while trying to debunk the legendary stories of a hard daily routine at the Engineer Training Centre. The men all marched into the 1st RAE Training Battalion prepared for whatever the Army was going to throw at them. In just two weeks time, the four mates found themselves huddled together in the ill-destined dugout. Three would die, and one would have his life changed forever.

Joyce Merritt was still pregnant during the moving funeral for her husband. She later inscribed Ivan's headstone with a personal message of farewell;

'... *Dearly Beloved Husband of Joyce. His Duty Nobly Done*'.

The wife of the 26-year-old soldier who survived the initial blast, but succumbed two hours later, posted the following message in the newspaper to mark the tragic death of her husband;

'... I have you in my memory, God has you in his care.'

On the 1st of August 1945, less than three months later, at the Unley Private Hospital, Joyce gave birth to a son. She named him Geoffrey Ivan. Sadly though, further tragedy plagued the Merritt family following Ivan's death. His younger brother Gilbert would prematurely lose his only child in 1950 and, by 1957, sadly, Gilbert, aged just 29, would also be dead. Ivan's serving brothers Dick and Frank both discharged from the Army in February and March of 1946 respectively.

Queenslanders

9—Q273551 Ernest 'Ernie' Frederick Poschalk (aged 18)
Plot B, Row B, Grave 12

Ernest Frederick Poschalk, or 'Ernie', was born in Townsville on 9 January 1927. He was the eldest child of Ernest and Cecilia Poschalk. At the time of Ernie's birth his father was a clerk in the Queensland State Government Insurance Office. His parents were married in 1925 and, with his father a successful insurance clerk, the young family survived the horrendous unemployment rate of more than 30% during Australia's Great Depression years. But following the Wall Street crash of 1929, the fiscal uncertainty of the country meant that his father's position resulted in the family living a somewhat 'nomadic' life. His father was busy taking up positions with the Insurance Office in posts around Queensland such as Townsville, Cairns, Roma, Brisbane and Maryborough. The family would also continue to grow over the years with the further addition of two brothers, Kevin and Gordon and a little sister, Beth.

As a 1-year-old, Ernie moved from Townsville to Roma, and then in 1932, at aged five, moved with his parents further north to Cairns. It's here that his father became a vocal and important representative 'Brother' of the Cairns' Branch of the Grand United Order of Oddfellows (GUOOF). After four short years, from 1936, the family was back living a little further south in the tropical city of Townsville. Knowing nothing of the GUOFF, cynically believing it to be some of cult or masonic brethren-based institution, I was somewhat embarrassingly surprised to learn that the GUOOF was an organisation that was established in Australia in 1871 as a friendly society. Open only to Australian-born natives, the organisation was formed by Australians who believed in the principle of mutual self-help to provide some of the medical and other essential services that weren't provided by governments of the time. Ernest's namesake father was one of those *types* of men. Each week his father made a small contribution to a common fund that paid benefits to those in the group who became ill, lost work or suffered hardship. It was a noble and spirited cause. By the late 1930s GUOOF had developed products such as hospital and medical insurance, household insurance, personal and housing loans and life insurance. As a member of the Government Insurance Office, Ernest Poschalk and his entire family were well-known in the financial and community circles. Well-respected, but more importantly, valued members of the United Order and the North Queensland communities, the Poschalk family were much admired and appreciated Australians.

Meanwhile, life for young Ernie and his family remained nomadic on the back of his father's vocation and calling. Ernie's brother Kevin became a successful academic, including Dux of the Cairns School whilst his younger sister Beth was herself doing well in school. By 1937 the young family moved to Brisbane and resided in Wooloowin, a working-class suburb on Brisbane's north side.

As Ernie grew up he became a charismatic fun-loving young man who continually had his mum in hysterics over his hijinks. He was relatively tall with a fantastic sense of humour—a real practical joker. But one of Ernie's greatest attributes were his lungs. As a schoolboy he put his qualities to good use in the school band. He was a magnificent musician, especially on most wind instruments. So good in fact, in 1939, aged 12, Ernie was crowned the Champion of Champions in a local contest playing the euphonium. He was the eldest Poschalk sibling adored by his younger brothers and sister. He was talented, charming and a great role model for his relatives.

In 1939, the arrival of war changed Australia's way of living. It was especially poignant for young and impressionable teenagers like Ernie, who thought the war was an entire world away with no need to worry. He wasn't to know that it would rage on for many more years and that he'd eventually get caught up in it. He was just happy being a teenager, playing practical jokes. But after learning the miraculous story of his uncle it seems Ernie was pre-destined to join the Army.

Herbert Poschalk was Ernie's uncle, his father's younger brother. Herbert enlisted in the 2nd AIF shortly after Menzies made the declaration of war. But Herbert's enlistment was to be no ordinary enlistment and certainly no ordinary service in the Australian Imperial Force. His uncle's story no doubt inspired the eldest Poschalk son to consider a life as a soldier. In April 1940 the Army's maximum enlistment age was 35. In Herbert's case he was already 35 years-old having been born in 1905 as the second of the three boys. Living in Chillagoe, 83 miles west of Cairns, he travelled the distance to the Cairns enlistment depot where he attested that his birthday was 29th of September 1914, making him 26 years and six months old and therefore eligible to enlist. This 'little white lie' continued to confuse military authorities in years to come. Herbert recorded his brother Ernest, as his Next of Kin, c/o the State Insurance Office. Little did he know the Army was due to increase the enlistment age to 40. It occurred the following month in May of 1940. After his enlistment Herbert joined one of the first Queensland units in the 6th Division to be sent to Greece. By October 1940 he was in Palestine and by April of 1941 he was on a ship bound for the fighting in Greece. However in June of 1941, following the Allied capitulation and subsequent retreat to Crete, Herbert eluded capture and escaped into the hills. Armed with a small pistol and nine rounds of ammunition, he became an Australian guerilla.

For twelve months he lived in Cretan villages and evaded capture from the routine German patrols that entered the villages and pillaged the villagers of fruits and harvest or anything else they could get their hands on. The Cretan villagers, living with the prospect of a death sentence for shielding Allied soldiers, provided Herbert with food, clothing and protection. After approximately twelve months he joined a band of Greek guerillas and essentially became a Greek soldier for the next year. He learnt their language, engaged in forays into German territory with his Greek counterparts, and as far as Australian authorities were concerned, was 'missing' in action in German-occupied territory; presumed killed. Meanwhile, back home in North Queensland, the military authorities reported Herbert missing—believed killed or captured by the Germans. There was no communication back to the Poschalk family for well over two years. His military papers, including his Service and Casualty Forms had all been stamped 'DECEASED'. But the family stayed strong, sensing and hoping he was in fact alive.

In September of 1943, the elusive Australian/Greek guerilla commando, reported himself to a New Zealand Discharge Depot in Egypt declaring himself an Australian soldier. Immediately, a 'SECRET' message was sent to Ernie's father back in Maryborough, where they were now living. The message notified the family that Herbert was alive and well and being repatriated back to Australia. Then, in November of 1943 Herbert finally marched into an AIF Reception Depot having been on the run for 830 days. His survival was a remarkable story. As far as the impressionable Ernie was concerned, uncle Herbert was a hero. The media attention he received in Brisbane on his return from overseas reinforced on the receptive Ernie that a career as soldier was now his mission in life.

As a member of the school band, playing the bass trombone and euphonium, although not as academic as his brother Kevin, Ernie completed his schooling with distinction. But approaching the age of conscription meant ditching the trombone for a rifle. It was time to become a man, much like his larger than life hero uncle. In order to complete his transition into adulthood, Ernie got a job as a junior motor mechanic in Maryborough. But his motor mechanic qualification allowed him to play many a practical joke; usually on his mum.

On one particular trip Ernie was driving his prized Maxwell utility south on a holiday with his mum in the passenger seat. But her constant nagging at his erratic driving forced Ernie to suggest a change of drivers. Mum agreed. As she unknowingly moved around the rear of the car, Ernie quickly went to work. Slipping into the driver's seat, Ernie's raucous laughter saw his mum seeing red. Meanwhile, Ernie was beside himself with laughter. He looked across to see his less than impressed mum sitting in the driver's seat missing the most important piece of driving equipment—the steering wheel. In the time it took them to both swap seats, he had quickly detached it and thrown it away.

When February 1945 rolled around, Ernie was listening to the Government's ludicrous plan to defend the country from Japanese invasion by establishing what was known at the time as the 'Brisbane Line'. Ernie soon realised that war was no longer a world away and that turning eighteen meant possible conscription. Suddenly things like the nonsense talk of this so-called 'Brisbane Line' was a reality and he could very well be one of those men defending the country from invading Japanese in his hometown. But turning eighteen also meant girls and it didn't take long for the funny and charismatic Ernie to be touched by romance. He soon met and fell in love with a girl by the name of Imelda Gott. Ernie and Imelda's relationship blossomed and the two 18-year-olds became engaged. Life for Ernie and Imelda was almost perfect but the constant possibility of conscription hung over their young heads.

Unable to dodge the inevitable, although perhaps not meeting the physical stereotypical mould of a herculean-built country sugar and banana farmer from country Queensland, Ernest 'Ernie' Poschalk the 18-year-old academic, became one of Queensland's conscripted sons. Devoid perhaps of physical strength like his Prussian-born relatives, Ernie's strength was found in his mental resilience to adversity. He was also strong in the bond of family and commitment. His relationship with Imelda would ultimately suffer as well. Without the requirement to do compulsory Militia training with the CMF, Ernie was conscripted straight into the 2nd AIF in 1945. He'd had a great upbringing but the excitement of his uncle's story triggered a response to now embrace life as a soldier. Hearing the exploits of his uncle in the wilds of Crete made his life as a junior motor mechanic seem a little ordinary but as a motor mechanic he knew he was perhaps destined for a job as an engineer.

Ernie's good sense of humour took a backwards step when his notice for conscription arrived. Enlisting and marching into the General Details Depot on the 17th of February 1945, he was no longer the practical joker, the funny and mischievous son; musician or mechanic, he'd somehow become *Q273551 Ernest Frederick Poschalk* the somewhat serious soldier.

On the 26th of February 1945, with his orders to attend recruit training, Ernie marched out of the General Details depot. Earlier in the day he'd walked out of the front door of the modest high-set Queenslander home at 261 John Street in Maryborough; his destination was the train station and the start of a fresh journey. Before he knew it he had marched out of one military unit and was on a south-bound train bound for Brisbane. Leaving his fiancée Imelda to ponder what might have been, his brothers and sister continued his legacy—playing practical jokes on their mum. It was a humorous memory that Ernie kept replaying in his head as he momentarily escaped the reality of what he was undertaking.

For Ernie it was onwards from Brisbane to Sydney and then a further train to Cowra where he would also undertake recruit training. He arrived at Cowra on the 2nd of March and was immediately and somewhat reluctantly introduced to a foreign world of discipline and physical activity. Marching in to the 2nd Australian Recruit Training Battalion was a world he wasn't used to. The role of the Recruit Battalion was to turn boys into men, and within the first two-weeks of his transition Ernie found himself in trouble with the chain of command. It was a baptism of fire. He'd committed a silly disciplinary offence, which meant getting charged by the hierarchy. Unlike many who make it through recruit training unnoticed, getting charged meant that Ernie was made to grow up just that little bit quicker. His humour wasn't going to help him at Cowra. On the 14th of March he faced a charge brought down by the Admin Commander for negligently losing his regimental equipment. Found guilty he was awarded a five-pound fine. Thankfully for Ernie the remainder of his recruit training was completed without further complications—or charges.

Completing the arduous regime, the young, and now extremely fit, Queenslander marched out of recruit training as a man brimming with confidence in his newfound profession as a soldier. Convincing allocation officers that his job as a motor mechanic would benefit the Royal Australian Engineers, Ernie was destined for the enormously busy Engineer Training Centre at Wagga Wagga.

He'd already replaced wind instruments with war implements—now, on the 7th of May, alongside other young men in the same situation, like Denby Grasby, Alf Witt and Allan Bartlett, the troop train to Wagga was full of budding engineers. For Ernie, as a previously unemployed junior motor mechanic, his journey into adulthood was almost complete. Inspired by his uncle's amazing stories of adventure he continued his transformation into adulthood amongst fellow trainees of the same age daring dreams of service and glory. He was a lovable rogue who no doubt made many of his tent mates at Kapooka laugh with his practical jokes and mischievous sense of humour. In two weeks, Ernie and his infectious humour were gone.

The Poschalk family inscribed 18-year-old Ernie's headstone with a simple soldier's inscription;

> '... *His Duty Fearlessly and Nobly Done Ever Remembered*'.

Although his inscription was humble, it symbolized Ernie's short military journey. His mother Celia died a few short years after his death on the 1st of November 1951. She was just 48 years of age. Ernie's father moved further south in the years following the death of his son and wife. He settled near Brisbane and remarried in May 1957 aged 55 to Mary Agnes (formerly Molloy, nee Christopherson). Ernest (or Ernie as he was also later called) and Mary lived a long life together. Ernie senior died at Redcliffe, Queensland on 26 September 1990 aged 88, whilst his wife lived on for a few more years. Sadly she too passed away in September 2003 aged 93.

Many thanks to Gordon Poschalk for the photo of his big brother Ernie; reportedly, it's the only one in existence.

✝

10—QX63309 *Frank Wilfred Platt (aged 20)*
Plot B, Row C, Grave 10

Frank Wilfred Platt was the second-born son of Frank and Martha Platt who arrived on the 27th of September 1924 in Brisbane. Their first-born son, James, was named after his dad's father who was a noted Inspector of State Schools in Queensland for more than 42 years around the turn of the century. Frank's namesake father was himself a Great War veteran with his own story to tell. Whilst his mother, well she was the Englishwoman swept off her feet by a young Australian soldier serving with the AIF in London. Frank's mum provided the requisite romantic component to his father's somewhat sheepishly described story of how the couple met.

According to his service records, Frank's father joined the AIF and was allotted to the Army Medical Corps in October of 1915. Then, by April 1916 was on a ship bound for Europe as a member of the 1st Reinforcements, 1st Australian Auxiliary Hospital. However, by October that year, the family hadn't heard from him. Deeply concerned they wrote numerous letters but never received a reply in several months of writing. Little did they know Frank senior had been admitted to hospital with something resembling muscular rheumatism. Spending some time in France, he returned to London as a member of the 16th Field Ambulance and was busy falling in love with Martha, a 25-year-old spinster, three years his senior. So, with the Commanding Officer's permission, Frank's parents were married in Dorset, in the west of England on 26 May 1917. Following a posting to the 2nd Australian General Hospital in London, the newly-married Frank and his slightly more experienced bride, returned to Australia allowing him to discharge from service in November 1919 where he settled into civilian life as a clerk.

Accompanied by his war-time bride, the couple moved around in the Brisbane south suburbs including a place at Rose Avenue, Yeronga, an inner suburb of Brisbane about three miles south of the city. James was born in 1920 and on the 27th of September 1924 in Brisbane, the young family was still living in Yeronga when Frank Wilfred 'junior' was born. The family remained living, working and schooling in the inner city area for many years. Surviving Australia's Depression years, by 1937 the approaching war years saw Frank's father, thankfully, still employed as a clerk for a large shipping firm. Meanwhile young Frank completed his formal education and sought work following in his father's footsteps as a clerk.

In May of 1940 Frank's older brother James completed his Militia training as a member of the Royal Australian Artillery. He was later sent to Port Moresby in New Guinea to man the large guns located at Paga Hill as a member of Paga Battery, one of Port Moresby's gun batteries, so named after its location and serving an important role as front-line security for the Islands. It was perhaps once again inevitable, much like Denby Grasby and

Allan Bartlett, where fathers and/or siblings had already volunteered, that Frank, at sixteen years, watching his brother go off to fight was himself always going to join the Army. Inspired by the colourful war stories from his father as a member of the Medical Corps in the Great War, teenager Frank could do one of two things. He could lie about his age and try and enlist without his parents' permission, or he could wait a few years until he was old enough—he wisley chose to wait it out.

Settling down for the next two years studying to be a clerk, like his father, Frank chose a path of further academic qualifications. Knowing that his brother had been conscripted, he knew there was a real possibility that he too might get conscripted and momentarily disrupting his academic pathway. In October of 1942, that's exactly what happened. Shortly after turning eighteen Frank got called up for service. By then he'd completed his initial schooling and his educational qualifications meant that he possessed academic aptitude and promise. With a scholarship to Brisbane Grammar and graduating from a Business Coaching College, Frank was nine months into a Bookkeeping and Arithmetic qualification when his academic pursuits were interrupted by conscription. He soon completed his Militia training and returned to his normal life, but remained painfully aware that at any moment he could be called up to serve in the 2nd AIF.

In 1945, at twenty years of age and with the war drawing to a close, Frank Platt couldn't wait. He still hadn't been called up and was now seeking a change from the books. Motivated by the propaganda saturation still encouraging young men to enlist, and although working to earn a living as a clerk alongside his father, the excitement and adventure of soldiering was too great an opportunity to miss. He ditched the books for a rifle and decided to voluntarily enlist. With his parent's approval, 20-year-old Frank enlisted on 21st of February at Redbank, a small suburb just south of Brisbane. Unlike his father who'd initially been rejected from joining the AIF, due to being flatfooted, Frank passed all the medical requirements.

Before too much longer, the brown-haired, brown-eyed junior clerk was receiving his movement orders. He was no longer living in the shadows of his father's occupation, he was now *QX63309 Frank Wilfred Platt,* a soldier. Completing the recruiting process he became a fully-fledged 2nd AIF man and prepared to take on recruit training.

Spending five or so days in the General Duties Depot, by the 26th of February, Frank was on a troop train making his way to the Australian Recruit Training Centre at Cowra. Beside him were many other fresh faces all displaying a mixed assortment of fear and excitement in their eyes. Marching in to the 2nd Recruit Training Battalion on the 2nd March, Frank applied himself to the eight weeks of rigorous recruit training. Like so many of the men we've met so far, Frank also marched out of Cowra on the 7th of May. And, much like the many young faces surrounding him on the train, throughout his training he displayed an aptitude for engineering and convinced allocation officers that he should become a Sapper. Frank was allocated to Engineers, and before too much longer, he too made the small train journey and marched into the ageing, yet still functional, Engineer Training Centre at Wagga Wagga.

He was now a member of the 1st RAE Training Battalion and certainly well on his way to fulfilling his vision of becoming a Sapper. It was an exciting time for the young former clerk. He was of course also looking forward to demolitions week.

Frank Platt senior and his wife Martha, with a saddened heart, farewelled their youngest serving soldier son. The simple inscription on his headstone reflected their prayers for his eternal safety, ' ... Fold Him In The Peace Of Heaven.'

11—Q273563 Stanley 'Stan' Robert Morphy (aged 18)
Plot B, Row C, Grave 12

A few years after Frank Platt was born, and certainly much further north, Wilfred and Elizabeth Morphy gave birth to their fourth child, a son, Stanley Robert. Stan, as he was known, arrived in the gold-rush town of Charters Towers on the 4th of January 1927.

Founded in the 1871, not by an explorer but by a 12-year-old Aboriginal boy, Charters Towers grew into a bustling North Queensland town and claims to also have a significant military connection with Australia's heritage. Many young men from the local grazing industry became Light Horsemen, probably because they were excellent horsemen with a keen eye for shooting. They volunteered for not only the AIF but also earlier for the Boer War in South Africa. One of those men connecting the town's early war history extends to the drover, horseman, poet, soldier and convicted war criminal, Harry 'Breaker' Harbord Morant. The charismatic roustabout made a name for himself as a hard-drinking, womanising bush poet who was friends with Henry Lawson and Banjo Patterson. Gaining renown as a fearless and expert horseman, 'Breaker' Morant was said to have traded horses in Charters Towers and married a local girl, Daisy Bates in 1884. As legend has it, Morant failed to pay for the wedding and stole some pigs and a saddle, forcing Daisy to kick him out. The couple separated soon after and were never officially divorced.

Meanwhile, Stan's parents, who were no way connected with Breaker Morant, were married in 1918 and the eldest child, Edward, was born in 1919. Thomas followed in 1922 and a sister Theresa two years later in 1924. Stan's arrival in 1927 made him the baby of the family. Although born a mere five days earlier than Ernie Poschalk, Stan arrived and was welcomed into the north Queensland town not far from Ernie's birthplace of Townsville. Ironically, although born only about sixty miles apart, Stan and Ernie didn't know each other growing up but would later develop a friendship and bond when they became soldiers together.

Stan's father was a cab proprietor when he married his mum when she was aged 23. His parents later moved to Sunburst Creek in Charters Towers where the family raised their four children for the next 25 years in the dairy industry. As dairy farmers, life for Stan and his siblings wasn't an easy upbringing. Early mornings and plenty of chores meant that hard work, which later developed a very disciplined, focused and serious young man, punctuated his childhood. For as long as he could probably remember Stan worked on the farm as a casual labourer.

But on Father's Day 1939, listening intently to the wireless up in the 'Towers', the Prime Minister delivered the declaration message which not only changed the Morphy family forever, but the landscape of the region. It was during the 1939-1945 war years that Charters Towers became home to a major American Air Force bomber base during the major naval battle, the Battle of the Coral Sea. At that time Charters Towers hosted 15,000 American troops. The Americans stored their small arms and large bombs under the town's historically significant Tower Hill. But the increased military presence in Townsville and Charters Towers hit particularly hard on food supplies, water, electricity, firewood and ice. Local residents were required to obtain ration coupons for daily essentials, including meat and other foods, but particularly meat. The control on meat became such a strictly-controlled food source that even hospital patients were required to surrender meat coupons.

Of Stan's siblings, Thomas and Theresa were influenced by the town's strong military presence. Both would have military careers during WWII and, much like Ernest Poschalk, Stan would become enamoured by the 'aura', and stories, of the US servicemen and the tales of adventure passed on from his older siblings. But military life offered an opportunity to get away from the dairy farm. Many of the region's men and women were volunteering. Of course it wasn't long before Theresa enlisted. At nineteen years of age, she volunteered for the AWAS in April of 1943. She managed to secure a prominent and prestigious posting working alongside General Blamey at the Advanced Land Headquarters in the Forgan Smith Building of the University of Queensland at St. Lucia in Brisbane. For Theresa, working in Land Headquarters, she was right in the middle of the war effort and she loved it. However, long before her younger brother had the opportunity to enlist, Theresa discharged in March of 1944 as a Lance Corporal.

Following his sister's discharge from Blamey's Brisbane Headquarters, Stan's brother Thomas, already a member of the CMF, enlisted at Camp Chermside north of Brisbane and became a member of the 69th Bulk Issue Petrol and Oil Depot Platoon (commonly referred as the BIPOD). The Platoon was located in Belgian Gardens in Townsville and was involved in the major repair and servicing of military vehicles.

In January of 1945, shortly after turning 18, the boy from the bush was conscripted before he could honourably volunteer himself. Attending the recruitment office in Charters Towers he became: *Q273563 Stanley Robert Morphy,* the accidental soldier. But conscription wasn't about to dampen his hard-working dairy farming ethos. By the 26th of February Stan had been processed and was inbound for Cowra to show off his hard-working attitude to instructors at the Recruit Training Battalion. Allocated to the 2nd Battalion, he marched-in on the 2nd of March but it wasn't long before the young man's inexperience in most things Army got him into trouble.

Within two weeks of arriving at Cowra, somehow Stan had managed to lose some of his equipment. Regrettably it was a case of Stan's neglect, and like his fellow Queenslander and 2nd Battalion mate Ernie Poschalk, he unfortunately had a date with the Administration Commander. Charged and found guilty on the 14th of March, he was awarded a five-pound fine. Needless to say he kept his nose clean and his equipment close by for the rest of his recruit training. Always in the back of his mind was the prospect of going to Wagga Wagga. He didn't want infantry, artillery, or logistics; he wanted engineers. So when allocation officers appointed him to engineers,

Stan was undoubtedly pleased—happy with his allocation he was still a serious young man.

On 07 May 1945, Stan travelled to Wagga Wagga, no doubt on the same troop train as Frank Platt, Ernie Poschalk, Denby Grasby, Ivan Merritt and a few others. Upon his arrival with his travelling companions, he marched straight into the 1st RAE Training Battalion. Finally his training and hard work had paid off as he settled into life and routine at Kapooka as a Sapper.

Inserted by his family on the first anniversary of Stan's death, 21 May 1946, the family placed a memorial notice in the newspaper;

> '... Your end was sad dear son, you made us weep and sigh; but the saddest part of all, you never said goodbye'.

The inscription on his headstone forever reminds all who gaze upon his final resting place just what type of young man Stan would have been,

> '... *A Great Lad Loved By All Who Knew Him—Ever Remembered*'.

New South Welshmen

12—NX141715 William 'Bill' (Billie) Barclay Cousins (aged 24)
Plot B, Row C, Grave 8

After reading about his involvement in this tragedy, looking upon the black and white studio portrait of a uniformed William 'Bill' Barclay Cousins, like me, you might perhaps start wondering about what type of man he would have been? Looking resplendent and charming under his Army slouch hat, he did wear a cautious cheeky smirk, which made me think he was nothing more than a tall strong laid back and laconic country boy! I visualized a young man who had a heart of gold, was unconditionally loved and adored by his parents and above all else was protective of his close-knit family. As I delved further into his short life I found a man loved and admired by his entire family and for that matter, anyone who knew him. Raised in the heat, dust and mayhem of outback cattle stations under the watchful eye of his disciplinarian but extremely proud father, Bill was a true 'character'. Affectionately referred to as 'Billie' by his grandmother, the easy-going country boy would grow into a strapping six-foot man whose cheeky grin appeared across his face as an almost 'goofy' smirk.

Born in Young, New South Wales on 29 October 1920 to William and Mabel Cousins, little William (or Bill as he was known) was raised on cattle stations along with his sister Jessie. As the only son, Bill was ultra-protective of his younger female sibling. But this task was made easy for him because his father William was not only a respected and hard cattleman, but a well-respected veteran of the Great War. A hero of the ANZAC Campaign, he was also the boss of the cattle station Bill was working on. But for Bill he wasn't the ANZAC legend or the cattle station boss, he was simply just Dad to him and Jessie.

Bill's hero father was a man he looked up to with great respect. He was a Trooper with the original 6th Light Horse Regiment. Enlisting in the AIF in September of 1914 at aged 22, Bill-senior was working as a farmer with *his* father (Bill's grandfather) at 'Myola' Station via Trundle in outback New South Wales. Trundle, located 262 miles west of Sydney, was a small town 'back of beyond' supporting surrounding cattle farms and stations. Also working at 'Myola', as a labourer at the time, was his dad's good mate, Tom Arnold. His father often told young Bill the amazing story of how he and Tom ran into each other thousands of miles away from 'Myola' in the most unfortunate of circumstances during the Great War. It was a story a young and impressionable Bill never forgot and perhaps convinced him later in life to join up.

The story goes that on the 1st of February 1915, William Cousins landed on the Gallipoli Peninsula. At Lone Pine in August of that year, an angry Turkish bullet managed to hit its target—his father's face. But Bill-senior was a big strong New South Wales country boy. Thankfully the bullet didn't kill him and William later made a full recovery in hospital. Within two weeks he had hastily rejoined his unit on the front line. Promoted to Lance Corporal at Anzac Cove in October 1915 and assigned as a temporary Corporal-In-Charge of some of his evacuating unit, Bill's father was one of the last of the Australians to leave the Gallipoli Peninsula when the 6th Light Horse evacuated on the 20th of December 1915. For his courage and dedication his father was later promoted to Sergeant.

A bout of mumps in February 1916 hospitalized his father in Cairo for a short period of time before being transferred from the Light Horse to the 51st Battalion. A promotion to 2nd Lieutenant meant that he was seconded for duty with the 13th Infantry Brigade Machine Gun Company. Sent to France with the British Expeditionary Force, his father was wounded for a second time on the 9th of September 1916. This time Bill-senior was shot in the back. Receiving a minor wound, he was treated at a field dressing station before once again being patched up. But instead of returning to the fighting, his Commanding Officer thought that the two previous strikes was enough and they weren't prepared to risk a third time; Bill was seconded to the Machine Gun Training Depot as an Instructor. It was in France where his father's story about his mate Tom Arnold from 'Myola' takes its incredible twist. The tale often had young Bill transfixed; such was the quality of his father's stories. You see Tom Arnold enlisted with the 45th Battalion, which was composed mostly of men from New South Wales, and sent to Egypt as reinforcements in September of 1916. Tom spent long periods alternating between duty in the trenches and training and rest behind the lines. First he was in and around Ypres in Belgium and then in the Somme Valley in France. When William Cousins was shot in the back and sent into the Machine Gun training depots, it was here he eventually caught up with his stockman mate. Swapping stories of Trundle and the home-front thousands of miles away, the two good mates eventually went in different directions. Sadly though the good mates never saw each other again—at least in Europe anyways.

By the end of 1919, his mother and father had returned to Trundle so that Bill's worried grandfather could once again look his hero son in the eye. Telegrams only told his grandfather part of the story, and, for many months, Bill's grandfather feared the worst. But his son was home before much longer. Then on a magical day for his father, he could make out the figure of

Tom Arnold walking the dusty track back into the cattle station. Carrying his kit bag confidently on his shoulder, he'd survived the war. The reunion was a magical moment on the outback station. The story of the reunion would be told over and over again and like a well-made suit, it never went out of fashion.

Thankfully the Australian Government was looking after old soldiers like William Cousins senior; especially when they were settling back into life via the soldier settlement scheme. The scheme was open to soldiers who'd served outside of Australia, either as a part of the AIF or as a part of the British Defence Service and who'd been honourably discharged. The scheme allowed all eligible soldiers to apply for Crown Lands. Settling into civilian life, Bill's parents decided to make a go for it themselves. Moving 100 miles south to Young in New South Wales, which was located eighty miles north of Canberra and more than 160 miles southwest of Sydney, they applied for the soldier settlement scheme. Filling out the appropriate paperwork, they gained financial assistance for the purpose of clearing, fencing, drainage, water supply and other improvements on their land and, with further government assistance, purchased stock, seeds, implements, plants and necessary material and settled on the land in Young.

No sooner had the young Cousins family moved to Young, Mabel gave birth to Bill on the 29th of October 1920. The family stayed in Young for a period of time but the cattle station 'Erragolia' in Ponto, about ten miles north of the inland New South Wales town of Wellington, needed an experienced station manager. Bill-senior seized the opportunity. Richard Jackson, a former AIF soldier had previously owned 'Erragolia Station' but he sold it and moved back to England to be with his wife. For the Cousins family, Erragolia was to be their home for the next ten or so years. William, Mabel, Bill and Jessie, moved to Wellington to allow his father to take up the important station manager's position.

Bill and Jessie grew up as carefree country kids. Living on the cattle station their lives consisted of school and helping out their parents and staff, including his father's enigmatic mate Tom Arnold. Bill would become a great asset for his ANZAC hero father. As a strapping lad he was employed as a farm hand and learnt to earn his keep the hard way. When Bill was barely a teenager, the town of Wellington unveiled its memorial to the dead of the Great War. In 1933, as an adventurous 13-year-old, Bill sat on the grass in Cameron Park, on the banks of the Bell River, with his little sister and watched as their father and his mate Tom Arnold took part in an unveiling ceremony of a memorial dedicated to local mates who never made it home. It was truly a moving ceremony and it became a poignant moment in the young Bill Cousins short but impressionable life. He dreamed of being a soldier just like his dad.

By the early 1940s, Bill was in his early 20s when his father had given up his position as station manager at Erragolia and moved to another property at Ibstock. It was near Maryvale just outside Wellington. Here his father was able to secure work as a farmer working alongside his now strong and reliable son Bill. As strong as a Maryvale bull, Bill now stood at around six foot tall. He was also the perfect protective big brother for Sister Jessie, often chaperoning her to the dances in the Wellington Town Hall. As a close-knit family in a large farming community, the two Cousins' siblings were inseparable. It wasn't until Jessie met her husband, Bill Morley, in 1941, that she finally tore herself away from the care of her older brother. For Bill his life took a different turn to that of his much-loved sister when the war arrived.

Bill was every bit as strong as his hardened father, and his strength was noted when he voluntarily enlisted for compulsory Militia training in May of 1941. Passing medically fit as Class I on his first go, Bill took his oath of enlistment to serve 'King and Commonwealth' at Wellington, on the 27th of August. It took until October for his military career to start when he became a member of the Militia's 3rd Army Troops at Narellan, about 250 miles southeast of his home at Wellington. From early November 1941 he was granted special leave to return home and wasn't expected back in camp until early January, the following year. Returning to camp as directed, Bill quickly applied himself to his duties and was promoted to Lance Corporal in late January of 1942. During February and March of that year he was detached from the unit to complete infantry training. And, by April, he was earmarked to go and do more training to become an engineer. Bill was soon detached to the School of Military Engineering in late October 1942 for initial engineer training, but not before he attended the recruitment office in Muswellbrook, New South Wales and registered for service with the AIF. Bill's service records indicate that he failed his engineer course. But it was clear that Bill, aged 21, was pursuing another career in the full-time AIF. He wanted to go to New Guinea to fight the Japanese—he desperately wanted a transfer to the AIF.

As a result of his defection, Bill was marched into the General Details Depot at SME on the 9th of December and about a week later he'd been granted special leave without pay to return home to Maryvale to inform his parents of his plans. Passing his medical examination and signing his oath of enlistment in January of 1943, he was no longer a part-time soldier; but a full-time soldier in the 2nd AIF ready to serve 'King and Commonwealth'. Within two weeks the extremely fit son of a Great War veteran was on his way to Townsville. It was his embarkation point for the long journey on-board a troop transport ship, the fast troop carrier *SS Taroona*, destined for action in New Guinea.

The *SS Taroona* would make almost 100 wartime journeys, but in January of 1943 it disembarked the young and inexperienced Bill Cousins at Port Moresby. As a member of the former 3rd Army Troop Coy of the Royal Australian Engineers (Militia), the unit had become an AIF unit and Bill had landed in New Guinea with his former mates. It was a world away from Maryvale and the farm but it's what he always wanted—to be a soldier on operations just like his dad. For the next sixteen months Bill served in the campaign during some of the most decisive actions of the war. Actions which included the stubborn rearguard action by his former Militia cohorts delaying the Japanese advance on the Kokoda Track long enough for reinforcements to arrive. Embarking on *SS Ormiston* from Moresby bound for Townsville in

May of 1944, a lucky and thankful Bill Cousins had survived the horrors of New Guinea. But a short stint of AWOL in July of 1944 cost him seven days' pay and he quickly realised that he was not only back in Australia, but he was a soldier and had a great career ahead of him. Having survived the horrors of the New Guinea campaign his unit headquarters recognised his engineering exploits in combat and decided he'd be of benefit at the Engineer Training Centre at Wagga Wagga. It was also an opportunity to get back to New South Wales to see his mum and dad after nearly two years away.

Posted to the 2nd Battalion at the RAE Training Centre at 24 years of age, Bill arrived in Wagga Wagga in September of 1944 and immediately took a week off without pay to return home to Wellington to see his mum and dad. Bill and Mabel were inconsolable when their strapping son walked through the gates at Maryvale. As far as reunions go it was right up there with his dad's legendary reunion with Tom Arnold after the Great War. Also waiting in the wings was his little sister Jessie. No more was she the little sister. When Bill gazed upon her he realised Jessie didn't need her big brother as much; in his absence she had become a mature young woman.

After a week of tales and tears, Bill returned to Wagga Wagga to take up his new position. Upon his return he was transferred, on paper, to the 4th RAE Training Battalion where he remained until the Christmas New Year period of 1944/45 working as an assistant instructor to some of the more experienced Sergeants teaching the art of military engineering. But a shortage of Corporal assistant instructors throughout the camp meant that Bill would regularly bounce amongst the training battalions assisting more qualified Sergeants where required. In May of 1945, Bill was working in the 2nd Battalion when he was summoned to assist a man he knew well—Sergeant Jack Pomeroy in the 1st Battalion. He knew Jack was an experienced demolitions instructor, having served in the Middle East and New Guinea and was enthusiastically looking forward to working alongside the experienced and well-respected Sergeant.

Back in Maryvale, Bill's war-weary parents and the entire extended Cousins family and friends were relieved that their son, brother and much loved local lad had secured an Australian-based position. At least for now, they thought—he was safe. Bill's enthusiasm during that fateful Monday afternoon lesson in some way may have contributed to tragedy.

Jessie was in hospital the day of the funeral. She wasn't able to accompany her parents on the journey to farewell her much-loved brother at the Wagga Wagga service. Instead, when the news report of the funeral came over the hospital radio, compassionate nurses turned it down. Jessie didn't have to hear the horrible broadcast. For her it was one of the hardest times of her life. Fittingly, the family inscribed Bill's headstone;

'... *Dearest Memories of Our Only Son and Brother Never Forgotten*.'

Wearing Bill's medals, Jessie made it to the 1995 Memorial Service. Overwhelmed with emotion when meeting up with other relatives of the victims, she also finally got a chance to thank the Titus family on behalf of her deceased parents for looking after her brother all those years ago. Throughout the service Jessie couldn't help but remember Bill as just a little boy. When she visited her brother resting in peace at the Wagga Cemetery she quickly realised that fifty years had since passed and she still missed him terribly.

13—N480870 Thomas 'Toddy' Woods (aged 25)
Plot A, Row c, Grave 6

Sydney and Dorothy Woods gave birth to a son they affectionately called 'Toddy' in December of 1919. Arriving in the Sydney suburb of Parramatta, Toddy was the third child for the young couple who'd married three short years earlier. But Sydney and Dorothy were amongst a growing number of parents who'd experienced great pain and loss in the early decades of Australia's defining 20th century. The couple's first-born child, a son (also named Sydney) was born in 1916, the year of their wedding. Bestowed the honour of carrying on his father and grandfather's name into the next generation, the newborn Sydney Thomas Woods made Sydney and Dorothy proud to be able to continue the family legacy. But the reality of the time would have a disastrous effect on this third generation of Sydney Woods. I've painfully discovered that infants growing up in Australia during the early 1900s experienced an alarming fatality rate. It was especially difficult for babies born into an era where the worldwide Spanish flu epidemic was building. It was a deadly epidemic eventually claiming almost 12,000 young and old Australian lives. Sadly, at less than 1-year-old, Sydney Thomas Woods, the elder brother Toddy would never know and the third generation namesake silently passed away.

It was with some trepidation then that Sydney and Dorothy welcomed their second child, another son Fred, born in Parramatta in 1917. Understandably anxious for Fred's survival as an infant, the family was excited when a third son was also born. With Fred thriving, the couple felt an uneasy family burden lift and in dedication to their lost first-born, they named their third son after Sydney's father and simply named this son Thomas. Born just after Christmas on the 28th of December 1919, he was a welcome distraction from his big brother, Fred who by this stage was nearly two.

In March of 1927, Thomas was seven and Fred just nine when their little sister Betty arrived. The family was living in the Sydney suburb of Alexandria and Thomas' father was an oil refiner at the nearby Kurnell Refinery on the shores of Botany Bay. It was hard hot work. By the late 1920s, following-on from the gloom and deprivation of lifestyle thanks to the Great War, Australia was approaching the brink of fiscal tragedy due to the 1930's Depression; at least the oil refinery industry provided stable and secure employment for the family man with the large brood.

Aged 14, Thomas' parents and siblings moved to a bigger residence in Ascot Avenue in the inner city suburb of Mascot. But a more suitable home was sought soon after. Before too much longer, the family moved about two miles west to Wilson Street. It was a more appropriate dwelling across the train line. Surviving the Depression the years passed by uneventfully for the Woods family. It was during his time living in Ascot that a young Thomas met and fell in love with Harriet Glover (who liked being called Ettie). Ettie was only about two years younger than he was, and it was during his relationship with Ettie that Thomas became affectionately known as 'Toddy'. Ettie became smitten by the colourfully handsome, debonair and most charming man who was an amateur piano player. It certainly helped that Toddy could tickle the ivories and crooned Ettie with the most charming of singing voices. Deeply in love with each other and the music they created together, the couple were inseparable over the next couple of years. Life for the courting couple became magical. Toddy, as the piano-playing crooner, was a charming ladies' man. Even Ettie's mum, Lavinia and her brothers and sisters, especially sister Blanche, were instantly attracted to the young musical man. They warmly welcomed Toddy into their large family.

Being the first of the Woods siblings to be married, Toddy, at aged 21, tied the knot with 19-year-old Ettie on the 23rd of March 1940 in Rockdale. Shortly after, replying to an advertisement in the Sydney Morning Herald, Toddy got himself a job at the ASCO Scales Company in Waterloo. Before they knew it life for the young married couple was almost perfect. So perfect in fact they were now planning for their first child. Moving into their own place in England Street, Brighton Le Sands, a southern Sydney suburb on the western shore of Botany Bay, life for Toddy and Ettie was a settled existence. Not quite utopian, but surely a baby would fix that?

But Australia during the mid-1940s was fast becoming defined by the actions of men like Toddy performing amazing perilous feats overseas as members of the Militia and the 2nd AIF. Without warning many men of conscript age were being called up. The war became a little closer to home when Toddy's older brother Fred, literally without any pre-warning or registration of intention to serve, was conscripted into the Permanent Military Forces in July of 1941. With Fred's call up Toddy knew then that his day would also come; but his life was ideal and he was planning to have a baby. For Toddy, and the rest of the world, the war got in the way of life. Perhaps, he naively thought, a new baby might save him from conscription.

13th of February 1942 was a defining date in the life of 22-year-old Toddy. It was the day of his conscription into the full-time military forces. Although he and Ettie had hoped for a baby, which may have prevented his call-up, there was still no baby by February 1942. Attending the Town Hall in Arncliffe on the 12th of March, Toddy witnessed the long queues of volunteer men stretching far down the aisles of the Hall. With so many men volunteering, he felt any attempt to dodge his obligation would be fruitless. He knew it was fast becoming his turn. In the end he felt a sense of obligation to wear a uniform to save the life and family he so desperately enjoyed. But, as he lined up amongst the throng of strangers, he couldn't help but think about the married life he and Ettie were only just beginning to enjoy. Especially now as they were busy planning for a family of their own. But Toddy harnessed a little personal secret that might have prevented him being accepted into the Military. With that in mind he proceeded through the enlistment process in the hope that his secret would go unnoticed.

Undergoing his medical assessment, the examining medical officer identified that toddy suffered from an 'organic heart disease'. Initially passed as temporarily unfit for Class I fitness, it didn't prevent Toddy from completing the remainder of the enlistment process. Perhaps his condition wasn't as bad as he suspected. Notwithstanding his medical setback, Toddy was versatile and sought after. He could drive a car, a motor lorry and ride a motorcycle and he could play the piano. Like the thousands of men before him, Toddy took the oath of enlistment and swore, ' ... to well and truly serve our Sovereign Lord, the King, in the Citizen Military Forces of the Commonwealth of Australia'. His heart condition wasn't serious enough to prevent his enlistment, but the Army weren't prepared to put him through vigorous recruit training just yet. Before he knew it, he was no longer only the assistant sprayer with the Australasian Scale Company, with the perfect life, wife and future; he was a now a soldier in waiting. Although the military had their man signed on the dotted line, it would take more than two years before Toddy would actually have to wear his uniform.

Returning to his life as Ettie's husband, working at the Asco Scale Company, he and Ettie had time enough to welcome their first and only child, a son—Alan. It took a further two-years, not until March of 1944 that Toddy was re-assessed for a second time by military medical authorities due to his earlier heart condition. His improved health became a double-edged sword. As he was now a lot fitter, his earlier heart disease didn't prevent medical authorities declaring him Class I fit this time around. He was now capable of being accepted as a general reinforcement. Satisfying the medical requirements for the rigours of becoming a soldier, Toddy was finally in a 2nd AIF uniform by September of 1944.

Upon his eventual enlistment, understandably, Ettie was distressed. Knowing so many men were already dying overseas, the thought of losing her true love was upsetting—especially now the couple had baby Alan to care for. Saturday the 9th of December 1944 was the day Toddy marched into the holding strength of the Recruit Reception and General Duties Depot at Paddington. Within three days he was at Cowra and a member of the 1st Recruit Training Battalion. Whilst there, for the next eight weeks undergoing hard physical recruit training, Toddy put his somewhat 'dodgy' heart to the test.

Recruit training for Toddy didn't go so well. But it wasn't his heart. On the 7th of February he was evacuated from Cowra and admitted to the 103rd Australian General Hospital at Baulkham Hills, Sydney, suffering from a benign bony outgrowth, or 'ostema', on his left foot. Making it difficult to march or run, thankfully he thought to himself, ' ... at least I won't be going to infantry'. Remaining in hospital until the 26th of March, he was finally sent back to Cowra to complete training. Although he'd missed about a month of training, he'd completed enough to convince allocation officers that he'd be better off in the engineers. So after a period of waiting for administration paperwork to catch up, Toddy marched into the Engineer Training Centre at Wagga, on the 23rd of April 1945, to begin training as an engineer with the 1st RAE Training Battalion.

Back in Brighton Le Sands, Ettie and baby Alan were concerned for Toddy's welfare, especially in light of his recent hospitalisation and they were now eagerly awaiting his completion of training at Wagga. Ettie and Alan wanted the return of their loving husband and devoted dad. It had taken a long time for Toddy to finally wear a uniform, but as far as Ettie was concerned she wanted him home so that they'd finally be a normal family. Waiting patiently for her husband to complete his difficult training to become a Sapper, Ettie was thankful that Toddy wasn't in the New Guinea jungles. At least, she thought, at Kapooka he'd be was relatively safe. But really Ettie knew little of the role of engineers. She certainly didn't know what went on at Kapooka and just how dangerous it was!

Ettie struggled with the death of her much loved and admired husband. Much like the other dads who also died in the explosion, it was always difficult for wives and mums to explain to children and grandchildren why their dad didn't come home. Thankfully Ettie was well-supported by family and friends. She personally thanked them for their support at such a terrible time by placing a notice in *The Sydney Morning Herald* the following Tuesday (29 May 1945). Ettie thanked relatives, friends and her spiritual support for their expressions of sympathy in the loss of her husband Toddy.

Toddy's headstone inscription simply reads;

'... *His Duty Nobly Done Ever Remembered.*'

14—N481536 Norman (Norm) Rourke John Dilley (aged 34)
Plot A, Row C, Grave 2

Born on the 15th of December 1910 in the Hunter Valley region, the life of Norman (Norm) Rourke John Dilley embodies the remarkable story of the young, and not so young, men hailing from New South Wales who answered the call and trained for war in the backyard of their home state. For many they were never really far from home. But unfolding Norm's story exposes a life punctuated by the love for his family but tragically cut short in the cruelest of circumstances.

Located on the Hunter River, approximately 103 miles North of Sydney, is the city of Maitland. Synonymous with the city is the large Dilley clan. The family earned the respect of the community for being one of the hardworking pioneering families in the Hunter River region of New South Wales. Dilley families had been populating this region since the 1800s. Albert Dilley (Norm's father) was born in 1879 and was one of thirteen children. With large burgeoning broods it's easy to see how the extended family grew into pioneering legends of the region. In 1902 Albert met and married Catherine Rourke. Between his father and his extended siblings, who were expert carpenters, builders, draymen and carters in the town, the entire family made significant contributions to the development of the Maitland/Hunter region at the turn of the century and beyond.

Shortly after their marriage Norm's parents had their first son. And, in 1904 they welcomed their second. Sadly, tragedy was not far away for the young couple. Their third child, also a boy, named James, was born in 1906 but the infant child died the following year in 1907. Following the heartbreaking tragedy Albert and Catherine decided to have another child. In memory of their departed infant son James, they called this next-born child, a boy, also James. With four sons already, the arrival of Norm, the fifth, was still a celebrated affair. Incorporating his wife's maiden name (Rourke) into Norm's name, their fifth child's full name became Norman Rourke John Dilley.

But Norm's parents didn't stop there. In the true descendant tradition of large family broods, the family grew by another six children; two boys and four girls between the years 1910 to 1923. Strangely it wasn't until 1914 that they had their first girl, Dorothy, and their last in 1923, also a girl, Margaret. In total Norm had six brothers and four sisters. It was fair to say that Norm Dilley, as the fifth child in a family of eleven siblings, grew up in an extremely large loving family. Much like the Pomeroys in Victoria or the Bartlett family of South Australia, being brought up in large Australian families with multiple siblings, life was comparatively tough. There was a distinct lack of material possessions and that meant more reliance on brothers and sisters, not only for 'hand-me-down' clothes but also for the love, support and family bond required to keep large families together and healthy.

The Dilley family of eleven surviving children and their vast network of aunts and uncles living in West Maitland supported each other through Australia's tumultuous Depression years of the early 1930s. Being of working age during the fiscally-challenged years, meant that Norm experienced the highs and lows of independence and hard work in his early adulthood—much earlier than most his age. It was an upbringing that made him appreciate what hard work can bring.

But compounding the hardships of early Australian life, especially for the Dilley family living in Maitland, was the fact that their hometown was flood-prone. Recording serious floods in the 1890s which killed nine locals leaving many more homeless and without any possessions, the region relied on large families, like the Dilleys, for manual labour. Rebuilding the town encouraged camaraderie during the bad times and the region relied on the large industrious Dilley family, not only for their manual labour, but for their sheer dedication to the community. It was the same spirit of community selflessness shown by the Grasby family of South Australia.

During his early twenties, Norm was living with his large family in Wallace Street. It was a modest residence for eleven children and close to the current day Maitland Showground. Norm, his father and many brothers, were all carpenters and busy building and repairing the town and surrounding districts following devastating seasonal flooding. However, even more family tragedy wasn't far away. Sadly Norm's father Albert, passed away in November 1930 aged 51; it was just before Norm was to celebrate his 20th birthday. With their father's passing, the family rallied together. It was always going to be that way in such a tight loving family. Suddenly with her youngest child only seven and her eldest 28, Catherine Dilley was being supported by her elder children.

Doing chores around the house was something that came naturally for large struggling Australian families. Catherine relied on her large family to rally around her to help raise the younger of the children. So when it came to supporting each other, it wasn't a task the family noted on a chores board hanging loosely on the kitchen dresser, it just came naturally. Surviving the worst financial conditions of their livelihood, and now the death of their father, the large family remained resilient and bonded strongly together with love and support. But the kids were always going to grow up and leave home; it would be Katherine's fear and joy all rolled into one.

Norm was an unskilled horse carter, but that didn't stop local girl Kathleen Marjorie (who preferred to be called Marjorie) O'Rourke falling for the dashing Dilley man. Married in May of 1937 in Maitland, Norm, then aged 27 and his bride Marjorie, aged 24, both moved to Sparke Street. It was a small residence adjacent to the train station in Maitland and it was where they set about starting their own family under the watchful eye of the extended family network. But sharing and helping other family members in need was a proud family legacy. Before settling into married life and raising children, Norm's elder brother Albert and his wife Annie moved in with him and Marjorie. It wasn't crowded, that's just what families did; they supported each other through the good and, all too often, bad times.

The hostilities of war meant that a heavy workload was placed on Australia's industrial resources. More than 900,000 men and women were involved in some way in the production of ships, planes, guns, mechanised military equipment or the material for making them during the war-years. At its peak, Australia was able to present itself as a highly efficient and largely self-contained arsenal, capable of turning out a great diversity of weapons and military equipment. Prime Minister Menzies had earlier returned from England in 1941 and proposed to establish three munitions factories in New South Wales. It included one munitions factory near Newcastle. Many existing industries including railways and fabric-making factories were co-opted for the war effort. As the war industry had now found its way to the Hunter Region, it goes without saying that the industrious Dilley family became heavily involved in the prominent industry—especially now a munitions factory had literally arrived on their nearby doorstep.

By 1941 Norm and Marjorie welcomed the arrival of their first child June. Their second child wasn't to arrive for a couple of years yet. In that time Norm, a tall lean man, with a fashionable part in his hair down the middle, set about establishing his professional career as the main income earner for his small family. In 1942 the Prime Minister's proposition eventuated and a large gun and ammunition factory was built at Rutherford; located approximately 26 miles west of Newcastle. Many tools, dies, and jigs required for the construction of war-fighting equipment were now being produced at the Rutherford plant. One of the local men caught up in the industrial push for more war equipment was of course Norm. By 1942, Norm was 32 when he took up a job as a tool setter at the nearby munitions factory. But for many of the Dilley family it wasn't enough to be working in the factories; they craved to be part of the real action.

On the 4th of January 1942, the youngest of the children, Norm's sister Margaret, by then aged 19, enlisted at the Maitland Town Hall, joining the Royal Australian Air Force. She was posted to the newly formed No 5 Aircraft Depot at Forest Hill near Wagga Wagga. It was about 288 miles south of the family home. Thanks to Margaret, thought of conscription often hung over the heads of the remaining Dilley children. Of all the siblings in the large home of volunteer and conscription age, it was 32-year-old Norm who would be called up first. For Norm it was a blessing in disguise.

Following the drawing-down of the war, the Rutherford munitions factory ceased production. As a consequence, Norm was terminated from his employment. The married man who had a family to provide for was now out of work. Considering a life as a soldier with a steady income and job security, Norm registered his interest at the Maitland Drill Hall. Conscripted and enlisted in Maitland on 27th of January 1942, Norm would become another of the so-called mature-age soldiers. They may have been novice soldiers, who were married with children, but they had life experiences before donning a uniform for the first time. With a hard work ethos it made the mature recruits a fantastic resource for the Army. Upon enlistment the father of two declared his occupation as a tool setter. It was a move that would ultimately prepare him for a later career as a military engineer.

Catching the train from Maitland to the Newcastle Hospital for a checkup, Norm passed the mandatory medical. Instructed to take the 'Oath' to serve 'King and Commonwealth', before too much longer he was allocated his official number and became: *N481536 Private Norman Rourke John Dilley.*

However, like most of the conscripted soldiers at the time, Norm had to wait to be called up for Service. Whilst waiting he did have to complete his initial military training at the nearby Rutherford military camp. It was here that Norm received his initial introduction to life as a soldier. Greeting newly-conscripted soldiers was a bare dusty and stark terrain. For most of the men, who were local lads, accommodated in less than inviting tents not far from home was hard to take. Climatic extremes at Rutherford made camp life unceasingly uncomfortable. For example, in the summer months, incessant flies were a constant unwelcome distraction for all the men who battled with the long hot days. For Norm, as a 32-year-old, he was mature and somewhat used to the harsh living conditions. Issued with a uniform, military equipment and the uncomfortable paillasse to sleep on, Norm accepted the conditions and soldiered-on. Sleeping in the cramped tents he endured the initial training of how to march and use a rifle. During these early days he was fortunate to have his family close. But it wasn't to last. Completing his compulsory training Norm did however have to wait almost two years before being called up for full time service. But joining the full-time Army meant that his life was to change dramatically.

By late 1944 Norm and Marjorie had already raised their daughter June and the arrival of Neil saw the young family move to a different residence in Lorn, Maitland. Their move came just as Norm was preparing to depart for military duty. In preparation for his full-time commitment, Norm was ordered to parade at the nearby Rutherford camp in early February of 1945. Within five or so days of being at Rutherford he was granted leave to go home and see his wife before being 'shipped' out for recruit training. For Norm, that meant eight weeks at Cowra away from his loving family.

On the 27th of February, leaving behind his wife and young children, including baby Neil who was only six months old, Norm departed for Cowra and wondered when he would see them all again. The following day, he arrived at Cowra and joined Bravo Company of the 2nd Australian Recruit Training

Battalion. Norm set about training for war amongst men almost half his age. Of course his constant distraction was the young family he'd left behind. He wrote to Marjorie every day and in return, his wife wrote back to him equally. Although it pained Norm to be away, the constant stream of letters made recruit training slightly bearable. It was just the tonic he needed to overcome his homesickness. He even wrote to June expressing his love for his daughter. In one poignant letter he made sure he checked on her pet bunny rabbit and also promised to send her a bag of lollies as soon as he could.

Completing recruit training and convincing allocation officers that he was well-suited to become an engineer, on the 30th of April Norm found himself on the train bound for Wagga Wagga. Again, amongst much younger-looking men heading in that same direction, Norm was on the cusp of a career as a combat engineer. Perhaps also sitting alongside him on the train was his 2nd Battalion colleague Bill Reid who also departed the camp on the same troop train that day! Although he missed his family terribly Norm quickly found comfort in the fact that he was only heading to the Engineer Training Centre at Wagga Wagga and he knew it would be an opportunity to catch up with his younger sister Margaret out at nearby Forest Hill. Sadly, he never got the chance to catch up with Margaret.

Many of the large Dilley family surrounded the mature Sapper's coffin as it was laid to rest in the Wagga Wagga War Cemetery. Catherine, their two young children, one of Norm's four sisters and three of his brothers, Bob, Wally and Keith, all made the journey to Wagga to farewell a much-loved husband, father and brother. Their simple inscription on his headstone reflected their pain of loss,

'... *God Rest His Soul*'.

Norm's wife Catherine and the two children returned to Maitland after Norm's passing. In later years, his son Neil lived with his wife just 100 feet away from the old family home at Belmore Road. Passing the house he was constantly reminded of his childhood and especially the day his father walked out the front door of the modest brick home to join the Army in 1944, and never returned. Catherine remarried in 1957 and passed away on the 25th of July 1986 at age 73. Neil now lives in the Rutherford area with his wife Susan. Together, they still bravely carry on the Dilley family legacy of community spirit.

15—NX204475 Colin Francis Boyd (aged 21)
Plot A, Row C, Grave 4

Born on the 13th of February 1924 in Cowra, Colin Francis Boyd was named after his father. His mother, Bridget, and namesake father, were unmarried when their second son arrived. Elder brother Jack, also born in Cowra, had arrived two-years earlier in January of 1922. It wasn't until 1930, when Colin was a sprightly 6-year-old that his parents eventually married. After their marriage the young Boyd family moved to Denison Street in Tamworth, New South Wales. It was a small unassuming home a mere block or so away from the local railway yards which provided employment options for Colin senior at a time when work was becoming increasingly rare.

As an engine driver with the New South Wales Railways, young Colin often watched his father go off to work and come home exhausted after a long day's work. Although it was hard hot work, the Boyd family was thankful for a steady income. But surviving the country's crippling Depression of the early 30s was becoming increasingly difficult. Suddenly, on the brink of financial insecurity, paid work around the country was becoming difficult to find. In their quest for stable employment and a decent life, the only option was to move to areas where work was available. Needless to say work in Tamworth dried up and the Boyd family was again on the move. During those hard times families relied heavily on family. For the Boyds, their saviour were Colin's grandparents. At the time they were living in Coonabarabran; a small town a further 100 miles west of Tamworth which could perhaps provide a haven for the young struggling family. So, appreciative of the opportunity to call on their support, the young Boyd family moved to Coonabarabran in search of work and much-needed financial support. Living with his grandparents Colin established a good relationship with his grandfather. But when young Colin learned that his grandfather was not only an ex-soldier, but also a hero of the 1st AIF he worshipped the man as more than just granddad. His grandfather's stories of the Great War impressed upon Colin the romanticism of being a soldier.

By 1933, aged nine, Colin and the whole family, including his grandparents, moved again. Located in the middle of dairy and timber country, Dungog was a small town 100 miles south of Coonabarabran and the extended Boyd family lived together in a small home in Fosterton Road. Much like their house in Tamworth, their latest home was once again conveniently located just down the road from the large train station. Colin's father gained employment as a locomotive driver. Living in a much smaller community, Tamworth and those early years seemed a lifetime away for the young Colin.

In 1937, with Dad still working for the railways, Colin was thirteen when the transient family made their greatest move. Packing up their belongings in Dungog, the entire family caught the train to their next home, this time 200 miles north in Grafton. Living in North Street during the late 1930s, Colin, now aged 15, left school and was looking for work. But the constant moving around was to have a detrimental effect on his health. Suffering from asthma as a child, he would later suffer from its nasty debilitating side effects such as the skin condition—eczema. It would become a problem for him in his important adult years.

The arrival of the war in 1939, when Colin was 15, brought industry and employment to the young men of similar age and circumstance. Before the war, rifles and machine-guns were already being produced at the government small-arms factory at Lithgow for the Army and were later assisted by smaller factories at Bathurst and Orange. But for Colin and his peers work became a priority over education. At the time Colin was transitioning from school to the workforce, New South Wales became a major contributor to the production of weapons and munitions to support the war. By June 1942 the government factory at Lithgow employed 5,058 men, at Bathurst 1,600 men and women had jobs, and a further 1,339 were employed at Orange. Smaller factories were also opened in 1942 in New South Wales country towns such as Mudgee, Forbes, Wellington, Cowra, Young, Dubbo, Parkes and Portland, all of which were connected by rail and for Colin, the son of a railway man, this meant plenty of work opportunities for him to pursue after leaving school.

In January 1940, Colin's younger brother, Jack, was called up for his universal service. Jack underwent his three months of compulsory training with the Grafton-based 41st Militia Battalion. By April of 1941, the eldest Boyd son, John, had also decided to join the full-time military. Enlisting in the Air Force as a canteen steward, John would remain in the Air Force for nearly five years.

Having filled out a National Service registration card shortly after turning 18, Colin Boyd got his chance to serve when he was called up for universal training in June of 1942. By this stage Australia's involvement in the war was at fever pitch. Colin had since moved away from his family and was now living in Sydney and sharing a house with a friend in Sloane Street, Haberfield. In the meantime he found work as a mechanic's assistant in a munitions factory in Sydney. He passed his medical examination at the recruiting depot at Ashfield in Leichardt and completed his compulsory training.

After leaving the munitions factory Colin got work soldering and assembling, and was back living with his parents in North Street, South Grafton. But like many men who got a brief taste of military life and found it appealing, Colin found a calling to enlist in the 2nd AIF. Marching into the Paddington Recruiting Office on the 13th of September 1944 aged 20, Colin listed his namesake father as his next of kin. He volunteered his service to truly serve his Sovereign Lord, the King. Successfully passing the doctor's examination, Colin was deemed physically fit for active service abroad. Issued with his equipment, like the thousands before him he was ushered into the Sydney showground. Amongst the busy scene of uniforms, marching and a lot of yelling, He joined the thousands of other recruits, many with similar stories and circumstances, all waiting to march-out to recruit training. The young man roughed it in the showground stalls for the next four days. Participating in basic military skills training he was being prepared for the next phase. Sleeping on concrete floors before finally marching out of the showground on the 19th of September, Colin was allowed a couple of days' leave to say his goodbyes to family and friends. He was bound for Cowra and the 1st Australian Recruit Training Battalion. Catching the troop train from Sydney to Cowra, he arrived on the 30th of September 1944 somewhat prepared for whatever it was the Army threw at him.

But Colin's early childhood affliction with asthma would later surface in the less than hygienic conditions at Cowra and eventually would have a marked effect on his health. Exacerbated by the primitive conditions found at the Cowra camp he developed severe dermatitis of the feet. Admitted to the 114 Australian General Hospital in Goulburn, the young man was suffering from a form of 'cystic abrasions' on his feet. His ailment made it difficult for him to wear boots, let alone walk or undergo rigorous training. Suddenly, Colin was bed-ridden. Evacuated to 113 Australian General Hospital (113 AGH) in Concord later in the month, he was also suffering from dermatitis on his face, especially his nose.

But Concord was too close to the city of Sydney for country boy Colin not to be tempted and enticed by Sydney's hypnotic culture. At one point, without authority, Colin skipped hospital. Determined to see what all the fuss was about he headed into town for a slice of Sydney's night-time action. He knew he'd be in trouble, but it was a small price to pay for an experience he would never forget. As expected his short stint of AWOL landed him in trouble with authorities. Although AWOL for less than two hours, it was enough for the Commanding Officer to charge him and award a small monetary fine.

Spending a couple of months in hospital, medical staff tried to stop the growing skin infection taking-over his face. Colin was discharged from 113 AGH on the 23rd of April 1945 suffering from what doctors diagnosed and indicated on his medical form as a 'congenital deformity of the nose'. During his hospital convalescence he was allocated to engineers after successful completion of his recruit training. So, within a week of discharging from the hospital, Colin was marching into the RAETC at Wagga Wagga. Once there he was posted to the 1st RAE Training Battalion and straight into the novel training program to become a combat engineer. At Kapooka his facial affliction was no longer going to hold him back, he was going to be busy and experience real danger for the first time.

It's unsure which members of the Boyd family attended the mass funeral at Wagga Wagga to farewell young Colin. His headstone inscription reflects the short life he experienced as a soldier. It simply reads:

'... *His Duty Fearlessly and Nobly Done, Ever Remembered.*'

16—NX205863 Joseph William Faull (aged 18)
Plot B, Row D, Grave 12

Roy and Kathleen Faull gave birth to Joseph William on the 30th of December 1926 in Parkes, New South Wales. Named after the 'Father of Australian Federation', Sir Henry Parkes, the township of Parkes, located nearly 200 miles west of Sydney, was a gold-rush town during the 1880s. And, much like the pioneering Dilley families of Maitland region, so it is in Parkes with the Faull family. The heroic Faull family pioneering legacy is so significant in the Parkes area that local streets, such as 'Faull Crescent', located south of the city centre, are dedicated to the family.

However, with pioneering comes inevitable heartbreak. For more than forty years tragic stories of suffering would befall the pioneering Faull family. In 1927, when Joseph was just one and his older brother Len, just four, the extended family, which included his uncle and his family, were all living in a small four-roomed weatherboard cottage in Kurrajong Street, Parkes. In the early hours of a warm December morning of that year, the family was asleep on the verandah of the small rented residence. Awoken by the sound of cracking timber, the home was well and truly alight when the Faull family had finally been roused from their sleep. The home and its contents were totally destroyed. Thankfully every member of the resilient and lucky family survived. Supported by family in their time of need, Roy and his young family moved into another residence of his brother Joseph's, his wife Margaret and their surviving children. Joseph's uncle's eldest son died after a fall from a cart in 1912. He was just 13. Then having survived the Great War, after serving in Europe with the artillery, his uncle Joseph returned safely but lost his fourth-born child Ken, aged just fourteen in 1921.

During the years 1930 to 1936, Joseph's father was a labourer for the Parkes Municipal Council and was working for his brother who was the foreman. With a young family, Joseph's parents provided a stable life raising their children through Australia's Depression years. It was tough, but the support of a significant and admirable family network in Parkes made it bearable. It was perhaps inevitable that Joseph himself would become a carpenter, much in the mould of his father. However, during post-Depression years, the family moved 65 miles north in 1936 to the town of Dubbo. At the time Dubbo was a growing country town meaning manual laboring work was aplenty. After arriving the family moved into a residence at Wingewarra Street, just around the corner from the Dubbo showground. For young Joseph, living in Dubbo gave him an opportunity to finally complete his formal education and he was soon able to find work as a casual labourer.

By December of 1940, Joseph's dad, Roy, now aged 40, was on the cusp of the conscription age limit. Believing he had evaded service, his 40th birthday brought the worst possible news—conscripted. Enlisting in Paddington in inner Sydney, Roy was posted to 1st Australian Railhead Supply Detachment. His enlistment meant that his young family living in Dubbo, including Joseph who by then was 14, would need to move to Sydney.

Moving in to Wordsworth Avenue in the city suburb of Concord, it was Joseph's elder brother Len who was the first sibling to be conscripted. Completing his compulsory Militia training, Len became interested in the Royal Australian Artillery and found a career as a Gunner in the 83rd Mobile Searchlight Battery, serving at Oro Bay, New Guinea. For Joseph he had other ideas about life; he was also slightly rebellious. Sydney-life was tempting for the former country boy. It wasn't long before the temptations of the big smoke caught up with the impressionable Joseph. It's here he met a streetwise bloke by the name of Terence Moore. Born five days apart in 1926, Joseph and Terence were destined to be buddies and best mates. Being the same age as Joseph, together the two young impressionable men tested the limits of police tolerance as sharp-witted larrikins. Later, having been convicted of stealing by the Sydney Children's Court, it wasn't the ideal start that his Joseph's family expected in their new life amongst the city slickers.

Meanwhile, no doubt pensive about his own contribution to the war effort and the prospect of conscription, and shortly after turning eighteen in December of 1944 Joseph, along with Terence of course, made a pact. To his parent's delight it wasn't a life of petty crime just to survive—he turned to soldiering. With his father and brother serving in the Army, Joseph decided it was time to enlist; after all he was unemployed and needed some stability. It was always his dream to work and serve alongside either his brother or father.

On the 9th of February 1945, with his parent's permission, Joseph or 'Joe' as he liked to be known as he grew up, attended the Paddington Recruiting Office in Creek Street and enlisted in the 2nd AIF. Passing the medical examination, he was now *NX205863 Joseph William Faull*. On the 15th of February, Joseph marched-in to the holding strength of the General Details Depot waiting for his movement orders. Issued with his latest identity and unfamiliar equipment, Joseph was sent out to the Sydney showground and into the 'infamous' holding pattern for fresh recruits. For about five days, like so many unfortunate recruits before and no doubt after him, Joseph endured the mundane training until sufficient recruits had been mustered before sending them onward to training locations in bulk.

He arrived at Cowra for initial recruit training on the 20th of February and was marched into the 1st Australian Recruit Training Battalion. Before too much longer he'd met and re-established his friendship with his Sydney mate Terence Moore. The pair became inseparable. Leaning on each other for support, the great mates completed the difficult Cowra regime. Both were hoping for the opportunity of being allocated as engineers and thereafter a posting to the Engineer Training Centre, Kapooka Camp where'd they become Sappers together.

On the 30th of April Joseph completed his recruit training. Along with other men from the 1st Battalion such as Alf Witt and Colin Boyd,

Joseph marched out of Cowra in the morning and by the afternoon was marching in to the RAE Training Centre and attached to the 1st RAE Training Battalion. No doubt Joseph was surrounded by many of his Cowra mates, including Terence when they bunked down in their tents on that first night in camp.

Roy and Kathleen inscribed their 18-year-old son Joseph's military headstone:

'*... God Doth His Own In Safety Keep He Give His Beloved Sleep.*'

Kathleen passed away in December 1963 aged 56. His father Roy followed almost four years later on the 3rd of April 1967, aged 66. His mum and dad are both buried together at Point Clare Cemetery, New South Wales.

17—NX205652 Terence Ronald Moore (aged 19)
Plot B, Row B, Grave 10

Growing up on the streets of the bustling Sydney metropolis during the troubled days of the Great Depression made city boys not only streetwise, but cagey, resourceful, resilient and for many extremely thick-skinned. Terence Ronald Moore was no exception. Born on Christmas Day 1926, the fair-haired blue-eyed was six years-old when three-quarters of a million people witnessed the official opening of the iconic Sydney Harbour Bridge on the 19th of March 1932. With celebrations that included an aerial display by the Air Force, a parade of floats and marching bands, fireworks and carnivals, it was a day no 6-year-old boy would ever forget.

His father, Edward, was a radio expert, but his natural mother, Beryl, was no longer on the scene. In 1933 Terence was aged seven when he finally realised that his father was living with his grandmother, Elizabeth Anne, who was not his real mother. Terence's grandfather, Arthur, had passed away just one year before he was born, leaving his grandmother and father as his guardians when the family settled into a house in Samuel Street, Tempe, an outer-Sydney suburb. When he was ten years of age, the family moved less than two miles away to a house at 92 Princess Highway, Arncliffe where the young Terence grew up using his streetwise city-boy skills.

Although Terence's father Edward was only 21 when he fathered him, the arrival of the war in 1939 meant that his father would be a prime candidate for conscription. So in June of 1940, when Terence was only 14, the same age Joseph Faull was when his father was conscripted, Edward was conscripted. Due to his technical expertise as a radio technician, his father found himself in the Air Force and was sent to West Melbourne where he was a trade trainer at the School of Technical Training. No doubt, influenced by his father's profession, Terence attended the Kogarah Technical School where he completed six months of technical training.

His father remained in Melbourne for a short period and eventually discharged from the Air Force in February 1941. Meanwhile, Terence had earlier met Joseph Faull and the two became good mates with plenty in common. The pair of hardened young men became involved with Charlie Company of the 18th Garrison Battalion. Providing coastal defence for the Sydney suburbs of Bronte, Dee Why, and Manly during 1942—1943, as young enthusiastic pretend soldiers they were eager to get involved in the war when they came of age.

Turning eighteen on Christmas Day 1944, it was the opportunity to enlist in the full-time 2nd AIF that Terence had been waiting for. His chance would come sooner rather than later. The day after another mate of his, Pete Jackman from Mosman, enlisted at Paddington, and coincidently the same day West Australian Alf Witt enlisted in Karrakatta; Terence Moore was standing in the recruiting line at Victoria Barracks in Sydney. On the 23rd of January 1945 he became *NX205652 Terence Ronald Moore*.

Three days later on, Australia Day, Terence marched into the General Details Depot and within the customary period of four to five days, he'd marched out and was on the troop train to Cowra for recruit training. Little did he know at the time, although separated by thousands of miles, he and Alf Witt had more than just the same enlistment date in common. The lives of Terence, Alf and Joseph were destined to collide at Cowra and then again at Wagga Wagga. By the 1st of February, Terence marched into Cowra and was allotted to the 3rd Australian Recruit Training Battalion. But all was not well with his training and more importantly, his health. Within two weeks he found himself admitted to the 11th Camp Hospital suffering from dermatitis. Unfortunately he had been temporarily separated from his mates. Evacuated to 114 General Hospital for treatment, it wouldn't be until early April when he'd arrive back at Cowra fit enough to continue. But he didn't return to the 3rd Training Battalion; instead, he was sent to the 1st Battalion. Whilst there he managed to complete his training alongside Alf Witt, Colin Boyd and his mate Joe Faull. Thankfully, by the 30th of April he'd managed to march-out and was now bound for a life as an engineer.

Surrounded by his new close mates and friends, he marched in to Kapooka Camp and the RAE Training Centre the very same day. No doubt the first couple of days at Wagga Wagga for Joseph and Terence would've been filled with a lot of nervous laughs and jokes but before too much longer the city boys would've gained a streetwise confidence in their training schedule to become effective soldiers. They took that confidence with them when they commenced demolitions training in week 4.

Edward, the single father of a loving only son, farewelled his boy at Wagga Wagga amongst other grieving parents. He left a simple but fitting inscription on his son's headstone:

> '... *Dearly Beloved Only Son of Edward A. Moore of Arncliffe, N.S.W.*'

18—NX205938 Edward Charles 'Ted' Robson (aged 18)
Plot B, Row C, Grave 6

Edward Charles 'Ted' Robson was born to parents Jack and Maude on 24 July 1926 in Tumut, New South Wales. He was the fourth son for the young couple who were married in Tumut in 1912. For Jack and Maude, aged 22 and nineteen respectively, it was not only their marriage in 1912 worth celebrating; the couple also welcomed the arrival of their first-born, a son, Ernest. They would eventually welcome a large brood of seven children, four boys and three girls (Ernest, Cecil, Norman, Edward, May, Joyce and Patricia). As you'll discover later, Patricia, was something akin to either a miracle or a mistake. But whichever of the two, she was a welcome addition to a more than loving large family.

Whilst father Jack originally hailed from nearby Junee, Maude was the daughter of Catherine Kershaw, who came from one of the more influential pioneering families of Tumut. Located in the Riverina region of New South Wales, the township of Tumut is idyllically located on the banks of the Tumut River. Found approximately 250 miles southwest of Sydney and 330miles from Melbourne, it also rests less than fifty miles east of the unofficial Riverina capital of Wagga Wagga.

Following the birth of Joyce in 1924, the family moved to Sydney and in particular the western Sydney suburb of Liverpool in search of greener pastures. With dreams and hopes of a better life, they'd hoped Sydney would offer them boundless opportunities to make money and a better life. But on the brink of Australia's financial peril, hard times in the big cities made life difficult. With some reluctance, the family returned to the sanctuary of Tumut. The journey was however undertaken with caution as Maude was pregnant carrying Ted. The family managed to return to Tumut in time for Ted to be born in the July of 1926. Upon their return Jack secured work as a painter.

Fast forwarding over the next three or four years and 1930 arrived. With their eldest Ernest almost a man at age 18, and Ted just four years of age, the family made a crucial decision to once again leave Tumut in search of work. Minus the two elder sons, Ernest and Cecil, who'd remained behind in Tumut having secured work, the remainder of the family again relocated to Sydney. Jack, having secured work as a storekeeper in the Sydney suburb of Darlinghurst, settled into city life and stability with Maude and the remaining Robson siblings. But the evil of Australia's great financial Depression caused Ted's father to lose his prized store. In an instant their less than ostentatious, but nonetheless comfortable, way of life, returned to a struggle; much like the struggle they experienced in the early years. Downtrodden and disillusioned with the competitive post-Depression retail industry, Jack Robson kissed his dreams goodbye. On the wrong side of forty he was forced to return to work in backbreaking manual labour positions.

In the foothills of the Kosciuszko National Park in New South Wales rests the Khancoban Forest camp. In 1936 the Forest camp required labourers. In the lingering shadows of Australia's revival from financial uncertainty, Jack was forced to drag his wife and young children to a remote isolated existence. It was the only work he could secure. Living in what was known as No 1 Forest camp, it was far cry from the hustle and bustle of city life, but for the family it was a well-paying job. The remoteness and gypsy-like lifestyle meant the youngest Robson children didn't receive adequate formal education. It was here, in the dark deep solitude of the National Park, where 10-year-old Edward did most of his growing up. Working under his father's tutelage, a young Edward Robson got an early introduction to hard work and toil. Helping his father at labouring work in the forest camp was a hard unforgiving life for a small boy and a man over 40.

1937 brought an unexpected surprise for the Robson family. Finally rounding out the large Robson family, 47-year-old Jack and 43-year-old Maude brought their final gift into the world. Almost 23 years after their first-born, the mature-aged parents gave birth to daughter Patricia. She arrived via the Corryong Hospital and first glimpsed life in the serenity of the forest camp.

Experiencing his teenage years, living and working in a remote National Park during the late 1930s and into the early 1940s, world events occurring outside the cocoon of the family's wooded existence would eventually lead to dramatic changes in young Edward's life. In 1939, at age 13, the declaration of Australia's involvement in the war significantly altered the lives of the entire Robson family. Much as it did with the hundreds and thousands of other Australian families, with four eligible sons, for Jack and Maude the arrival of the war and the Prime Minister's declaration amplified their horror. It was an especially poignant time for Edward who, as an impressionable teenager, heard that his older siblings back in Tumut were now responding to the government's call for available men to sign up.

Ernest, the eldest of the Robson boys, was the first to answer the appeal to take up arms in service for King and Commonwealth. On the 28th of May in Tumut, he volunteered and was sent to Sydney. Swearing his Oath in Paddington, he was allocated to the Electrical Mechanical Engineers where he was sent to an Advanced Watercraft Workshop. For little brother Edward, the letters sent to his mum and dad from his brother telegraphed stories of what life would be like as a soldier. From his brother's description of disciplined life, Edward no doubt pictured a more exciting world outside the confines of his current existence. '... Perhaps', he thought, the Army just might be his opportunity to get out of the forest—a real chance to go on a journey and to see the real world.

Following the outbreak of the Pacific War, the Government expanded the Australian part-time volunteer military force in February 1942 making membership open to men aged between eighteen and sixty. In August 1942, Jack Robson was working in Gwabegar in the Pilliga State Forest, a small village located in the middle of the southern hemisphere's largest naturally-forming cypress pine forest. Somehow, amongst the pine of this enormous forest, the War Office found Jack. On the 8th of August 1942, aged 52, Jack was conscripted into the 25th Battalion Volunteer Defence Corps for part-time duty.

Anticipating that other members of the family would also be conscripted, Cecil, another of Edwards's older brothers was the next Robson family member to answer the Army's call. Cecil had already completed his voluntary Militia training and naturally progressed into the 2nd AIF. Enlisting in Singleton, New South Wales, he left his wife behind in November 1942 and saw overseas operational service with the 2/25 Field Park Company. Norm, the third eldest, followed Cecil's approach by completing his Militia training and service before volunteering for the 2nd AIF. Enlisting in Dubbo in July of 1943, he listed his father as his next-of-kin and became an engineer. He was later posted to 11 Small Ships Company, which saw action in Borneo.

By 1944 Edward, or 'Ted' as he was better known as he got older, being the youngest of the Robson's sons, was now eighteen and living in Illawarra Road, Moorebank working as a dairy hand at a nearby farm. He'd broken away from the forest camps and life amongst the pines where he and his father were doing backbreaking manual laboring work to seek his own future. Moorebank was a suburb abutting the thriving military base at Holsworthy in New South Wales. At the start of the war there was a huge increase in troop operations in the Sydney camp, with troop numbers exploding to somewhere in the vicinity of 40,000 men. The facilities at the site provided operational training for the troops before they were sent overseas. For Ted, living and working so close to this thriving military territory provided great exposure to life as a soldier. His close proximity to everything Army, and the fact that his three eldest brothers, and father, were serving, perhaps convinced Ted it was his turn to volunteer.

Although he was working as a dairy hand, perhaps at the large Glenfield Dairy Farm nearby, he was impatiently hoping to serve alongside his brothers. He'd earlier submitted a National Register Card but hadn't heard anything, so, on the 7th of February 1945, at aged eighteen years and eight months old, the farm hand decided to make his move. Attending the recruitment depot on the Hume Highway at Liverpool, Ted volunteered for the 2nd AIF. On the 16th of February he travelled into the city where he attended the Victoria Barracks recruiting depot at Paddington. The fit labourer easily passed the medical examination and signed his Oath of Enlistment. Allocated his official number he was now *NX205938 Edward Charles Robson* and like the other New South Welshmen he was bundled off to the Sydney showground for his rudimentary training.

Marching into the General Details Depot on the 21st of February, Ted endured a six-day wait at the Sydney showground for more volunteers to make up the numbers in the holding strength before being moved to Cowra. Finally, after tolerating the primitive conditions in the cold and damp showground pigpens, he was on the troop train to Cowra and recruit training. Marching in to the 2nd Australian Recruit Training Battalion he would eventually meet up and befriend fellow New South Welshmen also in the 2nd Battalion, Norm Dilley, Kevin Hurst and Joe Collins.

Uneventfully completing recruit training in the allocated time frame, along with his latest comrades, Ted marched out of Cowra on the 30th of April. Much like his 2nd Battalion mates on the troop train bound for Wagga Wagga, Ted convinced Corps allocation officers and passed aptitude testing to become an engineer. Well-suited to become a Sapper and wear the mantle with honour, for Ted it was onwards and upwards to Wagga Wagga and the RAE Training Centre. The hard-working son of a forest worker and the youngest of Maude Robson's four soldiering sons had the world at his feet. For Jack and Maude they couldn't have been any prouder of their youngest son.

Embarrassingly the telegram delivering the grevious news of their sons death incorrectly listed Cecil, not Ted, as one of Kapooka's victims. After the family realised Ted was the actual victim, the large Robson family, including dad Jack, sister May, Cecil's wife Lorna and grandmother Catherine, attended Ted's funeral. The 18-year-old was the only Robson soldier to die during WWII. His headstone inscription relayed a simple message signaling his short life as a soldier:

'... *His Duty Fearlessly and Nobly Done—Ever Remembered*'.

19—NX205951 Kevin 'Kev' Alexander Hurst (aged 18)
Plot B, Row D, Grave 3

On the 27th of January 1927, Kevin Alexander Hurst was born in the New South Wales town of Narrandera to Richard (or Charlie as he was known) and Jessie Hurst. The couple was married in nearby Wagga Wagga in 1923 and by the time Kevin (or Kev as he was later referred to by family) arrived, Charlie and Jessie had already welcomed their first daughter Edna in December of 1923. The couple also welcomed their first son, Mervyn in February of 1925. Kevin's parents would eventually give birth to seven children—four boys and three girls. But it's the family's military history which sets the scene for Kevin's destiny in the Army. It's a history certainly worth recalling.

Long before his parents met, Kevin's dad Richard was born in the remote central New South Wales town of Hillston in 1894. In 1916 Richard Hurst-senior and his wife were living at 'Glen Ayron' Homestead in Hillston Road, a remote property via Carrathool, when two of their sons, William and Kevin's dad Richard, enlisted in the AIF. The brothers Hurst saw combat in Belgium and France, but their war experiences told two completely different tales of adventure. Kevin's uncle was wounded in action in Belgium in September of 1917. Whilst recovering from those initial minor wounds, he was wounded a second time. This time, exactly one year later, he was in France but unlike his previous recovery this was significantly more serious. Receiving shrapnel wounds to his right leg and left arm, his injuries were too severe to rehabilitate with normal procedures—the military doctors made the life-changing decision to save his life. To do so they amputated his left arm. He was soon returned to Australia and back to Carrathool and his loving family.

The older of the Hurst soldiers, Richard, Kevin's father, had a different tale of the war. Enlisting at age 22 and single, his operational service overseas was a great adventure. After all he was the son of a publican and a hard-working farmer from a small New South Wales country town. When he hit the big time city of London he was dazzled by the many opportunities offering selfish pleasures to lonely soldiers. But before he could taste the adult feasts found in the debauched city, he first had to do what he was there to do—fight. Soon, exposed to the prolonged damp, unsanitary and cold conditions found in the trenches, a severe case of trench foot developed and he was consequently evacuated for treatment. It was during his hospitalisation and recovery that he managed to get himself into more serious trouble with the military authorities for a 'non-combat' related illness.

Apparently, so the records tell us, during his medical treatment for trench foot in London Charlie had a brief, but nonetheless, telling encounter with a lady plying the 'oldest profession'. The end result of his happenstance was the contraction of a debilitating disease. It was an affliction which generally accompanied brief lapses of personal judgment—no doubt alcohol had altered his common sense. But this type of affliction wasn't an isolated problem for the soldiers of the AIF in the Great War. The AIF took drastic action to prevent such conditions decimating combat soldier capabilities. According to the official records of the Great War the Australian Forces Command took a heavy stand on such behavior. The first line of action taken by the Australian Command and Commonwealth Government was to embody in Australian Finance and Allowance Regulations a special military order dated 1st February 1915. Taking a hard stand on the problem, authorities were determined to hit soldiers where it hurt most; their pay. Needless to say Kevin's father was out of pocket for the duration of his 'undisclosed' treatment. But more importantly, he was cured and sent home in one piece. It was a stark contrast to the fate suffered by his now limbless brother.

It was following the Great War, when the Hillston area, located approximately 155 miles northwest of Wagga Wagga on the Lachlan River in western Riverina, got divided into relatively small rural properties onto which returning soldiers were repatriated under the soldier settlement scheme. One of those repatriated soldiers was Richard. He'd arrived back in Australia in August of 1919 and returned to Carrathool where he made use of the soldier settlement scheme at nearby Narrandera and worked as a farmer. He started his fresh approach to life and in 1923, aged 29, met and married Jessie McLean in a ceremony in Wagga Wagga. Jessie was a local girl from the town of Hay, about nine years his junior. Edna was born in December of that year. Sometime after her birth the family moved to Griffith, about forty miles north of Narrandera where Mervyn was born. Finally, January of 1927 saw the arrival of Kevin. With a shock of curly red hair, just like his mother, his appearance was in stark contrast to his bald-headed father. Shortly after Kevin's arrival the young family was back in Carrathool where Richard, working with a horse and cart, was employed as a carrier making deliveries.

The family remained in Carrathool whilst the young Hurst children attended schools in the region. It was at school where Kevin's nickname 'Brickie' manifested itself. With curly fire-coloured hair, which was 'as red as a house brick', it was inevitable that one of the eighty or so kids attending school in Carrathool would get creative. Most likely given to him by one of the local boys, the schoolyard name stuck.

But Australia's misery years of the 1930s affected the small town of Carrathool. Needless to say the family survived the fiscally gloomy years with Kevin's father toiling away as a carrier, now father to seven children.

Meanwhile his mother kept house and successfully raised Kevin and his six siblings. But it was the arrival of WWII which prompted the Hurst family to respond to Australia's needs. Suddenly Kevin was paying more attention to his father and uncles' stories about their time in the Army. Briefly catching snippets of his father's conversations with his mum, Kevin became educated in some of the horrors of the Great War.

But as a veteran of the Great War it was experienced men like Richard who the 2nd AIF was chasing. But they weren't pursuing the older veterans for combat roles; instead they needed experienced soldiers to guard prisoners of war in the Garrison Battalions. For Richard Hurst there was a renewed sense of obligation toward battalions like the 16th Australian Garrison Battalion, which was formed at Sydney; their role was to guard three high-security internment camps in rural New South Wales at Hay where Kevin's mum was born. Constructed in 1940 shortly after the start of the war, these were internment camps known as 6, 7 and 8. Joining more than 200 residents of Hay enlisting during WWII was Kevin's father Richard. Enlisting in Carrathool on the 13th of April 1942, he was once again a soldier. By this time he was 48 and father to seven kids. Serving for the next four years Kev's father saw the strength of the Garrison increase to more than 688 old mostly retired men just like him guarding dangerous prisoners of war and civilian internees. Some of the first arrivals into the camp were more than two thousand refugees from Nazi Germany. They were interned in camps 7 and 8 near the Hay showground whilst Italian civilian internees were housed in a camp near the Hay Hospital. For the Hurst and MacLean families, the impact of Australia's contribution to the war was real and occurring in their own backyard. For Richard Hurst it was a bittersweet taste of revenge for the pain the Germans had inflicted on his maimed brother. Richard was promoted to Warrant Officer Class Two and remained at the camp until the end of the war

The impact of the internment camps on the region and the local men guarding prisoners of war provided the incentive for Kevin's brother, Mervyn, to enlist in the Air Force in March of 1943. Mervyn set off to New Guinea and would not see his family again for the next two and a half years. Then, in the following month, Kevin's eldest sister, Edna, the eldest of the family, made her way to Victoria Barracks at Paddington in Sydney to become an AWAS. No doubt itching for his opportunity, Kevin the larrikin 16-year-old who just happened to be an excellent horse rider, loved a beer and a 'scrap' at the local on a Saturday afternoon. For him the war and enlistment was a world away. But whilst working various jobs as a station hand, and at many of the town's stores delivering groceries, he watched helplessly as his father, brother and now his sister went off to join the fighting. Eager to enlist 'Brickie' still had a couple of years to wait.

No sooner had Kev registered his intention to enlist, shortly after turning eighteen in January of 1945, he was called up. Saturday the 17th of February, the very same day Ernest Poschalk was enlisting in Queensland, Kevin caught the train and was in Sydney at Victoria Barracks Paddington taking his oath of enlistment upon enlisting in the 2nd AIF. Proud as punch, the recruit they called Brickie became *NX205951 Kevin Alexander Hurst*.

Following in the footsteps of his siblings, his father and his brave uncle, whose amputated arm was a constant reminder to the entire Hurst family just what war was about, Kevin's turn to be a soldier had arrived. The bar fights and larrikin lifestyle he was used to in Carrathool was temporarily put on hold—until at least he got into the Army. Marched off to the Sydney showground and into the General Details Depot on the 21st of February, Kevin soon found himself amongst many other 18-year-olds with similar dreams of service. No doubt he would have met up with other Sydneysider's Ted Robson and Allan Flood. For six days the men toiled away doing preparatory training at the Showground until they were finally placed on the troop train on the 27th of February; destination Cowra.

Loaded down with equipment (all inscribed with their personal details), it wasn't too long before Kev and other men, including new mate Allan Flood, were herded out of the showground stalls and sent off to Cowra. On the 28th of February Kevin marched into Cowra and was assigned to the 2nd Australian Recruit Battalion where his new mates would include men he'd come to know as Bill Reid, Ernie Poschalk and Norm Dilley.

The bustling recruit-training centre saw the back of excellent recruit Kev Hurst on the 30th of April. Experiencing little disruption during training, the fit 18-year-old completed the first step to becoming a soldier in the allocated eight weeks. Allotted to the Corps of Engineers, along with his latest companions, Kev boarded the recruit troop train for Wagga Wagga bound for the RAE Training Centre. At the 1st RAE Training Battalion he was introduced to a different group of men, which included Alf Witt, Colin Boyd, Joe Faull and Terence Moore. The men had all marched out of Cowra, albeit from the other training battalions, and now were bunking down together in the six-man tents.

When Kev arrived at Kapooka he met many new faces. Already bunking down in the tent he was to spend the next three months in was 19-year-old Allan McPaul from Kyogle in New South Wales. Allan was in B Company, which meant he had already completed a couple of weeks of training, before Kev and his new tent mates arrived. About a week earlier Allan had celebrated his 19th birthday in about week two or three of his training and now welcomed some fresh blood to the tent in the form of Kev and his Cowra buddies. Introducing themselves and becoming acquainted the men would talk about anything and everything. At one point the conversation centred on the instructional staff where Allan proceeded to tell his new tent mates about one of the Sergeant demolition instructors who was a really nice bloke to know. That man was none other than Jack Pomeroy.

Meanwhile, Allan and his new tent mate with the flaming red hair, Kev 'Brickie' Hurst, would soon get to know each other. But they didn't necessarily become fond mates. Their training was a week or so apart, but needless to say Kev and his new-found tent mates all got along, especially the guys he'd just marched in with. Thanks to Allan and his mates, it wasn't too long before the young man with hair as red as a house brick had his new moniker solidified by his humoured tent mates. His endearing name, like most defining features of a new recruit at Wagga Wagga, somehow stuck. The instructors would soon identify him as just Brickie!

Richard and Jessie Hurst made the trip from Carathool to Wagga Wagga to attend their son's funeral. They inscribed their 18-year-old son's headstone with a personal message of farewell:

'... *Fond Thoughts Of "Kev" Too Dearly Loved To Ever Be Forgotten*'.

Mervyn Hurst was in New Guinea with the Air Force when he got word that his younger brother Kevin had been killed. He was unable to make the journey. Kevin's younger brother, Bobby, was too young to make the 200-mile trip and to the best of his knowledge it was just his parents who were present to say goodbye at Wagga Wagga. Richard Hurst discharged from his position as a Prisoner of War guard at Hay in January of 1946. He died in 1961 aged 66. Jessie, Kevin's mum, died in July of 1969 also aged 66.

Mervyn returned from New Guinea and discharged from the Air Force in March of 1946 whilst Edna was discharged from the Army in July the same year.

20—NX205969 Allan Flood (aged 18)
Plot B, Row B, Grave 14

Being born in Sydney in 1927 meant Allan Flood arrived in Australia's largest city during an increase in 'organised' crime. Allegedly caused by restrictions on so-called 'recreational' pursuits, Sydney descended into a dark chaos on dimly-lit streets. Although carrying a pistol or weapon was commonplace in Sydney at the turn of the century, the introduction of the 1927 *Pistol Licensing Act (New South Wales)* by the New South Wales State Parliament imposed severe penalties on persons carrying concealed firearms and handguns. As a result Sydney gangland figures chose razors as their preferred weapons. It was the capacity to inflict disfiguring scars, without breaching laws, which confirmed the razor as the popular weapon of choice for the criminal elements controlling Sydney streets during the era. In scenes reminiscent of lawless London streets at the turn of the century, the country's major city may not have been the ideal 'nursery' in which Allan Flood would learn life skills.

Alfred and Ethel Flood gave birth to their son Allan on the 31st of January 1927 in the Sydney suburb of Neutral Bay. Born just four days after Kev Hurst, albeit Kev was born in the country, and living in the city of Sydney during the 1920s, it was perhaps a more dangerous beginning for the young Flood. After all 'Razor Gangs' not only dominated the Sydney crime scene during this time, their existence introduced an unnecessary fear into the lives of ordinary men women and children trying to forge a life in the bustling city.

Allan's parents were both born in the late 1890s. His father, in conditions far-removed from the city, was born at Billabong Creek in the New South Wales Riverina in 1892, whilst his mother was born in Sydney in 1894. At aged 22 his father married his 20-year-old bride in a ceremony in Sydney in 1914 and the following year the first of their five children were born.

Controversially, according to the *New South Wales Police Gazette* of 1921, Allan's father was no saint. Sydney's rampant crime wave caught up with the Flood family. No doubt trying to provide for his young family, on the 1st of July 1921 Alfred was convicted of embezzlement in the North Sydney Police Court. Alfred Flood left his pregnant and distraught wife and brood of young Flood children alone when he was sentenced to six-month's imprisonment with hard labour. He was also placed on a 12-month good behavior bond.

Without the support of her imprisoned husband, Allan's mother delivered Allan's older brother Jack on her own.

Allan was the last of the Flood children to be born. His mother was 33 when she gave birth to him and by 1930 the young family had settled into a residence at 60 Albany Street, Crows Nest. His father would soon secure work as a chainman; no doubt working on the Sydney Harbour Bridge construction along with thousands of other blue-collar workers. It was secure employment, at least for the next fifteen years as the bridge developed.

Tragically Allan was just ten years old when his mum passed away in 1937. She was only 42. The youngest of the Flood siblings, Allan's father and older siblings were left to raise him in Sydney's troubled streets. But it wasn't long before the elder Flood siblings had children of their own and were quickly leaving the family home. Abruptly, there weren't many Flood siblings left at home to look after an impressionable teenager. Apart from his hard-working father, Allan was left to grow up in his teen years without a sibling role model on the home front.

In 1939, within two quick years of his mother's death, Allan was nearly thirteen when the Prime Minister declared that Australia was at war. At the time it didn't mean much to him but expectantly other members of the Flood family reacted to the nation's call. Regardless of their current stations in life they either voluntarily, or via conscription, found themselves in the armed Services.

Shortly after marrying in 1941, Allan's brother, Jack, was conscripted and served with the Defence Platoon at 28 Brigade HQ. Discharging less than twelve months later, he found work as a carpenter's labourer. Meanwhile, Allan's brother-in-law (his sister Helen's husband William Harper), volunteered for the 2nd AIF, enlisting in Paddington on the 24th of April 1941. Joining the 2/19th Battalion, he trained at Wallgrove, Ingleburn and Bathurst before embarking to Singapore to support his fellow battalion members fighting the invading Japanese in the Malaysian jungles and Singapore. Like most Australian units involved in the 'Fall of Singapore' his brother in law's 2/19th Battalion fell into a desperate retreat that ended with surrender on the outskirts of Singapore city on the night of 15 February. Initially imprisoned in the sprawling Changi prisoner-of-war camp, it was not long before members of the 2/19th were allocated to external work parties. William Harper, Allan's brother in law, was now listed as missing in action.

Back in Crows Nest, the lack of information in the Harper and Flood families concerning William's welfare caused heartbreaking years. It was perhaps inevitable that shortly after turning 18, Allan himself would register his interest to enlist in the 2nd AIF. Submitting his National Register Card whilst still living at home all he had to do was wait. Allan dealt with his brothers'-in-law absences with some difficulty; he wanted desperately to join up to do his part. After leaving school and registering his interest he found work as an apprentice engineer. But with his sister's husband missing in action Allan decided to move in to his sister's residence in Talus Street, in the nearby suburb of Naremburn just near the Royal North Shore Hospital. Here he was able to provide first-hand support and care for his big sister whilst the whereabouts of her husband William were still unknown.

Whilst waiting for his call up, Allan secured work as a process worker with Bryant Brothers around the corner from his house in Herbert Street, St Leonards. Before too much longer his registered interest in the Army was answered. Allan found himself in Victoria Barracks, Paddington on Monday the 19th of February 1945. Passing his medical examination and taking his oath of enlistment, Allan was just another impressionable wide-eyed 18-year-old who turned up at Paddington to enlist that day and in doing so joined the band of New South Welshmen waiting in the pens. Similarly, the following day, many more young impressionable Sydneysiders would do the same; men like Joseph James Collins, who would follow the same enlistment procedure that Allan undertook the very next day. All men enlisting around the same time would've been shepherded into the same stalls together at the showground for a number of introductory days. Allan, the baby of the Flood family was now a soldier given his official number: *NX205969 Allan Flood*.

Marching-in and taken on the holding strength at the recruit reception depot on the 22nd of February 1945, Allan was transported to the nearby showgrounds for processing. Here he would meet other men with the visions of service and sacrifice—young men like Kev Hurst, who was not only making himself comfortable at the showground pens; his 'Brikie' nickname was beginning to stick. His curly red hair had been cropped short even more so now closely resembled a house brick. Within five days, Allan and his latest mates were being marched out of Paddington enroute to Cowra.

Allan left the hustle and bustle of the big city behind him as he marched into the 2nd Australian Recruit Training Battalion on the 28th of February 1945. Replacing city buildings with bunkers and obstacle courses, the young soldier embraced the challenges of training for war head-on. After a little more than two months of intensive training, he completed his indoctrination and convinced allocation officers he was best suited to engineers. So, on the 30th of April 1945, Allan marched out of Cowra with fresh new mates, including Joe Collins, and, after catching the troop train for the 120-mile journey south, they arrived at Wagga Wagga and the RAE Training Centre. Immediately the men were marched in to the 1st RAE Training Battalion. In less than a month the training would escalate to include the dangerous fourth week of demolitions training

Surviving perilous years as a prisoner-of-war along with his comrades, Allan's brother-in-law William Harper and members of the 2/19th Battalion were liberated in late-August 1945. Almost immediately they began returning to Australia. The Battalion was formally disbanded later in 1945, having suffered the highest casualties of any Australian Army unit during war. Thankfully William returned to the waiting arms of Helen. It was a moving reunion. For Helen it was an empty feeling as her younger brother Allan, her saviour in times of need during William's absence, was never coming home and was now resting peacefully in the Wagga Wagga Cemetery.

☩

21—NX205981 Joseph 'James' Collins (aged 18)
Plot B, Row C, Grave 14

The fourth generation of Joseph Collins was born on the 8th of January 1927 in Nowra, New South Wales. His parents, Joseph and Lavinia, were married six years earlier in 1921 at a beautiful ceremony in the same southern coastal regional town. By the following April, the couple's first-born, a son, John arrived. Over the next decade the couple would have three more children, Joseph (known more commonly as James, so as not to get confused with his father), Hedley and their only daughter, Dorothy. But it was James, the 4th generation of Joseph Collins who proudly bore the name of his forefathers and continued a family legacy. It was a tradition that had started more than a century earlier with James' great grandfather. His son, Joseph, gave birth to James' father, another Joseph Collins, who was born in 1883 in Nowra. Now it was the fourth generation of the namesake who was destined to carry the family tradition into the next generation—and hopefully beyond.

Sadly and in mysterious circumstances, Joseph's brother Hedley passed away in the Sydney suburb of Annandale on the 9th of November 1927. Little brother James was barely eight months old. Sadness ripped through the entire Collins family for a long time.

Less than three years later James' father was a farmer and living with his young family in East Street, Nowra. It was a nondescript street just on the outskirts of the town centre. Whilst his mother was content tending house, it was the uncertainty of Australia's fiscal demise suffered in the 1930s which forced Joseph's father to change professions. Farming was quickly becoming a liability and unable to adequately provide, his father was forced to move to nearby Browns Hill, just south of town. Here he turned his hands to manual laboring as a timber cutter. Young James was ten when two of his father's sisters also moved into the family home in 1937. Although crowding the children's upbringing, having the support network of family made life a little easier on the young Collins' family.

Over the next few years the family remained in Nowra. Living in the rural setting allowed the teenager James to earn a decent wage thanks largely to an introduced Australian pest. Australia in the late 1930s experienced plagues of wild rabbits, especially in rural New South Wales. In some regions such as Tamworth and Armidale, landowners were failing to control the pests, which normally included employing trappers, poisoning the rodents or even digging them out. In some situations landowners themselves turned their hands to trapping, rather than work the land, rabbits were being trapped for their skin and meat and rabbit skinners were paid per rabbit. It was reported that a top skinner could skin seven rabbits a minute, six hours a day while standing knee deep in rabbit heads. It was a lucrative venture of the time; sufficiently so that it became a profession for many.

The arrival of the war brought changes to the livelihood of the man who now went by the name of Joe Collins. Ditching the names James and Joseph, he preferred the shortened version and felt comfortable with the moniker. It seemed perhaps to be a little more grown up. After all, the young man was earning a keep in a profession that demanded some adult respect. Rabbit trapping became Joe's method of earning a decent wage; especially as a young man growing up in a competitive labour market. But the arrival of the war years in 1939 brought changes to the laconic country living experienced by the large pioneering families.

Jack, the eldest of the Collins boys at 21, was the first to be conscripted into the Militia and during his soldiering duties in Atherton, Queensland; he transferred his skills into permanent soldiering by enlisting into the

2nd AIF in January of 1943. Meanwhile Joe's sister Dorothy had married Tom O'Brien in 1938 just before the outbreak of the war, but like Jack, soon after conscription into the Militia, he was soldiering in Geelong, Victoria, when he too was recruited into the 2nd AIF. Seeing his brother, and now his brother-in-law, step into the world of full-time soldiering, Joe waited patiently for his turn at the big adventure. As a rabbit trapper wasn't the career or adventure that a young boy could be satisfied with, he eagerly waited for his 18th birthday, impatiently plying his trade until his time was due. His chance came in February of 1945.

Living at Browns Hill, near Nowra, Joe noted on his enlistment that his father Joseph was his next-of-kin. After gaining his father's permission to enlist he listed his occupation on his Attestation Form as 'Rabbit Trapper'. With brown hair and piercing blue eyes, the doctors took no notice of the two tattoos on his right arm and passed him medically fit for service. Signing his oath of enlistment on the 20th of February 1945 at Victoria Barracks Paddington, it was a long way from rabbit trapping but finally he was *NX205981 Joseph James Collins*—the soldier. Sent home to organize his affairs he was due to return to the General Details Depot at Paddington on the 26th of February.

Returning on the due date he marched-in and was taken on the holding strength at the showground; the very next day he was on his way to Cowra. Arriving at the 2nd Australian Recruit Training Battalion the following day, sitting amongst and cracking nervous jokes on the transport with Ted Robson, Kev Hurst and Allan Flood, the jokes quickly dried up as they spied their training base for the next few months. But Joe was an experienced bushman and brimming with some confidence; perhaps at Cowra he'd feel right at home. Completing training relatively unscathed, by Monday the 30th of April, he convinced allocation officers he was well-suited to become an engineer, and by the end of the day he was on the train bound for Wagga Wagga along with new mates Ted Robson, Kev Hurst and Allan Flood.

Marching-in to the RAE Training Centre he was ready to start his combat engineer training. It was his chance to become a genuine soldier; just like his brother and brother-in-law. He was nervously excited about the dangerous training he was about to undertake; but now surrounded and supported by different mates all feeling those same emotions he was experiencing, he was ready. Cautiously confident of surviving the training, especially the obstacle course and the realistic battle inoculation training—Joe was fit and prepared. He was on the cusp of a bigger and better life. A life far removed from the perils and indignity of ripping skins off rabbits—at least that's what he thought!

Joe never got to meet up with his brothers also serving during WWII. Their younger brother's headstone reflected his short time as a soldier:

'... *His Duty Nobly Done*.'

22—NX81964 Stanley Ernest Ross (aged 28)
Plot A, Row C, Grave 12

One of five children to Francis and Elizabeth Ross, Stanley Ernest Ross was born in Mungindi, New South Wales on the 30th of January 1917. Before the less-educated amongst us, including yours truly, ask where the 'bloody' hell is Mungindi, I can add that it lays claim to being the only border town in the Southern Hemisphere to carry the same name on both sides of a State/Territory border. Well, when you look at a map of Australia, specifically the border between New South Wales and Queensland, there is a straight horizontal borderline before it becomes 'squiggly'. Well, roughly, where the squiggly meets the straight; that's Mungindi and Stanley Ross' birthplace.

His parents were married in March of 1914 in Queensland and both seemed destined for a life on outback livestock stations. The outback couple experienced the pain of loss when their 4-year-old son, Neil Francis died in August 1917—the couple would however cherish their three daughters, Ethel, Shelia and Thelma and surviving son Stanley who became affectionately known as just Stan. By 1919, his father Francis was the Station Manager at Moogan Cattle Station, whilst his mother kept house at the nearby Mungan Station. In the early years after school pastimes of ball sports such as football and fishing provided normal highlights for kids growing up in small communities, but Mungindi was a tiny town with not a great deal of new people arriving. Often this meant family siblings would become more than just family—they were also good friends.

Maintaining the status quo of life in early country Australia of the 1920s and 1930s, it was an unusual shift away from the rural occupations on cattle stations, which saw the family move to the 'big smoke' of Sydney in the late 1930s. Life for the Ross children attending local schools consisted of regular air raid drills. They'd often find themselves lying in slit trenches for protection against imaginary Japanese planes conducting strafing runs overhead. Living at 35 Morris Street, Summerhill, a suburb about five miles southwest of the city, Stanley got his first job as a motor mechanic around 1939. At that same time, his brother was a labourer whilst his sister was busy getting married. But such was the strong bond of family, that Stan, his brother, recently married sister and brother-in-law were all living in the Morris Street household together. But 1939 also brought the uncertainty of life and created compulsory heroes courtesy of the war.

At 28 years of age, in comparison to the younger volunteers, Stan was no spring chicken. No doubt the bombing of Pearl Harbour, and the subsequent call to arms following the strategically unsuccessful infiltration of Sydney by Japanese midget subs, motivated and incensed New South Welshmen of all ages to enlist in the military. Stan's mature life experiences meant he was a man full of state and national pride when he considered volunteering himself in January of

1942. A motor mechanic by trade, he knew it was a skill that would benefit the Services in defending his country. So, with patriotic dedication whilst seeking out his own adventure, Stan found himself in the enlistment queue at the Randwick Racecourse in Sydney; he'd made his decision. Enlisting into the 2nd AIF, by early January 1942 he was a full-time soldier, *NX 81964 Stanley Ernest Ross.*

Marching out from the General Details Depot in Sydney, Stan was posted to 1st Armoured Brigade Company on 16 January 1942. By the following April he was being sent to Queensland in preparation for a deployment to the Northern Territory, in particular Alice Springs. Spending almost two months in the Territory with the 1st Anti-Aircraft Brigade he returned to Greta in New South Wales in July that year. However, something happened to Stan at Greta which sent him on a path not appreciated by the Army.

In September of 1942 Stan went AWOL. A warrant was issued for his arrest. He'd had been AWOL for 21 days when the authorities finally caught up with him. Punishment was a period of detention in one of the purpose-built detention centres. Serving his time manfully and wearing his punishment well, Stan was back in his unit by December and being deployed to Western Australia where he arrived in February of 1943. After serving a further five months in the field without incident, Stan received some well-earned rest and was sent to the General Details Depot in Claremont for seven days. Returning to the field as directed, it wasn't until the following February where Stan once again felt the full wrath of his Commanding Officer.

This time he failed to appear on parade and, with his past track record, he was awarded the most unwanted punishment—Confined to Barracks for seven days. This meant no alcohol, no socialising and perhaps he was even living with the Provost in the makeshift guardroom. Needless to say Stan didn't like being confined or stuck out in the middle of nowhere at Greta. He did the only thing he could do; he went AWOL again. So from the 28th of February Stan was again on the run evading the Provosts and anything that resembled Army authority.

He remained on the run until the 2nd of March when he was recaptured. This time his punishment was a little more severe. The Commanding Officer was clearly trying to send a message to the rebellious and reluctant soldier. Fined a total of five pounds, Stan received more severe punishment in the form of a further 28 day's loss of pay and an additional forfeiture of an additional three days' pay. In the end the unwilling and reluctant soldier wasn't going to be paid for a whole month. But if you thought the mature soldier had learnt his final valuable lesson, it wasn't to be. Stan's military indiscretions were far from over.

January 1945 Stan went AWOL again. This time it may have only been a quick three-day sojourn from his soldiering obligations but strangely, considering his past record, Stan's punishment was becoming a little lenient. Awarded a three-pound fine, the harshest part of the punishment was a further fourteen days Confined to Barracks. Stan was fast becoming one of 'those' soldiers with significant administrative issues. The type of soldier who may have been everyone's mate but with a poor military discipline record, he was not the hierarchy's favourite soldier. He was essentially an administrative nightmare and they had to do something with him. In the end the Army had to act.

Reluctantly, Stan was posted to the 1st Australian Base Depot Personnel in March of 1945 perhaps so the authorities could keep an eye on him and figure out what to do next. By the 9th of April Stan got his posting orders and was sent to the 1st Battalion at the RAE Training Centre in Wagga Wagga. Perhaps the engineers could sort him out? On the 2nd of May 1945, Stan marched out to the 3rd Army Transport Training Depot to do a two-week Mechanical Fitters Course; returning back to Wagga on 15 May 1945. Within a couple of days Stan found himself amongst some younger men at the 1st Battalion. Although it was a little unusual for a motor mechanic and fitter to be participating in the engineer training schedule, it was even more unusual that the wayward soldier was now working amongst the younger more impressionable men. He joined in with the men who were in about week three of training and working towards demolitions week; week four in G Coy.

There's a lack of creditable information surrounding Stan's family both in his life and in his death. It's unclear whom if any of Stan's family attended the mass service. His beautifully-crafted headstone probably doesn't reflect the gruff character of Stan, however his headstone is properly inscribed with a soldier's farewell:

'... *His Duty Fearlessly and Nobly Done Ever Remembered*'.

23—NX205833 Jack Clinton Nixon (aged 30)
Plot B, Row B, Grave 16

The son of William 'Bill' and Elsie Nixon, Jack Clinton was born in Cobar, New South Wales, on the 3rd of November 1914. Located 442 miles northwest of Sydney, Cobar was a significant major mining community with a rich history going back as far as the early 1800s. It's a remote town located at the crossroads of two major highways, one leading to Queensland, the other to South Australia.

Although mining had long been the lifeblood of Cobar, by the time Jack Clinton was born in 1914 and into his early years, difficult and lean times had struck the region. Copper mining operations ceased in 1920, and by the 1930s, the town's population had dropped to a little over 1,000. During this time, Jack's parents and his two brothers, Robert and William (aka Billie and Bob), were carving a life for themselves with his father hard at work in the mines as a labourer. In 1936 Jack, at just twelve years of age and his elder brother William, had moved about a block away from their mother into a house in Becker Street. His big brother was working as a blacksmith and Jack found work in the mines as a labourer. Taking up a position as a station hand at a nearby property, Jack's father left Jack's mum at home, alone.

In the early 1940s Jack met and fell in love with Alice Nicholson; she was a woman six years his senior. His life with Alice became busy, so busy in fact that he hardly saw his parents and hadn't seen his father for a couple of years. By 1943, Jack was 29 and Alice 35 when the couple decided to make the trip to Sydney to tie the knot. Following their nuptials the couple returned to Cobar and into the Becker Street address where Jack continued his work as a miner.

Like many other families of the time, much like those we've already met, the arrival of the war brought a change of fortunes. It was especially hard on the families in small struggling mining communities like Cobar. It was inevitable then that the Nixon men (Bob, Bill and Jack) along with many men of recruiting age in the Cobar community would eventually be summoned to join the Permanent Military Force. If not volunteering, alternate conscription into the part-time Militia was also beckoning.

War was furthest from Jack's mind when he married his sweetheart Alice in 1943 and welcomed their first child Neryl in 1944. Neryl was Jack's little princess. But much like Jack Pomeroy, struggling between life as a father to his young family and his patriotic commitment in replying to the country's call to arms, Jack found it difficult to tear himself away from his princess. But his little princess was less than twelve months old when her father decided to make a life-changing career decision. Perhaps influenced by his wayward ways and conviction in the Cobar court for drunkenness, Jack changed the course of his life when he decided to put his family first. Pursuing his obligatory duty in the Army, he knew that temporarily he'd be out of his little princess's life, but deep down he knew it was for the better, after all he was joining the Army for security and to establish a future for her.

Deciding it was better to do it now before Neryl got a lot older, Jack decided to register his intention to enlist in the 2nd AIF and go off to fight. But at the time he was far from the normal demographic for those men enlisting in the 2nd AIF. Having been born in November 1914, Jack was already thirty years of age. He was married, the father of a small child and now forging his own life away from his parents when he decided to volunteer. But amongst the throng of much younger and fitter volunteers, Jack sought out the challenge. A little portly, due in part to a content married life, the stodgy 30-year-old was prepared to do his part.

Leaving his wife, young baby girl and extended family behind, Jack clambered up onto the train a little nervous, anxious, excited and concerned. As he set out on his journey he was wondering just when he would see his wife and baby again. But at the same time Jack knew it was something he had to do; it was hopefully his family's future he could secure. As a slightly overweight 30-year-old, he was also a little concerned that he might not make the grade physically. But he was going to give it a go nonetheless. Leaving his young family, his job in the mines and his ageing parents behind in Cobar, it was a life-changing decision. He was immensely proud of himself making the move. His trepidation and nerves were masked by a beaming smile towards his mature wife and angelic daughter as he sat inside the train waiting for that first shunt out of the station. It was going to be a long lonely journey.

But as the train departed Jack was travelling toward a different future. Leaving Cobar, Jack had plenty of opportunities to change his mind, but he didn't. He remained committed. He left the town that Tuesday morning bound for Sydney and the recruitment offices. With a photo of his adored baby Neryl tucked tightly in his trouser pocket, Jack landed in Sydney and negotiated his way to the enlistment office. Standing and being processed in the recruitment line before Jack at Paddington on the 9th of February 1945 was Eric Phillips of Newtown, Billy Cox and an 18-year-old from country New South Wales he would later get to know well by the name of Stan Emery. Jack found out that Eric was off to join the infantry and that Billy was off to work with the Army War Graves Unit. As for Stan, he and the much older Jack would eventually become good mates a little later down the track as engineers. For now they were just another group of volunteers trying to make a difference. Before all that was to occur Jack was processed at the recruiting office. Passing the medical examination Jack signed his oath of enlistment and became *NX 205833 Jack Clinton Nixon*, a soldier. Amongst the throng of volunteers he did wonder to himself just how many of them were also married and had kids? Were they also feeling the same sense of guilt?

New South Welshmen were transported to the Sydney showground for holding until they were ready to be dispatched to their respective recruit training units. For Jack, like so many of his New South Welshmen counterparts, the 15th of February 1945 was the day he was sent to the dry and dusty fly-infested Cowra where he was allotted to the 1st Recruit Training Battalion. With his mining background he was hopeful that a career as a combat engineer was there for the taking. But first things first—he had to survive recruit training. Courageously completing the physically grueling recruit regime designed for much younger and fitter men, Jack Nixon displayed physical fortitude beyond his portly frame. Immensely pleased with himself, he'd completed the first step of his career change. A letter of congratulations from Alice and Neryl was all the motivation he needed as he progressed to the next phase; Kapooka Camp

Alongside fresh mates like the much-younger but likeable Stan Morphy, by the 7th of May 1945 Jack marched out of Cowra. Holding his head perhaps a little higher than his younger counterparts, he was leaving Cowra an extremely proud and remarkably fitter man. Caught up in the disciplined life of soldiering, he soon found himself on the train bound for the extremely busy Engineer Training Centre for his next phase of career transition.

No sooner had he arrived at Kapooka he was allocated to the 1st RAE Training Battalion and one of the six-man tents dotted amongst the landscape of the Pomingalarna Range. Much to his surprise Jack was astounded at how fast he was becoming a Sapper. He did miss his wife and their little princess and was looking forward to completing his training to be with them once again. His time at Kapooka was made a little easier though when on the very first night a familiar face bunked down beside him in their tent. It was none other than the young man he met at recruiting nearly three months ago, Stan Emery.

Participating and receiving experienced instruction through the engineer training regime, Jack soon found a remarkable confidence and felt highly capable even amongst the much younger brigade. Within two weeks Jack was dead.

By May of 1945 Bill Nixon hadn't seen his son for about three years. It was heartbreaking therefore for him to hear that his son was one of the 26 men killed at Kapooka. Immediately, with the heaviest of hearts, Bill travelled the 400 miles from Cobar to Wagga Wagga for his son's funeral. After three years without his son in his life he barely had time to say goodbye as he arrived in Wagga Wagga late but thankfully just in time to see his son being lowered into the ground. Bill later inscribed his son's headstone with a simple final message of love:

'... *Until The Dawn Breaks*'.

Following the death of her adored husband, for many years later Porkie's beloved Alice pined for her lost love. Although raising baby Neryl allowed her to focus on being a mum, after eight lonely years Alice re-married. Her new husband was also a widower having also lost his wife too early. Together the couple experienced a happy life together, but they also exhausted a lifetime pining for their lost loves.

Sadly Alice died in 1995 aged 87. But thanks to daughter Neryl, who has kept a family secret close to her heart to this very day about the location of her mum's ashes, she took it upon herself to reunite her mum and dad in body and soul. Fifty years after the tragic explosion robbed Alice and Neryl of their much-loved husband and father, Neryl made sure her mum Alice now rests eternally at peace alongside Jack; her one true love. Wrapped firmly in the arms of the charming man she first married back in 1943 who was abruptly taken from her just two short years later, the couple continue to proudly look down over their adored daughter.

24—NX180219 Leslie John Mather (aged 20)
Plot A, Row C, Grave 10

On the 31st of January 1925, Leslie John Mather, the second son of Joseph Mather and Margaret Canty, was born in Nyngan, New South Wales. A small town situated on the Bogan River, Nyngan lies lazily on the Mitchell Highway between the inland towns of Narromine and Bourke. Although it rests approximately eighty miles east of Cobar on the Barrier Highway, needless to say being located more than 230 miles northwest of Sydney and about 130 miles from the 'back of Bourke' (a well-known Australian colloquialism describing the isolation of being 'in the middle of bloody nowhere'), for the young Mather family remote Nyngan simply meant work opportunities. For Joseph, working with the New South Wales Railways, it was hard hot work but was there a reason for an 'escape' to the isolation and anonymity of Nyngan because of a darkened past?

As a 21-year-old, Joseph enlisted in the AIF on 11 December 1915. He followed in the footsteps of his older brother David who'd earlier enlisted in November of 1915. Allocated as a Gunner in the Field Artillery Reinforcements at Enoggera Camp in Brisbane, it's fair to say that Les' father and uncle David had 'indifferent' military experiences, none of which you would say equated to a career. The Mather boys proved a handful for the AIF authorities. Both Joseph, and his big brother by about five years, David, absented themselves without leave from Enoggera for more than 120 hours in February of 1916. Both men pleaded guilty to their respective charges. Following a short trial, an equal retributive punishment of 120 hours detention was awarded for Joseph's indiscretion whilst the older David was Confined to Barracks for twelve days. Completing his punishment in late February, Joseph was either rebellious or he sought out and took a likeness to the Brisbane nightlife. On the 18th of March he re-offended; another case of AWOL, but with an additional compounding charge of drunkenness. Again arrested and charged by the Enoggera Provosts, his punishment on this occasion was a little more severe—seven day's detention at Enoggera. Hopefully this was going to convince Joseph that AWOL was not the solution; however, his big brother had other plans. This time it was David who went AWOL. The day after his little brother was released from detention David was AWOL again. Arrested and given a further six days confined to barracks, it was clear soldiering was not what the Mather boys expected when they signed up. By now it was clear that both David and Joseph Mather didn't fancy life in the military and together they concocted a plan. It was all about completing their service without following formal discharge procedures.

On the 7th of April 1916, both men went AWOL together for the final time. When the Sergeant Major checked their billets, Joseph and David, including all of their equipment, were gone. This time the implications were a little more serious. Under the provisions of the *1903-15 Defence Act,* Joseph and David Mather had deserted their Corps. Under Section 72 of the *Army Act,* a military Court of Inquiry was held on the 2nd of May 1916 and found, that because the Mather brothers were absent with almost £17 worth of issued military equipment, the military wanted the men—and their equipment— back. As a consequence of the Court's finding, on the 12th of May 1916 the Camp Commandant at Enoggera signed two Australian Military Forces Warrants for their arrest. Warrants 253 and 254 were raised; one for the arrest of David, the other for the arrest of Joseph. Military authorities alerted all members of the Defence Force and Police Forces of the Commonwealth that 27-year-old David and 21-year-old Joseph were on the run and were wanted men. They weren't the only ones though. Thousands of similar Warrants were being raised every year. Needless to say the brothers Mather never returned to the Army.

Having evaded authorities for a number of years and managing to put his unfavourable military experiences behind, Joseph got on with life. Meeting and eventually marrying his true love Margaret in a small ceremony in Sydney in 1920, over the next five untroubled years the couple would welcome the arrival of three sons. With a small manageable brood of their own, for Joseph it was a far cry from the mayhem he'd experienced growing up in such a large family in Toowoomba. For the young country couple their first-born child, a son Raymond, was also born in Nyngan in February 1922, whilst Leslie arrived in 1925.

Shortly after the arrival of the brown-haired blue-eyed Leslie, before the family really had time to settle, forced by the constant pursuit of work they were on the move again. Settling into their latest home at Bourke in July of 1926 the couple's third son, Robert, was born. However, at the time Australia's economic climate was rocking on the brink of disaster. Work and suitable incomes to support growing families, like the Mathers, were quickly becoming scarce. Forecast to take a downward turn, with the pending demise of his income stream Joseph Mather found life a little unfair. Perhaps it was karma for escaping the responsibility of service in the Great War but it was especially tough on him as he had three hungry boys to feed—and all under 10. Assuming gypsy-like lifestyles, the family were once again on the move northeast in search of paying work.

Leslie was just five when the wind blew them into the agriculture-rich town of Goondiwindi. Resting on the New South Wales/Queensland border, Goondiwindi was almost 300 miles away from Bourke. Getting work as a labourer, Joseph and the family finally settled into a humble residence in Francis Street. Catching the cooling breeze whilst fishing from the banks of the nearby McIntyre River, his prior Army indiscretions and the difficulties of raising a young family now lapped unconsciously at his feet; life was turning for the better. But many years later, with trouble in Europe forcing the country's leaders to ponder the inclusion of Australian men in the conflict, 44-year-old Joseph was able to rest a little easier knowing that his age and his almost embarrassing first stint as a soldier wouldn't catch up with him.

Relocating yet again, this time a little further into the State, they found themselves living in the security of Joseph's parents Joe and Annie, in Bolton Street, Toowoomba in 1939. By now Leslie was fourteen years of age and the entire country was talking about the new war. Working as a shop assistant in November of 1941, Leslie's older brother Raymond was the first of the Mather sons to be conscripted into the CMF. He was subsequently called up for full-time duty in January of 1942. Working with ammunition as a member of the Advanced Ammunition Depot, Raymond would see service in North Borneo, Labuan and was serving in Moratai when his youngest brother, Robert enlisted immediately after turning eighteen in August of 1944.

By this time the family was back living in Sydney in Garden Avenue in inner-city Glebe. Living in Sydney allowed the youngest son, Robert to enlist at Victoria Barracks, Paddington. Robert would eventually go on to work at the 19 Australian Infantry Training Battalion at the Wallgrove Army Camp. Coincidently, used primarily as a staging and training area during the war, many years after the war ended the Wallgrove Army camp became Eastern Creek and is now home to Sydney's most famous motorsport racetrack.

With his older and younger brothers already serving, Leslie expected to be conscripted into the CMF. And, shortly after turning twenty in January of 1945, that's exactly what happened. He was working as a moulder and press operator when, on the 12th of February, he chose to travel to Cowra and enlist directly at the barracks of the Recruit Training Battalion. Striking up a conversation with a young man standing in front of him in the recruitment line, the two would later become good friends. The young man anxiously waiting in front of him was Geoff Partridge from Kempsey. Three days later he marched into the Recruit Reception and was taken on the holding strength to complete his CMF training. His military record reflects that he was a Private in the 2nd Australian Recruit Training Battalion. However, it appeared as if Les Mather was destined for full-time service with the 2nd AIF. Because, after just five days, he marched out of Cowra's holding strength and formally marched-into the 2nd Australian Recruit Training Battalion ready to undergo his recruit training in the full-time Army. It wasn't until Thursday the 1st of March 1945, nearly two weeks after marching in to the Recruit Training Battalion, that his CMF number was scratched from his Attestation Form and replaced with his AIF identity *NX180219 Leslie John Mather.*

Leslie completed the physically rigorous and challenging training regime designed to turn him into a fighting soldier. Without any setbacks or complications he completed the training in the customary eight weeks. Persuasively convincing allocation officers of his suitability to become a combat engineer, Leslie marched out of Cowra on the 07th May 1945 along with other members of his battalion, which included Denby Grasby, Ivan Merritt, Ernie Poschalk, Colin Hurley and Frank Platt. Together the six men took the troop train to Wagga Wagga the very same day and marched directly into the 1st Battalion RAE. Bunked down in uninviting tents that appeared to the men to have been spasmodically erected around the Pomingalarna Range, the previous strangers were fast becoming good mates. Together, along with the other 8000 or so men making the camp home, they were looking forward to taking on the 'Mad Mile'. As young men they were certainly looking forward to the 'Lady Blamey'. For Leslie, after years of moving around with his family as a kid, he was happy just to be staying in the one place for a period of time, even if it was a little daunting due to the unknown perils of Kapooka.

Joseph and Margaret were present at the funeral and with deep emotion farewelled their young son. The family left a moving inscription on the 20-year-old's headstone:

'... *In Memory Of Our Loving Son And Brother His Duty Nobly Done*'.

Joseph passed away in December of 1970 in Sydney. Sadly big brother Raymond was in Moratai when his little brother was killed. He never made it back for the funeral. He survived the fighting and lived a full life until his death in Brisbane in 2003 aged 80.

25—NX180545 Colin Leslie Hurley (aged 20)
Plot B, Row B, Grave 7

Born on the 10th of December 1924 via a sleepy hollow known as Herons Creek, Colin Leslie Hurley was the ninth and final child of Mark and Elsie Hurley. The large family consisted of seven girls and two boys. The two youngest, twin girls Ella and Zelma, were born in May of 1921. With more than a decade and a half separating the entire Hurley siblings, much like the Pomeroy's, the Dilley's, and the Bartlett's, living in large families meant there was never any shortage of help. For Colin's mum who kept home to a husband and nine children, it was no different. Whilst mum kept the household ticking along, Mark Hurley was off working as a labourer, most probably in the thriving local timber industry, desperately trying to support his large brood. Like all young men of the time who'd left school early to find work to help support large families, the arrival of the war in 1939 changed the future of 15-year-old Colin. Although he had registered to attend secondary school, he was much more comfortable driving a car and a lorry than sitting at a desk.

Herons Creek was a small but important township on the New South Wales north coast approximately 180 miles north of Sydney. The early settlement of Herons Creek consisted of a series of grasslands, which were maintained to support timber bullocking teams. For more than fifty years bullock teams criss-crossed forests hauling logs to local mills set up beside the roadways. A sawmill had been established in the town in 1915 and by 1920, two existing saw mills had amalgamated to form the Herons Creek Timber Mill. With timber mill sites also located in nearby Wauchope, Bellangry, Kendall, Telegraph Point and Sancrox, there was plenty of laboring work to see the local families through Australia's financial Depression and pre-war years. The arrival of the northern railway, and the opening of a local station in 1915, heralded a period of innovative prosperity within the regional timber industry and growing dairy industry. A local butter factory was established in nearby Kendall adjacent to the railway station. It was here at the factory where a young Colin Hurley originally found work as a factory hand before he volunteered for the 2nd AIF.

At eighteen years and one month-old, Colin was conscripted under the *Defence Act* for compulsory Militia training. Enlisting in Kendall, nearby to where he was living at Herons Creek, he was passed medically fit 'A1'. But unlike some of the men we've met so far who found their way to Wagga Wagga and the Engineer Training Centre, Colin was a late inclusion into the ranks of the permanent 2nd AIF. Whilst his official military record incorrectly lists his enlistment date as 30th of April 1945, it may not have been his enlistment date but it was however a date which determined his destiny.

Colin actually registered for enlistment into the CMF in January of 1943. Traveling to Cowra, he marched into the 2 Australian Recruit Training Battalion for his compulsory training. Completing it successfully and concluding his national service obligation, he returned to Herons Creek and his job at the Butter Factory. But unlike other 18-year-olds, who acquired the taste for military service and enlisted directly into the 2nd AIF after experiencing Militia training, Colin didn't automatically transfer; instead he returned to Herons Creek to work. But something wasn't right; he'd gotten the taste of life as a soldier more than two years ago and for some unknown reason he wanted more. Perhaps it was to escape the clutches of the butter factory but whatever it was Colin returned to Sydney to enlist in the 2nd AIF.

Arriving at the Sydney showground on the 14th of February 1945, the brown-haired, blue-eyed 20-year-old officially enlisted into the Army. However he was yet to be transferred from the CMF to the AIF. For a short while he continued his identity under his CMF number of *N481442* Colin Leslie Hurley. Marching into the holding strength at the showground, he remained there for a number of days meeting other previously trained Militiamen with the CMF, and the 2nd AIF GR recruits. No doubt the confident Colin boasted about his previous experience, almost expecting to be fast-tracked through recruit training due to his prior military qualification. On Wednesday the 21st of February 1945 he marched into the 2nd Australian Recruit Training Battalion for a second time. But Colin unfortunately expected too much and he was required to undertake the obligatory eight weeks of further instruction. It wasn't until he was almost finished that he was given his up-to-the-minute official 2nd AIF military number: *NX180545.*

Much like his new mates in the 2nd Battalion, Denby Grasby, Ivan Merritt and company, Colin also had the pleasure of marching-out of Cowra on the 7th of May. Much like his latest new friends he too performed well in aptitude-testing and, even though he was merely a factory hand in a butter factory, he convinced allocation officers that he was suitable for engineers.

As the only direct member of the Hurley family to serve in either the Great War or WWII, Colin was a man proud of his achievements. No doubt his older siblings were envious of the baby of the family as he set out to adventure the hills and valleys outside of the family retreat at Herons Creek. But it was a dangerous environment which Colin had propelled himself towards. Wagga Wagga and the RAE Training Centre were busy unforgiving and hard places to conquer.

On Monday the 7th of May 1945, the very same day he left Cowra, Colin and the men on the troop train pulled into the Kapooka Loop and marched straight into the 1st RAE Training Battalion. Kapooka would test the former butter factory worker, as the transition from factory worker to combat engineer was a big undertaking. His transition still required some serious application to training in order to transform himself into the soldier he wanted to be. It was a notoriously dangerous place, but one which offered enormous challenges. Surviving Cowra, Colin bunked down for the night in the six-man tents somewhere deep within the belly of the bustling Kapooka Camp. Trying desperately to get off to sleep was proving difficult; the young man's mind wondered—'... what's day one gonna be like?'

Colin may have been the only Hurley family member to serve in either of the wars; but his loving family recognised his short military service and sacrifice. They inscribed his headstone with the soldier's farewell:
'... *His Duty Nobly Done Ever Remembered*'.

26—NX180218 Geoffrey Wilton Partridge (aged 18)
Plot B, Row D, Grave 6

Famous for its floods—due to the fact that the town stretches out around a long loop of the Macleay River, the town of Kempsey is located about 220 miles northeast of Sydney and approximately half-way between Sydney and Brisbane. It was the birthplace of Geoffrey Wilton Partridge. Geoff, as he was known, was born on the 17th of October 1926 to Stanley and Laurel Partridge (nee Jeffrey) just nine months before one of Kempsey's most famous sons was also born in the city; Australian balladeer and legend Slim Dusty.

At the turn of the century and into the pre-Depression years, Kempsey initially flourished as a logging and sawmilling centre. But, as the red cedar resource was exhausted, the town focused efforts towards dairying as its main source of commerce. The impact of the rich dairy industry resulted in a large Nestlé factory being established at nearby Smithtown. Geoffrey's father, Stanley, the sixth of ten children himself, was 22 when he married Laurel in 1925 and she was just 21. The couple gave birth to their first-born, Geoffrey, the following year. Although originally establishing himself as a labourer in Kempsey, his father and his family soon moved to nearby Smithtown, approximately ten miles upriver past the town of Fredrickson. Coming from such a large family Geoffrey's father, and his nine other siblings, expanded the Partridge family throughout the region. It was a rich family history, which commenced three generations prior to Geoffrey's father, and in fact the first Partridge descendants arrived from England as early as 1818. The family was soon synonymous and well-known amongst the Macleay River communities. Their names stretched from Kempsey to Port Macquarie on the east coast. But it was during the Great War when the community received their greatest contribution from the extended Partridge clan.

During the Great War, Mouquet Farm stood in a dominating position on a ridge that extended northwest from the ruined, and much fought over, village of Pozières. Mouquet Farm was the site of nine separate attacks by three Australian divisions of the AIF between the 8th of August and the 3rd of September 1916. Included in the fighting with the 13th Battalion were three Partridge men from the Macleay River district, brothers Thomas and Patrick and their cousin, Charles. Regrettably, all three were wounded during the battle. Thankfully they survived, but sadly, their 1st, 2nd and 4th Australian Division comrades sustained over 11,000 casualties. It was during later fighting in Belgium in 1917 that Thomas sustained a gunshot wound to his stomach. He later died of his wounds and was buried in France. His brother Patrick and their cousin Charles returned to Smithtown. It was a most difficult journey and homecoming without him and it would take another war for Patrick to forget the horrors of the Great War, which took his brother. But the Partridges were resilient soldiers. Patrick's son Frank Partridge, Geoffrey's cousin, would later go on to receive the highest military honour—the Victoria Cross, when the next war was fought.

Back in Smithtown of the 1930s the Partridge family settled into a somewhat routine life albeit ably supported by a large family network. After taking over a dairying factory site, the Nestlé factory became operational in Smithtown in 1921 and was a staple source of employment for the small town, especially through Australia's Depression years. It was here at the factory that Geoffrey's father found work and perhaps a place in history. Apparently in 1933 the popular Australian beverage, MILO, was developed and manufactured and Stan Partridge was there to make it happen.

Meanwhile the young family saw the arrival of baby Shirley who entered the world in June of 1932. Being raised in the small New South Wales town, Geoff and Shirley attended the local schools and regularly caught the punt across the river to the neighboring town of Gladstone. Growing up on a farm, with wide-open spaces, the Partridge children enjoyed their freedom. Stan, Laurel and their young children worked the small dairy farm on which they lived until the government made it compulsory for those living and working farms to stay on the land. That was about 1940. It suited Geoff. He became a proficient horseman and tractor driver, and assisted by his parents, the family could manually milk up to sixty cows, morning and evening. Needless to say Geoff and his family were hard workers. But Geoff's talents weren't restricted to back-breaking farming work. As an accomplished musician, he displayed an artistic flair not often seen in the milking sheds.

Along with a couple of his mates, Geoff would play the violin whist mates played piano at the local dances. Not only was he as charismatic as he was genuine, and musically talented, Geoff was also thoughtful and considerate. He was fast becoming a responsible young man with a big heart. He was especially thoughtful and considerate to his only sibling, his little sister. He was constantly looking out for her. On one particular occasion Geoff and his little sister Shirley were in town when Geoff's mates tried to coax him into the local for a beer. But this charming man knew too well the responsibility he had for his sister's welfare. Rather than enjoy a beer, he declined the offer preferring instead to walk his little sister home safely. He even built her a tree house on the farm for her to play in. As far as Shirley was concerned, he was the ideal big brother.

Much like the Partridge contribution to the Great War, the second war was no different. The family had little reservation answering the bugle call to arms and the honour of service to King and Commonwealth. For his parents, they were thankful that Geoff was barely a teenager and hoped the war would pass before their son was called for Service. But conscription and the Volunteer Defence Corps of the CMF prevented the family from escaping responsibility. After all, Partridge cousins had served with distinction and paid the ultimate sacrifice in the Great War; deep down they knew it was now their turn.

Although war was declared on Father's Day of 1939, it wasn't until May of 1942 when Geoff's father was called out for part-time duty with the 30th Battalion of the Volunteer Defence Corps. By this time Geoff had moved to Tamworth and was living with an aunt where he was working a farm growing potatoes and other fruits and vegetables, and working as an assistant in a local fruit market. Turning eighteen in October of 1944, with 10% of Australia's population already in uniform, and a victory in Europe inevitable after heavy strategic bombing of Germany, Geoff knew it was his time to support the Australian forces in the battle against the Japanese invaders. In January of 1945 he returned to the Macleay River one last time during the school holidays where he took Shirley down the street for an ice-cream. She didn't realise it at the time but it would be the last time she would see her charismatic and fun-loving big brother.

With his parents' permission Geoff set off on the 320-mile journey to Cowra to enlist. Leaving 12-year-old Shirley behind, he wasn't sure when he'd see any of his family again. Passing the medical and aptitude testing he swore his oath of enlistment just after his mate Leslie Mather. Given his official number, he was now *NX180218 Geoffrey Wilton Partridge*. Marching into recruit training, Geoff was allotted to the 2nd Battalion alongside men like Denby Grasby, Ivan Merritt, Ernie Poschalk and Leslie Mather. The men trained hard and experienced the rigours and hard work of recruit training together. Unfortunately for Geoff his early training days were a little distracted.

Word had got to him that his mum was ill and required major surgery. However, convinced their son was safe whilst training in Australia, Geoff's parents, Stan and Laurel, insisted he stay at Cowra whilst Laurel got her treatment. With news from the home front that all was well, Geoff threw himself into the training regime. But like any good son, he was forever mindful of his mum's health and his little sister's welfare. Engaging himself in the routine, albeit with his family constantly on his mind, Geoff and his latest colleagues continued with the daily grind of being transformed from civilians into soldiers. In between exhausting days and nights of physical and mental challenges, Geoff found time to buy a small present for his sister's upcoming 13th birthday. It was a small gesture, but one which meant the world to him. Amongst the thousands of soldiers, many displaying entrepreneurship qualities, he found one particular soldier in the camp who was well-known for making bracelets out of three-pence coins (colloquially known as a 'thruppence'). The loving big brother sent one home to his little sister. Shirley was missing him terribly.

Upon marching-out of recruit training on the 7th of May, Geoff had already convinced allocation officers that he was better suited to engineering than infantry, artillery, transport or even medical corps. Now, accompanied by his cheeky but reliant new mates, which also included Les Mather, the fruit and vegetable salesman, accomplished violin player, sound horseman and big brother to Shirley, was happy he'd made it. But he knew bigger challenges were to come now that he was nervously racing towards Wagga Wagga to become of all thing one of the Australian Army's toughest of the toughest—a Sapper.

Owing to Laurel's operation and hospitalisation, the family was unable to travel to Wagga Wagga to attend the funeral of 18-year-old Geoff. Sadly he was buried without his parents or little sister to farewell him. Following the tragedy Stan never spoke of the painful day he lost his son. Mum Laurel never read the newspaper reports and decided for the remainder of her life she'd never watch a war movie or read a book about the war ever again.

It wasn't until August of 1945 when the family finally made it to the Wagga Wagga cemetery to pay their respects to their son and brother. Staff from the RAE Training Centre took Stan Partridge out to the explosion site; however, Laurel and Shirley refused to go. As a family though they did visit Geoff's grave together. A little white cross, amongst a small sea of white crosses, indicated to his parents and sister just where their son and brother was resting. It was a sad forlorn reunion and an equally emotional and lengthy goodbye. They did get the chance to organise his inscription on a large white headstone some years later, which now reads:

'... *His Duty Nobly Done Ever Remembered By His Loved Ones*'.

Although he never really knew his cousin Frank Partridge well, before he died Geoff knew he was a member of the Militia from nearby Mackville who'd enlisted in 1942. He perhaps may not have known that Frank joined the 8th Battalion and was sent to Bougainville. In his untimely death, Geoff didn't have the chance to learn that his cousin would become the last Australian soldier to win a Victoria Cross in WWII. For outstanding fortitude and heroism on the 24th of July 1945, his cousin received the nation's highest honour.

Back in Smithtown on the New South Wales north coast, they named many streets after victims of World War II. In his memory and honour, the town now boasts 'Geoffrey Partridge Place'. His name also lives on in his sister Shirley's son—and her grandsons.

Chapter Twenty-two
A Fond Farewell

Plus 1—SX34062 Allan Raymond Bartlett (aged 18)

Allan Raymond Bartlett was the middle child of 13 children to Henry and Ivy Bartlett. Born in Adelaide on the 7th of January 1927, Allan was a rather shy child. Coming from such a large family, his quietness allowed him to retreat within himself. As a result he became scrupulously studious and loved working creatively with his hands. Although he wore a cheeky mischievous grin, he was known to be a very compassionate young man towards everything and everyone who knew him or had the pleasure to meet him. With more than a passing interest in horses, Allan would also grow into an accomplished and skillful horseman. In a post-Great Depression South Australia, Allan made the most of his skillful hands and would later become a talented carpenter—understandable considering the family home in Hart Street, Semaphore, was directly across the road from the Le Fevre Peninsula Central School, which would later become a Boys' Technical School during the war years.

I can't help but think that the life in a bustling home of thirteen children wasn't without its dramas. So when I read a report from the *Adelaide Advertiser* of July 1935 about the Bartlett family of Semaphore, it quickly validated my thoughts about how hectic the large household would've been. About 1:30 pm on 15 July, 3-year-old Barbara Bartlett was playing in the backyard of the family home in Hart Street. Apparently Barbara, and her four year-old brother, was trying to chop some wood with a tomahawk. However the young industrious pair only succeeded in severing little Barbara's right index finger right down to her second knuckle and lacerating her right thumb. Right then I knew that life in the Bartlett house would have been absolutely disordered—but no doubt a hell of a lot of dangerous fun.

Of all the family members in the large Bartlett family, Allan formed a close childhood and later adulthood bond with his older brother, Alfred. As the family grew, so did their relationship. Two and a half years senior to Allan, Alfred was the 'big brother' that Allan looked up too. But with a shock of red hair, Alfred looked nothing like his more sedate and 'refined' younger brother who had brown hair and hazel eyes. Because his head looked like a red house brick, Alfred was given the nickname 'Brick', a nickname that continued into the Army when, at 19, he enlisted into the 2nd AIF in July of 1943.

For Allan, his 18th birthday in January of 1945 was especially important. He'd finally reached recruitment age, and with Alfred already serving, it was inevitable that he found his way to the Wayville showground. With his parents' consent he enlisted on Thursday the 15th of February 1945. Working as a guillotine operator and still residing in the family home in Hart Street, Allan was roughly about the same age and possessed the same sense of adventure as his fellow South Australian, Denby Grasby. A mere 36 days separated the birthdays of Denby and Allan and, much like Denby, Allan had only turned 18-years-of-age just over four weeks prior to him being in the volunteer line at Wayville. Allan watched the dashing Denby, a man he would later get to know really well, get put through his enlistment process before him. In between Allan and Denby stood Walter Platten and Stanley Thomas. Stanley would later go on to work in Port Operations for about a year whilst Walter served for some four years in the Army as a driver in a transport platoon.

But it was Denby, standing three men ahead of him, that Allan would later travel to Cowra with and work together with during their recruit training. It somehow felt right that that these two men found themselves in the same recruitment line bound for Kapooka and hopefully a fight with the Japanese. Coming from similar backgrounds it was fitting that Allan was given his official Army number *SX34062 Allan Raymond Bartlett;* just three more official numbers later than Denby. The pair was very much alike—not only in age, but also character. They were nice, polite and family-oriented young men. They both shared the dream of sacrifice in service and making their families proud. Importantly, both young men had other family members already in uniform. Allan didn't know it at the time, but he, Denby and a more mature man the young men would later meet at Wayville by the name of Ivan Merritt, were all heading to Cowra together and their appointment at one of the Australian Recruit Training Battalions. Meanwhile Allan reported for duty at Wayville the following day and was granted three days leave without pay to go home and say his final goodbyes to family and friends. He was due back at Wayville at 0800 hours on the 20th of February.

Staying overnight at the Wayville Showground, Allan and his new mates Denby and Ivan departed South Australia the following day. Leaving Adelaide and his eleven siblings behind, Allan got to know his new mates even further on the four-day trip to Cowra. Living, eating and sleeping all things Army, the three men would soon get to know everything about each other. Allan had a lot to tell his newfound friends, especially about his large family and his elder bother. Marching in to the 2nd Army Recruit Training Battalion on the 24th of February, the somewhat shy and humble Allan gravitated more towards the more mature Ivan Merritt than the boyhood hijinks and charms of Denby during more than two months of recruit training. Ivan would help Allan through the challenging recruit-training regime in the dry dustbowl of the fly-infested Cowra. Denby meanwhile had the type of gregariousness that Allan needed but couldn't find in his humble self; he turned to Ivan who perhaps played the big-brother role that his real brother Alfred normally played. Despite their differences the three men would become good mates especially considering that Allan was also going to the RAE Training Centre, Wagga Wagga to become an engineer.

On the 7th of May 1945, Allan marched out of Cowra and on the same day marched into the 1st Battalion of the RAE Training Centre alongside Ivan and Denby. After more than two months together at Cowra together the men would have to put up with each other for a lot longer, especially now they were bunked-down in the same uninviting six-man tents on the windswept Pomingalarna Range at Kapooka.

Meanwhile, for Allan, in May of 1945, big brother 'Brick' wasn't too far away. He was in Bonegilla, Victoria attached to a transport unit. '…Perhaps,' he thought, 'by chance they might catch up with each other?'

Allan miraculously survived the Kapooka explosion by mere fate and the helping hand of a savior. But it was Alfred Bartlett who provided the most moving story of human spirit and brotherly love as he helped Allan through the immediate and lifelong consequences of the tragedy. It's a poignant story that will never be forgotten. One which reminds us of the power of family, of the special sibling love and bond of brothers and the legacy of mateship in times of trouble. The large Bartlett family was all that—and much more. They were blessed with two true heroes in their lives.

Nursed back to health by his loving big brother Alfred, Allan was too injured mentally and deemed unsuitable to continue serving in the Army any further. Allan was discharged from the 2nd AIF later that same year. But much like the men of his generation, those who witnessed the atrocities of war, Allan never talked about the explosion. Silently though he wept for the damage done.

Allan's photo became the photographic verification of the tragedy. The moving images of him openly weeping during ceremonial services provided the requisite story behind the scenes. Allan's story was the sad tale of an old WWII soldier who lost his mates in the most horrific of circumstances.

During my research I had the most humbling of experiences talking to Allan's son, Bruce. The way Bruce described his father; I sensed a real bond of powerful love and respect between the two men. Bruce advised me that his father was in his words ' ... a true gentleman.' Working with horses as a stable hand, he later put himself through night school to become a carpenter to provide for his family. Bruce supplied me with a photo of his aging father taken just a few short years before his death. Bearing the physical scars of the Kapooka events of 1945 on his left ear, the humble, no-fuss soldier got on with life. But for Bruce, the day he accompanied his ageing father to Wagga Wagga to commemorate the 50th anniversary in 1995, was a day he'll never forget.

Assisting his father in unveiling the very first commemoration on the explosion site—a plaque symbolizing public recognition of the tragedy—finally, for the humble man who carried a heavy burden of guilt for a long time, it felt a little lighter that day. Watching his father reminisce in a flood of emotion, it was a proud Bruce Bartlett who supported the arm of his father that day. Accompanied on the other arm by Alfred, the three Bartlett men paid their respects to fallen comrades, good mates and men they barely knew. They were all heroes in Allan's eyes. It was befitting that he could finally say goodbye to the mates he left behind after he returned to South Australia following his discharge. With a tearful parting, albeit fifty years too late, he bid his tent mates Denby Grasby and Ivan Merritt a fond farewell.

Regrettably, and with deep sadness, Allan passed away on the 29th of August 2005 aged 78. His death closed the door on the enduring legacy as the only eyewitness to the Australian Army's worst training disaster. Finally, his tormented memories of that fateful day sixty years ago were at peace. His brother Alfred 'Brick' Bartlett pre-deceased him in May of 1999 aged 75. Both Allan and Alfred are interred alongside their mum and dad, at the Cheltenham Cemetery, South Australia.

Conclusion

Kapooka temporarily ceased operations as a military camp after the war years, eventually closing for military duty in 1946. What was at best a temporary solution to a military problem, the camp quickly developed into the largest military camp in the Commonwealth. For the township of Wagga Wagga, their support was also a meaningful contribution and achievement. But the end of the war didn't mean the building and facilities at Kapooka would go to waste. Its second life was as a migrant camp. Most migrants arriving in Australia after WWII seeking a new life arrived by ship. They were disembarked in the major cities and transported to migration hostels in rural areas, often in former military barracks. Kapooka was an ideal location. Administered by the Department of Immigration until 1951, it became a long-term holding centre in rural NSW for the influx of immigrants arriving into Sydney. Unlike other military camps in the area, such as Greta, Bathurst, Cowra and Uranquinty, all of which became private property, Kapooka Camp was returned to the Army after it served its time as a transit migrant camp.

In 1951 Kapooka commenced its third life. Resuming its military status, the 1st Recruit Training Battalion was established on the camp grounds. During the Korean War, Kapooka was joined by the 2nd Recruit Training Battalion which occupied temporary buildings on the ridge south of the main camp. But the 2nd Battalion subsequently disbanded in 1953 and reformed in Puckapunyal many years later. In 1965—66, a more permanent camp was built on the site to train in excess of 10,000 national servicemen, as well as regular Army soldiers during the Vietnam War period. And when the Women's Royal Australian Army Corps School closed its gates in Georges Heights in 1985, Kapooka entered the challenge of mixed gender training. No longer were female soldiers imprisoned behind barbed wire fences, away from male recruits. Male and female Army recruits were finally being trained together as Kapooka became a co-ed training centre for all of Australia's Army recruits.

But long before migrants, Korean and Vietnam servicemen, and modern day recruits marched once again across the camp, authorities cleared up the area. At the end of its life as the Engineer Training Centre in 1946, traces of engineer habitation was subtly removed. Adding to the mystery of how this tragedy was forgotten by the nation, the barren land that was the expansive demolition area was returned to the owner, and the camp's spare land reverted to private farmland, just as it was when the Army first took control in 1941. This included filling-in the dugouts on the demolitions area. Underneath less than fertile grazing land, the scene of the country's worst military training accident simply disappeared—meaning, apart from a single gnarled box eucalypt tree standing lonely in a barren grazing paddock on San Isadore Road, the exact location of Jack Pomeroy's ill-fated dugout was quickly relegated to local memory.

Over the years, the former demolitions area was ploughed and farmed until it no longer represented its war-years' identity. In later years, a farming shed was erected on the site totally changing the identity of the paddock. In the end, the only real archeological remnants of its engineer origins were large concrete buttresses left behind when the engineers departed. And, of course, dominating the landscape was the scraggly lone box eucalypt which the G Coy squad enjoyed their lunch in the scarce shade of 21 May 1945. The site of Australia's worst military tragedy was fast becoming lost. But for how long?

It would take a further 45 years before the story of the Kapooka tragedy and the location of the dangerous demolitions area would re-surface. The tragedy needed an advocate. Unexpectedly it found two. One was a former WWII AWAS woman; the other, a Vietnam veteran and the then Commandant of Kapooka.

In the late months of 1991 into the early months of 1992, Commanding Officer of the 1st Recruit Training Battalion, Colonel Gordon Hurford, a larger-than-life Vietnam veteran (coincidently my CO when I was a young Army Recruit in 1991), had a chance meeting with a local Wagga Wagga woman at an AWAS reunion being held at Kapooka Camp. The woman just so happened to be a strong advocate for local community causes, including WWII affairs in the town. That woman was former AWAS driver, Sheila Oehm. During their meetings, Sheila advised Colonel Hurford that she was a munitions driver in World War II and was present when 26 Sappers were killed in 1945 after a horrific explosion at an area of Kapooka camp that had largely been untouched by modern-day development. She took the Colonel on a tour down San Isadore Road. Sheila stopped on the side of the road and using the gnarled box eucalypt and concrete buttresses as a reference, explained to the Commandant with razor-sharp precision her very location when the dugout exploded in front of her eyes. With emotion in her every word, she told the Colonel how the explosion killed many men she knew well, including Jack Pomeroy. She briefly told the Colonel how she was travelling back with the intention of going into the dugout at Jack's suggestion.

It was through Sheila's impassioned perseverance to modern day military authorities, including Gordon, that the tragedy at Kapooka should no longer be a quietly-kept shameful secret resigned only to a handful of local memories. She was determined to remember that Jack and the other 25 soldiers, many of whom she shared lunch with nearly 45 years earlier, had since been forgotten and she was determined to do something about it. She had been suffering in silence for a long time and called upon the influential Vietnam veteran to initiate some form of military recognition of the site to preserve the memory of the men she called friends. Not only that; it was time for Wagga Wagga to remember the fateful tragedy which tore open the heart of their town. Gordon Hurford was clearly more than obliging.

Using Sheila's memory and Gordon's pursuance of the tragedy's details, the actual location of the ill-fated dugout was still problematic for historians to pinpoint accurately. A thick layer of farmed land and tall grass hid any real reference points. They were concerned that the actual site would be lost forever. Vigilant military historians became involved in the project.

They turned their attention to the photographs taken during the experimental demolitions by the Court of Inquiry in May of 1945. Victorian historian David Mitchelhill-Green identified the positioning of the two large concrete buttresses in the 1945 images. During visits to San Isadore Road he studied the lay of the local land and discovered a small part of one of the two concrete buttresses remained in the same location as it was in the photographs all those years ago. Although the landowner had placed a shed in front of the remaining buttress, inadvertently masking their identity, dedicated historians were soon zeroing in on the actual dugout site. Then, using photos of the demolished dugout taken at the time of the explosion, it was clear that the same prominent eucalypt tree could be seen in the background of the 46 year-old black and white image. It gave historians, including David, Gordon, and Sheila, the most likely positive location. The tree that shaded the Sappers from the hot afternoon sun on 21 May was suspected of surviving the last 46 years. They stood staring at what they agreed was the very same tree. Combined with Sheila's 46-year-old version of events, historians were quietly whispering, with an air of confidence, that the site had been confirmed. The eucalypt was the same one which was closest to the demolished number one dugout. After serious consideration of the modern-day layout of the land, which now had significantly more trees, and relying on Sheila's memory of the fateful day, including her direction of travel in her munitions truck, the location of the ill-fated dugout was identified. Gazing at the tree, even today in photographs, there's an eerie calmness about it.

The 50th Anniversary of the opening of Kapooka Camp in 1992 brought many influential military officials and dignitaries to Wagga. Due to Sheila's initiative, Gordon, and the many families who supported a push to conduct a remembrance of the tragedy, gathered there on the 6th of April 1992. And, with the approval of the landowner, a granite plaque was mounted on one of the concrete buttresses. Commemorating the 26 victims and dedicating the site as significant for Army's history, the ceremony was unremarkable as far as military ceremonies go. Unveiled in a low key ceremony by the then Chief of the Defence Force, General Peter Gration, at 1400 hours, in the presence of Wagga Mayor Pat Brassil, Gordon Hurford and a small group of dignitaries, the tragedy now had a place where families could visit. There was however still one problem. The plaque and concrete block and ultimately the explosion site, was still on private property. As it was still a grazing paddock, public access to the site was prohibited. The actual site, for now, was unreachable.

Three years later in 1995, marking fifty years since the explosion, the Army held another ceremony. This time a local committee, working with a Federal program 'Australia Remembers', combined with the local historical society organised a Commemorative Service. Several hundred people attended. One of the curious visitors inconspicuously dressed in a smart suit wearing a cheeky 68-year-old smile that day was Allan Bartlett. Travelling with his own son Bruce and his savior brother Alf ('Brick'), it was Allan's first return to Wagga Wagga since he was discharged from the Army in 1945. For Allan, it was an emotionally overwhelming visit. At 68 he was getting on and he knew this might be the last time. In a moment of quiet reflection, he said his own special goodbyes. Wiping tears from his tired eyes he remembered the laughter of Denby Grasby and the maturity of Ivan Merritt, his two South Australian mates. He remembered Bill Cousins; remembering the moment the tall Corporal walked towards Jack Pomeroy that fateful afternoon and the horrific aftermath. He was taken back to the pain of hospital where nurses stripped his burnt skin from his body. But Allan held no grudges against anyone for his pain. Placing a wreath at the plaque he was able to say his final goodbyes.

During his requiem at the service, Padre Gary Kennedy said the Kapooka tragedy would've '... generated doubt about the Army and doubt about God in the minds of the victims' families'. Addressing the gathered onlookers, Kapooka's Chief Instructor, Lieutenant Colonel Mike Taarnby added, '... let men and women be reminded of the tragedy surrounding war lest all the suffering be in vain.' Many families of the victims, including Jack Pomeroy's daughter Maureen (now Mrs. Raunic), Bill Cousin's little sister Jessie, (now Mrs. Morley), Shirley Booth (nee Partridge) Geoff Partridge's sister and 'Toddy' Woods' sister Betty Murphy (nee Woods), proudly addressed the media after the ceremony. Maureen was quoted as saying that 'Jack' Pomeroy was the father she never knew. '... I have never been able to have that mourning process, but this service is something where I have been able to express my grief,' she said. Maureen mentioned that the service was difficult '... but it was one of the most important things I have done, it's been a healing process for me.'

Bill Cousin's sister Jessie was in hospital the day of the funeral in 1945 and regrettably wasn't able to farewell her big brother. Her attendance in 1995, fifty years later, allowed her to catch up with other families and the Titus family where her brother sought refuge on Sunday nights. '... I'm just overwhelmed by it all,' she said, '... I just feel I'm back to when he was a little boy.' Wearing her brother's medals at the ceremony she remembered her time in hospital when the names of the victims were being read over the radio, ' ... they turned the radio down so I did not hear much about it; they did not want to upset me.'

Like Jessie, Shirley Booth wasn't able to attend her brother Geoff's funeral in 1945, she was only 12, and her mother had been bedridden after a serious operation. She remarked, ' ... it was devastating for Mum and Dad, I don't think they ever got over it.' 'Toddy' Woods' sister Betty explained that the entire family was devastated by Toddy's death. ' ... My mother could never come to terms with it (his death), but over the years I have.' But the majority of the media attention during the 1995 Commemorative Service was focused on one person—Allan Bartlett. As the sole survivor, even fifty years later, people were still asking him what happened. Rooted in his lost memory, to the frustration of authorities over the years, the few seconds he remembered before the explosion will never provide a reason for the tragedy. For such a significant turning point in Allan's life, he lived a life of frustration that he wasn't able to tell them more.

In 2007, thanks to the agreement of all parties, the granite plaque was moved off the private property. It was moved to a more accessible roadside position in the shade of a new eucalypt, and now overlooked the cow paddock which was once the demolitions area. From its new roadside position visitors to the site could still peer into the former demolitions area to see the one real monument. The prominent eucalypt hauntingly dominates the landscape and provides a natural reference point in a field where so many dreams were lost. Battle-scarred, but resolutely standing upright strong and proud, the tree not only provided shade for the thousands of trainees who sought refuge from the sun whilst having a smoke on the barren demolitions area in 1945, it was in the bosom of its very roots where the fateful No 1 dugout was first dug. From the gum's prominent position, overlooking the ill-fated dugout, its leaves quivered when, in the most horrific underground hell deep within her roots, 26 men met their maker. Littered with flying debris of timber soil and human remains, its very foundations were weakened. But in the aftermath of such hell, the tree also witnessed the extraordinary power of the human spirit during the rescue and recovery of the helpless victims. Even to this day the grand old eucalyptus stands proud and provides a resting place for weary crows.

In April 2008, after being advised of the significance of the site and the details of the tragedy by David Mitchelhill-Green, *The Australian* newspaper contacted local Wagga Wagga historian Dr Peter Rushbrook, of Charles Sturt University. The collaboration of three men, which now included journalist Cameron Stewart who wrote the article '*Buried beyond history*', resulted in the newspaper printing the story and explaining how the event has since been ' ... airbrushed from official histories of World War II', and was, ' ... lost to popular memory outside of Wagga'. The story prompted the Army to respond in more detail. A military guard of honour commemorated the 63rd anniversary. It was a low-key service attended by a little more than eighty people. Regrettably, one key man was missing. Having said his goodbyes ten years earlier, Allan Bartlett had since passed away. He didn't get to witness this newfound interest in the tragedy and his incredible story of survival.

But amongst the appreciative public who read the 2008 newspaper coverage was Beth Neilson, the sister of Ernie Poschalk. Beth graciously thanked *The Australian* newspaper for its fitting coverage of the ceremony and remembering her brother so fondly. Many individuals vicariously connected to the Army's worst training debacle, whether by virtue of parents who were at Kapooka that fateful day or were seeking some understanding of just why 26 men buried in the Wagga Wagga War Cemetery all died on the same day, now continue to attend commemorative services. Whether they were the siblings of pall bearer, padres, munitions drivers, medical orderlies or just relatives of Sappers who were there at the time, the work of *The Australian* newspaper and the two historians has now given them a spiritual place once again connecting them with their loved ones and the tragic story of their final resting place.

By 2010, the Army injected much needed attention into the tragedy. It promised the families and all those affected by this incredible saga that every year the memory of the soldiers would be honored. Beginning at 1430 hours—symbolically marking the few minutes before the explosion—every year a special service at the memorial site would honour all those involved. Observing a minute of reflective silence at the actual time of the detonation, remembrance of loved ones lost in the eerily stillness of a May afternoon provides a deeply emotional experience. The 65th Commemorative Service included the unveiling of a dedicated permanent Kapooka Tragedy Memorial. Major General Steve Day and Major Craig Williams of Kapooka unveiled a fitting memorial stone. The stone was blessed by the Army Padre, and in doing so, spiritually declared the site a sacred military monument. Finally, the men and the Army's greatest accident were afforded the respect of spiritual relevance.

Also attending the 2010 memorial dedication were more than forty family members of the victims, some of which included Maureen Raunic and 'Porkie' Nixon's much-talked about daughter Neryl (now Neryl Hogan). The day was highlighted by the performance of the song '*21 May 1945*'. Written and performed by violinist David O'Neil and RAAF College Group Captain Bob Rodgers, it's a solemn dedication to Jack Pomeroy and the day it all went horribly wrong. Its haunting and heartfelt lyrics reduced many onlookers to uncontrollable sobs of pain. You can find the lyrics in the opening pages of this book.

In 2012, the Army built a commemorative enclosure across the road from the former demolitions range. Allowing for car parking and a more accessible opportunity for the public to visit and pay their respects for the 26 men, the site is an important part of the nation's military story. But they're not just any '26 men'. Now their names, their identities and their lives are there finally correct and for all to see. A tree has also been planted for each soldier killed and a name plaque placed in front of each of the trees representing their identity and life. Today the trees grow strong and the new memorial stone allows visitors to the site to reflect on the horror of 21 May 1945. The stone reads:

> It was the Australian Army's worst accident, a tragedy so grim and gruesome it tore open the heart of a country town.
>
> 'On the 21st of May 1945, in a single blinding flash of gelignite, 26 young lives were snuffed out in an underground bunker.
>
> When they buried the victims three days later, half of the population of Wagga Wagga—7000 men, women and children-lined the streets to bow their heads at the passing parade of coffins. It remains to this day the Nation's largest military funeral.
>
> But then something strange happened: Australia Forgot.
>
> They will remain in our hearts and minds forever.
>
> Lest We Forget.

Author's Post Script

'Roll Call'

In the wake of World War II, hundreds of thousands of Servicemen and Servicewomen had to adapt to life as civilians albeit in a changed world. They knew the years ahead would be challenging, especially for the many families of men, women and children who'd lost loved ones during Australia's six years of fighting. Of the almost 40,000 killed, and the many more thousands injured, maimed, or mentally impaired, life was never the same. For those families living in the aftermath of the Kapooka tragedy, much like the lives of others who lost loved ones, life was extremely empty and always disrupted with those lingering unanswered questions. Whilst many were able to put the horrors of Kapooka behind them, the events of that day left immovable scars on so many others.

Although I haven't been able to establish the post-war stories of every man and woman involved in this horrific event, I've included some of the more poignant reflections on their lives following the tragedy. Titled 'Roll Call' it's my summary of the precious lives of some of the important figures dealing with the aftermath of Australia's greatest wartime training accident.

Sergeant Colin Kendall: Following the explosion, Colin was admitted to the 54th Camp Hospital having sustained minor burns to his face, hair and eyebrows. Treating doctors also diagnosed Colin to have sustained severe nervous shock. Medical staff established that the Sergeant would be incapacitated for a period of four to six weeks, thus preventing him from returning to duty. According to doctors, Colin would have permanent ill effects of a 'mental nature'. However, less than two months later, Colin's mental anguish was somewhat delayed and it was replaced with the gift of life. His wife Phyllis gave birth to their daughter, Carol.

At 31, totally and bravely disregarding doctors' advice and the horrors he'd already experienced, Colin wasn't ready to leave the Army. Instead, he transferred to the Dental Corps. Colin continued to serve the country for another 24 years. Fittingly the Dental Corps motto was, *Honour the Work*. On the 1st of January 1969, honouring his work, the 55-year-old's name appeared in the *London Gazette* on page 38 in the New Year's Honours List. For his meritorious military service it read:

> THE QUEEN has been graciously pleased, on the advice of Her Majesty's Australian Ministers, to approve the award of the British Empire Medal (Military Division) to the under mentioned *28374 Sergeant Colin James Kendall*, Royal Australian Army Dental Corps.

Sergeant Roy Tafe: The heroic Sergeant was discharged from the 2nd AIF on the 6th of December 1945. He lived a full and rich life as a devoted husband to Grace, his two children, five grandchildren, and nine great-grandchildren. Roy passed away at the Pittwood Nursing Home, Ashfield, on the 12th of July 2011 aged 93, more than 65 years after the day he pulled Allan Bartlett from the demolished dugout. His caring spirit was still active in his passing. In line with his wishes, in lieu of flowers, donations were made to the Heart Foundation.

Captain Edward Merry: Worldly and experienced, the resilient Captain enlisted in 1940 after working as a young man in Tasmania where he met his wife Lillian. In 1928, when Ed was just 23, the couple celebrated the birth of their first-born son John. But Ed and Lillian Merry quickly matured as a couple and learnt that life was tough as their young son only survived for a couple of weeks after birth. It was a tragedy that created a steely resolve in the young, soon-to-be soldier. Prior to his joining the Army Lillian gave birth to three girls (Heather, Helen and Delba between 1930-1938), so when Ed took up his position at the RAE Training Centre and was in charge of young men like Kevin Pierce and Allan Bartlett, these type of boys reminded him of his eldest daughter, Heather. He'd no doubt looked upon the younger soldiers with that same sense of paternal concern for their welfare as he did with his own offspring.

The brave and well-organized OC of Golf Company 1st RAE Training Battalion finally left Kapooka camp and was posted to the Headquarters of the 7th Australian Division. Eventually he discharged from the Army as a Major. Later moving to Queensland with his young family, Ed Merry set up a grocery store in Cooroy, inland from Noosa on Queensland's Sunshine Coast and worked the store into retirement. Ed Merry passed away long before his time. At aged 64, he passed away living in Mitchelton, in north Brisbane on the 10th of December 1969. Ed Merry is buried at the Pinaroo Cemetery, Albany Creek.

Researching this book I had the pleasure of talking to Ed's second eldest daughter, Helen, and his granddaughter Fiona. Helen and Fiona both described Ed as a lovable and humble man who rarely spoke of his Army career. I had the absolute pleasure of advising them that their father and grandfather organized the recovery, with immense dignity and honour, of the 26 deceased and survivors from the barren demolitions area that fateful day. In addition, I also took immense pride in advising them that Ed Merry was responsible for coordinating and assisting Archie Smith in organising the largest multi-denominational military funeral in Australia's history. I was humbled by the opportunity of meeting the surviving family of this man; in my eyes and no doubt the eyes of a nation, Ed Merry was a national hero.

Captain Archibald William Leslie Smith: The former shopkeeper responsible for one of Australia's most horrific identification parades transferred from the 4th Bn RAE to the 1st Bn RAE in June 1945 where he was appointed as the Adjutant. Within two months he was hospitalized at the camp with a severe case of bronchitis, and, finally, after 1482 days in uniform,

Archie Smith was discharged from service in November of 1945. After putting the horrors of the war behind him Archie got on with life. He silently passed away on 24 October 1989 in Endeavour Hills, Victoria, aged 75. In contrast to the chaotic scenes he witnessed at Kapooka's makeshift morgue 44 years earlier, Archie was cremated and his ashes buried amongst Springvale's 'Garden of No Distant Place'. It's a fittingly beautiful, serene resting place where he's surrounded by plants, flowers and bridges over gentle streams.

Brigadier Alexander Forbes: His military career ended when he retired on 1 December 1946. In 1961 to mark the 50th anniversary of the Royal Military College, Brigadier Forbes' original Sword of Honour was presented to the Corps of Staff Cadets by his son, the Honourable Doctor Alexander James Forbes, CMG, MC. His son was himself a graduate of the RMC, holder of the Military Cross for gallantry and Minister for the Army during a distinguished military and political career. Currently, on all ceremonial occasions at the Royal Military College, Duntroon, the Battalion Sergeant Major carries Brigadier Forbes' original sword as the first Sword of Honour.

Sheila Oehm: The munitions driver became a long-standing member of the Wagga Wagga community, in particular the Enamellers Association, until she found a passion for growing orchids. She passed away in March of 2009 and was laid to rest beside her late husband William, who died in 1963 of injuries related to his World War II service. The couple are buried in the Wagga Wagga Cemetery, a short distance away from the victims of the tragedy she helped recover 64 years earlier.

Thomas Musto: 1944 and 1945 were bad years for the 41-year-old mature soldier. Sadly his father passed away in 1944 followed by his mother who died just one month after the horrific Kapooka tragedy which scarred her son, in June of 1945. For Thomas, the munitions storeman discharged from the Army as a Sapper from 2/13 Australian Infantry Battalion on 20 November 1945. After his discharge he resumed working as a postal worker in the Wagga Wagga district for many years later. He remained married to Mary and was still living and working in Wagga in the 1980s. Thomas died in April 1988 aged 84 and just over ten years later his wife passed away in October of 1998 aged 91. They're buried together at the Wagga Wagga Lawn Cemetery.

Gordon Hurford: The retired Colonel and Vietnam Veteran is alive and well and living on the Sunshine Coast in Queensland. My former Commanding Officer, as a career military man, is never 'not' working. Currently, in supposed retirement, he's the Chairman of the 2nd Battalion Royal Australian Regiment Reunion Committee. In 2012 when we met at the Maroochydore RSL, we shared a coffee and a laugh at the coincidences of life. Based on his earlier work, and a chance meeting with Sheila Oehm, he told me the amazing story of how the accident site was rediscovered. It was a part of the story that stood to correct my earlier erroneous assumptions about the details of the tragedy and where it took place. Once again, the wise words of my former CO had me captivated. He not only steered me in the right direction but also encouraged my endeavours to write about the forgotten events.

Keith Kuhn: Keith is now vision-impaired and has difficulty writing, but I had the honour of speaking to the 92-year-old ex-Digger and his wife Antoinette via telephone in late 2012. The couple lives in the tropical heat of Darwin. Keith was more than glad to talk about his days at Wagga, especially during the tragedy when he remembers ' ... people were running around everywhere in a panic.' I was certainly humbled to speak to this ageing Digger. He's a true character and a legendary soldier.

'Hank' Keenan: The 87-year-old former Sapper responded to a request I'd placed in the *Vetaffairs* Newspaper seeking information about the tragedy. Hank wrote a short, scratchy, one-page letter explaining that he was there on the day and sent it to the Office of the Department of Veterans' Affairs. It was graciously provided to me and I finally made a phone call to the colourful and charming man in early 2013. My phone call found him living in Reservoir, Victoria. According to Hank, the explosion occurred because the trainees were smoking.

Des Surkitt: In early 2013 I also caught up, via telephone, with Des who now lives in Sale, Victoria. The 89-year-old former Sapper 'bent my ear' for about an hour. It was the most sobering and memorable hour I'd had for a long time. Lucid and very well aware of the tragedy, Des was on parade the following day when the Sergeant Major read out the names of the victims during the Company morning parade. Looking for volunteers who knew the victims, it was Des who packed up Toddy Woods' equipment in the tent. Humbled by his strength and resolve after so many years living with the events of that day playing over and over in his mind, in the finish I was disappointed the phone call had to end.

Des moved to Sale in 1953 where he took over a small engineering business. After a heart scare in 1990, a pacemaker momentarily slowed him down. Then, on New Year's Day 2010, he had to say goodbye to his wartime bride of 65 years, his darling wife Patricia. Thankfully, Des has been one of my greatest confidants along this journey; even proofreading the first manuscript. Des recently attended the Memorial Service for the 69th Anniversary of the tragedy at Kapooka during which he was introduced to and formed a special bond with another Kapooka Sapper Paddy Cranswick. Well into their 80s, the two men made a promise to each other to make sure they made it to the next memorial service in one year's time. They left Wagga Wagga with high hopes they could hang on for another year.

Today, on his third pacemaker/defibrillator, Des still drives a car and rides a motorbike, much of which, I might add, is to the chagrin of his daughters.

Stan Emery: It was 86-year-old Stan Emery who told me about Jack Nixon and his nickname 'Porkie'. After speaking with Stan over the telephone I wouldn't be surprised though if he was the cheeky bugger who gave Jack his unwanted title. Once again I was left humbled in amazement by the clarity of the memory of men like Stan Emery. He was sharp, concise and kept me on my toes. Almost 68 years after a particular demolitions lesson given by Jack Pomeroy in early May 1945 an alert Stan could still vividly remember his instructor that day. Describing Jack as a tall, fit, good-looking man, he said he was well-spoken, a true professional of his craft. He further labelled him as a man who was easy to get along with and a real 'decent bloke' whom he trusted implicitly with his life during dangerous lessons.

Stan did however seek my help. He wanted to know if baby Neryl, Jack's little princess was still alive and well. In a sense he too wanted some closure on this tragedy. Jack was always flashing this photo around of baby Neryl and Stan could see and hear the pride in Jack's voice when he spoke of his little girl. He essentially left me with a task to find Neryl quickly and put her in contact with Stan before he too passed away. He's spent everyday of his life since the disastrous events worrying about this little girl who'd lost her father so very young. I was of course proud to help out this old Digger. He'd spent the last 68 years worrying, it was the least I could do.

Paddy Cranswick: Not long after the explosion and formal identification of the victims Paddy was advised that Geoff Partridge didn't heed his advice—instead he returned to the demolitions area and became one of the victims. Paddy refused to go to the funeral. On the day of the funeral he was on a train bound for Adelaide to see his ill sister. When he arrived back home, less than a week after the explosion, he claims no one in South Australia knew anything about it. That's the period in his life when Paddy Cranswick shut out the tragedy. He never heard another word about it—ever.

I had the honour of speaking with the former Sapper via telephone in July 2013. I was put in contact with 87-year-old Paddy via Tom Locke representing the Department of Veterans' Affairs in Perth. Tom warned me to be gentle with Paddy; his age was fast catching up with him. My instructions were to speak slowly and carefully. Bloody hell Tom—you were wrong. Paddy Cranswick could talk the leg of a wooden chair. For 45 minutes I listened to him, mesmerised by his stories of a bygone era.

Although I introduced myself to Paddy at the start of our conversation, by the end of the 45-minute phone call, he couldn't recall my name.

But like most soldiers of his age, Paddy has significant and lasting memories of that fateful day and his run-in with Geoff Partridge in the tents. For well over fifty years Paddy's memory never went back to Kapooka. However, only more recently, he woke from his sleep drenched in a pool of sweat and startled by thoughts of Kapooka. Finally he was back. His recent subconscious return to Kapooka had heightened his memory, which was now extremely sharp in the details of 'that' day. He even knew there were apples and oranges available outside the canteen on the morning of 21 May. ' ... Not a day goes by when my memory doesn't take me back to Kapooka and that day.' ' ... I remember exactly where I was standing and what I saw.' Everyday it's the same painful vision—walking to Headquarters, hearing the explosion and the crackled message for help.

Much like the troubles Sheila Oehm experienced in reconciling her involvement at Kapooka before she passed away, Paddy is still battling with DVA for help. His recent reunion with Des Surkitt at the 69th Anniversary Memorial Service, and his pact with Des to be there in 2015, has given Paddy a sense of healing. Maybe he can finally let go? I look forward to meeting him in 2015.

Allan McPaul: Allan responded to my post on the noticeboard of the *Vetaffairs* newspaper almost ten months later. He managed to send me an e-mail where he explained that he was a pallbearer on that fateful day. Along with almost 155 other young soldiers, he'd managed to lock away the memory of the nameless wooden coffin atop his young shoulders being herded towards a hole in the Riverina soil. Time had eroded his memory a little, and he didn't really recall whose coffin he carried that day but after he told me that he knew the guy had red hair, straight away I knew he was talking about Kevin Hurst. The mere mention of his name allowed Allan to go back more than sixty years to the day he met Kevin in the tent at Kapooka.

Today 86-year-old Allan lives in a retirement village in South Brisbane.

Sir Thomas Blamey: It was the insistence of Prime Minister Menzies, who'd only returned to the Prime Ministership in 1949, that an ailing Sir Thomas Blamey was restored to active duty in order to be promoted to the rank of Field Marshal. Prior recommendations by Menzies to recognised Blamey's contribution had been mooted. This time the bedridden ex-Commander-In-Chief was promoted in the King's Birthday Honours of 8 June 1950. Presented his Field Marshal Baton by the Governor-General in a bedside ceremony, Blamey became the only Australian serviceman to achieve the highest military rank. Sir Thomas died of a cerebral hemorrhage at the Heidelberg Repatriation Hospital, Melbourne Victoria on 27 May 1951. Needless to say Blamey's lifetime achievements, which included his family's enduring legacy within the Riverina district, fittingly declared Sir Thomas Blamey one of Wagga Wagga's favourite sons. In years to come, honouring the legacy of an iconic local figure, Kapooka Camp was officially opened on the 6th of December 1966, ironically the anniversary of his son Dolf's death, where the Camp was named in his honour—Blamey Barracks. At the official opening ceremony, Sir Thomas Blamey's wife and his second son, Lieutenant-Colonel Blamey, were guests and witnessed the seven million dollar camp become the home of the 1st Recruit Training Battalion. Today Kapooka is known as 'The Home of the Soldier'. Recruits are taught the skills of weapon handling, field craft, physical fitness, drill, soldierly qualities, navigation, first aid, character guidance, development of self-discipline and the standards expected of a soldier representing everything that embodied the life of Thomas Blamey.

Jack Roach: After watching the development of Kapooka Camp, including the thousands of soldiers training on his leased land, on 12 April 1951, aged 85, Jack Roach, the last of the original owners of the Kapooka site passed away. His brother David had previously died in 1949. Showing generosity in death, thankful and appreciative of his good fortune at the hands of a needy government and the Army in a time of crisis during the war years, Jack Roach left his entire fortune, a total of 149,000 pounds, to charity. One year later, the station and land that had provided Kapooka's initial home base, following the generosity of one of Wagga Wagga prominent families, was soon to be relinquished. The enviable relationship between Kapooka and the Roach family was over. By 1952, following Jack's death, Moorong Station had seen the last of the pioneering Roach family as the land was listed for public auction.

Author's Footnote

Little did I know at the time of starting this project that my time as a soldier undergoing recruit training at Kapooka in 1991 had merely connected me *physically* to this story. By the time I'd finished researching and writing this story I'd discovered an unexpected *emotional* and *personal* connection. This story and its deeply-hidden family heartbreak was soon taking me to emotional places I hadn't expected or for that matter experienced before. It was a connection that came to me out of the blue, and as a result, had stunned me due to the unbelievable coincidence of it all. All along I felt an unusual linking to the story. As a former soldier there's always someone who knows someone who's a soldier. What I didn't expect was a family connection to one of the victims. In the most extraordinary of circumstances, my emotional and personal connection related me to none other than the larger than life Jack Pomeroy.

Pomeroy Family: I was born in the regional Victorian former gold-rush town of Ballarat in 1967. My family origins on my mother's side all hail from Ballarat North. Coincidently, it was the same side of town where the family of Jack Pomeroy settled shortly after burying their husband and father in Wagga Wagga. Intrigued by what I thought was some nonsensical sixth-sense, something motivated me to find out more about Jack's wife Dorothy and the four surviving Pomeroy children. Their whereabouts and welfare became something I *had* to know to close off that part of my research.

I began believing and hoping that by some miracle of fate, my family or I knew the Pomeroy family growing up in Ballarat. After all Ballarat North wasn't a thriving metropolis; surely my mum at least knew this family?

Previously I had read Jack's military file and later established the whereabouts of his three sons, but he had a 10-month old daughter when he was killed. During my primary research I couldn't find her. Quite strangely, it was a simple twist of fate that had me believing my nonsensical sixth-sense was somehow real.

It was on Christmas Day 2012 when I finally learnt what it was that interested me about this family, in particular, Jack's young daughter. It came to me as I was sitting with my family on the soft beach sand of Mudjimba; a small surfing town flanked by Maroochydore and Coolum on Queensland's Sunshine Coast. Seated beside me on my right was my wife and children, and on the left, my mother and father, Les and Eileen. Basking in the beautiful warm December morning sunshine of yet another hot Australian Christmas Day, we were all watching the small surf touch the beach, casually engaged in family conversation. My parents had travelled up from Victoria for the week to spend some time with my wife and I and our three adult children. Following the idle pleasant two-way conversation about how we were all doing, my mother's conversation turned into an inquisitive one. ' ... What are you up to these days?' she asked. It was then I began discussing my research and the progress of this book. I wanted the conversation to lean towards the unexplained attraction that left me wanting to know more about the Pomeroy family. Cautiously, I tested my hypothesis about my mother and her knowledge of the story and the family. The conversation soon lead towards Jack and the fact that his widow and their children moved to Ballarat following the tragedy.

My mother was born, raised, and schooled in Ballarat during the 1950s. I mentioned what I knew about the family. By then I was just hoping to get some positive feedback that I could work with to try and discover the whereabouts of Jack's baby girl, Maureen. Considering the Pomeroy children probably lived close by to mum's home in Howard Street, Ballarat North, perhaps the Pomeroy family knew the Muller family, and vice versa! After all Jack's daughter was about the same age as my mum.

As soon as I mentioned the Pomeroy surname something in my mother's eyes ignited. I'd stirred something. Quizzing me she asked with a sense of trepidation ' ... what was his daughter's name?' Hearing the expectation in her voice the hairs on my neck began to stand up and emotion was soon welling in my eyes. ' ... There were four children, three boys and a young daughter Maureen,' I said. Suddenly, it was as if Moses had parted the Mudjimba surf. My mother looked at me with *that* look; the one that instantly grabs your attention producing a stunned open mouth that struggled to get words out. Right then I knew the tragedy at Kapooka was a little bit more personal. Perhaps my sixth sense was real after all?

Sensing my mother was about to enlighten me with a snippet of coincidence, the type you automatically think ' ... geez it's a small world', she managed to close her mouth and inform me that her best friend at St Columba's School in North Ballarat in 1955 was none other than Maureen Pomeroy (now Maureen Raunic). It was something I didn't know about my mother when I began this journey. Apart from the remarkable coincidence, the unexpected news hit me hard. My bottom lip was now sitting in the warming sand. I was not only gob-smacked, but dare I say it, a little touched by sadness. As I gazed outwards towards Mudjimba Island, the reality of the tragedy hit me. This couldn't be true. Surely it's not the same woman! Deep down I didn't want it to be. Only because I didn't want this story to be a sad moment in my mum's life.

When we all realised it *was* the same woman, I was once again a little emotional. This time though I tried to hide my emotion. Why? I didn't believe my family would understand what this discovery meant to me. For months and months I'd been researching this family. Desperately trying to find a family point of contact, more importantly I needed to know just what happened to Jack Pomeroy's little 10-month-old baby girl after his death. As the youngest child, I knew she'd still be alive and I wanted to talk to her to learn more about the heartbreaking story of the responsible father and family man who went off to work on his 31st birthday and sadly never returned. Now, by sheer fate, on a hot Christmas Day, the whereabouts and story of baby Maureen was delivered to me—by my own mother. From the unimaginable source, I now had a living breathing witness to the family's life in the aftermath of tragedy.

Spending years together at school, my mother knew that her best friend's father had died in the war, but over the years, as a large family we never really sat down to discuss my mum's early schooling and who her good friends were. So, on this particular Christmas Day, I went from researcher to storyteller. For the very first time I told my mother the tragic story of Maureen's father—Jack Pomeroy, and the gruesome circumstances of his tragic death along with 25 other soldiers in his care. It was the first time in nearly sixty years my mother heard the truth about her friend. I told the story of a special man they called 'Jack', who was not only a loving father, but also a dedicated soldier and exceptional demolitions instructor. It was the sad story of the loving father who went off to work on his 31st birthday, but never came home to a family all waiting to share birthday cake with him. Mum went on to explain that as school kids she and Maureen regularly spoke about her father being deceased and that her brothers Barry, Les and Francis were well-known at the local YCW along with my uncles. As we all sat in the sand dissecting what I'd just shared, I think my family finally realised why this story renders me emotionally fragile.

By sheer coincidence, amazingly, the Johnston and Pomeroy families knew each other. Here I was writing an emotional book about the tragedy that claimed the life of my mother's childhood friend. Bloody hell—it really is a small world.

In researching and writing this book I didn't think I could trump the emotional closeness of my mother's involvement. But when I was responsible for an emotional reunion, almost seventy years in the making, suddenly my therapeutic journey of purpose and discovery I'd been looking for was fulfilled. It had to do with Jack Nixon's daughter and a promise I made to a WWII veteran.

Neryl Hogan (nee Nixon) and Stan Emery: In March 2013, once again I got personally caught up in the emotion of painful memories of that fatal day in May of 45. When Stan Emery contacted me and advised me that his tent-mate at the camp in 1945 was one of the victims—Jack Nixon, I made a promise to him that I couldn't, and didn't want to, ever back down until it was fulfilled.

You see Stan explained that Jack incessantly carried a photo and spoke feverishly to anyone who'd listen of his baby daughter, Neryl. Understandably, as his only child, Neryl was Jack's pride and joy. But when Jack was killed in the 1945 explosion Stan was burdened with the uncertainty of just what happened to Neryl. Every day for 68 years the face of that baby girl interrupted Stan's life. Accepting his twilight years were upon him, Stan expressed to me in a moving conversation that he wasn't prepared to pass away without first knowing that Jack Nixon's baby girl, who was so proudly impressed upon him, was, in his words, '... okay'.

When I finally found baby Neryl Hogan (nee Nixon) 68 years after the events which claimed her father, she was a former schoolteacher living a quiet life in Canberra. I was determined to honour Stan's wishes in passing on his heartfelt message to Neryl before he passed. I set out to reunite the former soldier and his mate's baby daughter. For this moment at least I wasn't just writing a book. I was scared of the responsibility, but at the same time humbled and deeply proud of the fact that I was about to make a significant difference in the lives of two families, both of whom I've never met face-to-face. But I knew they weren't just two ordinary families, they were two families that had been tormented by this unbelievable military tragedy for nearly seventy years.

After speaking with Stan, and later finding Neryl, I accepted that I had an important, and no doubt deeply moving, phone call to make.

Throughout my research I had always pictured Neryl as a 10-month-old baby girl who'd lost her dad in the tragedy. I never pictured her as grown women, not just any woman, but a survivor who'd lived a full life despite her tragic start. Emotionally for me it was always going to be a difficult task—so difficult in fact I kept putting it off. For days I was reluctant to dial the number. I knew how significant it would be for the two families and the emotion it would stir up. I knew just speaking to Neryl and passing on a 68-year-old message would affect me. It was hard not to be caught up in the emotion of a much-anticipated reunion in such tragic circumstances. I knew I'd break down into an emotional wreck on one end of the phone and therefore struggle to get the message out through my quivering chin.

After I eventually made the call I began to think a little more rationally. 68 years had since passed and Neryl was now the mature woman on the other end of the phone. We exchanged pleasantries and I introduced myself. I proceeded to explain to her what it was I was trying to achieve in writing this book and recognising the contribution of the 26 men including her father. Inevitably we briefly spoke about his tragic death but Neryl couldn't tell me much about her father, as she was so young. But that didn't dismiss the emotion I sensed in her voice. He was still her dad and although 68 years had passed she missed him terribly. I was nervously reluctant to tell her about Stan and his message. Not because of Neryl's emotive reaction, but more about mine. I knew then that telling Neryl of Stan's message of love and support would not only upset her but would turn me into an emotional wreck. As expected my emotional dam had been breached. Overcome with sentiment I advised Neryl that after nearly seventy years of worry, Stan Emery was looking for her. Thinking about her every day for seven decades, he just wanted to make sure she was okay. He needed to talk to the woman he knew so much about as a baby, even though they'd never met. In all my years as a military investigator I've never delivered such a moving message. I provided Neryl with Stan's details and told her I'd be more than happy to hear about her reunion and she could get back to me anytime day or night. Pensively, I waited for Neryl's response. Deep down I hoped the two families could finally meet face-to-face to look into each other's eyes and find a newfound inner strength and peace in each other's company.

Not long after, in fact only a couple of days, I received an e-mail from Neryl. Their telephone reunion took place. She described Stan Emery as a most charming gentleman. Stan began filling her in on some of those details about her dad especially the good times they had together at Kapooka. After reading her e-mail I was left feeling a sense of indescribable humility at being able to unite these two kindred people. Two people whom I've never met face to face, they've never met in person, yet are connected through the most precious of circumstances—Jack Nixon.

As I concluded my research into the circumstances leading to this historically discomfiting tragedy, I knew it was time to bring the tragic events of 21 May 1945 out of the dark recesses of Australia's forgotten past as soon as possible. I became more convinced that I *wanted* to tell this story and do it sooner rather than later. Then, after speaking with the families of the victims, like Neryl Hogan and Maureen Raunic who've lived with the painful memory of loved ones killed in a manner not fitting for a soldier; dishonorably and unceremoniously forgotten, I was touched with a profound sense of obligation. The tragedy at Kapooka instantly became the one story I *needed* to tell. It quickly became that nonfictional event I'd been looking for to research, analyse and importantly, write about. What I didn't know when I set out on the journey was just how much the story would affect me personally.

I also found some valuable quiet time to reflect on what I had achieved in instigating Neryl and Stan's emotional reunion. It was all about family and loss; and now a future.

Acknowledgements

For me, this story is a sad reminder of the human cost, the pain of loss and the sacrifices made by men and women who've served, or are still loyally and faithfully serving the nation, as members of Australia's military forces. But acknowledging their selfless sacrifices as sailors, soldiers, airmen and airwomen, a nation should also salute and remember the uncalled-for sacrifices foisted upon their most loyal and loving supporters—families. For them, the cost is always immeasurable.

For my part, bringing the story of the Kapooka tragedy back to life, including the identity and lives of its 26 victims, I carefully, humbly but always proudly and compassionately, listened to the tales and warm, heartfelt memories offered by families, friends and old 'Diggers' who frequently declared ' … I was there that horrible day.' Describing some of the victims as forgotten heroes, mates and adventurous larrikins, as nothing other than *cheeky buggers*, their short lives left an indelible mark in the hearts and souls of friends, family and Army mates for eternity. Many of my ageing contributors to this story are well into their twilight years. Reminiscences brought laughter and marked very fond memories of family members and former colleagues, and were a reminder to me of a life well-lived. Sadly though, our conversations had to focus on the inevitable in this story. In doing so, I forced you to conjure up horrible examples of news and events, which quickly extinguished your jovial spirits and often brought an uncomfortable silence. Painfully inducing memories of a most dreadful time in your lives, the reality of the tragedy which occurred almost two generations ago, meant your smiles and friendly banter were hastily traded for emotional quivering in often failing *tired* voices. For your strength and courage I can't thank you enough.

On this note, I'd like to make special mention to the brave men who opened a 'portal in time' for me. Your participation allowed me to reminisce with you about your experiences at Kapooka Camp during the war years nearly seventy years ago. Your humorous tales of Australian life in a bygone era, such as the 'Lady Blamey' or the hijinks experienced in the 'Entertainment Zone', were delivered with the cheekiness of a soldier and the boyhood wit of a young larrikin. You had the knack of making me laugh one minute, but such was the gravity and impact of your loss in this seemingly preventable heartbreak, shortly thereafter I'm reduced to wiping away my own emotions. It was clear that the dramatic and unexplainable circumstances of losing these, and other mates, hadn't eased your sorrowfulness. The glassy pain of loss in your eyes, and the constant clearing of throats, became a stark reminder for me that this type of pain aches deep down inside your heart. I know showing emotion is sometimes difficult to do for a hardened WWII veteran, but rest assured when you display your emotions, Australia feels for your pain and loss. You've seen the worst of life and thereby deserve the right to shed a tear or two for a fallen mate. Wear your emotions with pride as we, the younger generations, appreciate that your tears flow as a sign of a job done well under horrendous and unimaginable conditions.

To Des Surkitt, Paddy Cranswick, Stan Emery, Bill Rhodes, Keith Kuhn, Hank Keenan, Allan McPaul, and Des Martin, at times our conversations were tough, and for your ongoing courage and strength, I can't thank you enough. For 87-year-old Paddy Cranswick, who was pruning roses and cut his finger in racing to answer my call—I hope your finger healed well. It was a pleasure being introduced to you all. I look forward to meeting up one day.

Uncritically we look upon all of our ageing veterans and often forget where they've been and what they did to ensure Australia's survival. Embarrassingly we forget what we can't see and the majority of Australians aren't even aware that a tragedy of such significance occurred at Kapooka all those years ago. In our ignorance we've contributed to resigning the accident to mere folklore. Doing so, we've unfortunately dishonoured many families who have stoically lived with this indignity for a long time. These remarkably resilient wartime families, and their modern day descendants, aren't angry at the forgetting or demanding explanations as to why the accident occurred, they simply want recognition. They're seeking respect and acknowledgement that their loved ones played a small yet simple and unenviable place in Australia's military history, but nonetheless were part of Australia's courageous contribution during the 1939-1945 war years. Interviewing them and observing misty eyes and hearing quivering voices, it's all about appreciation that no one died in vain. In their voices, I sensed obvious pain in their disregarded hearts when reflecting on the treasured memories of loved ones and long-lost friends. But their pain and outpouring of grief is often silent and withdrawn. It's a unique hurt. Not even the sands of time can cure this type of internal torment. Talking, laughing, reminiscing and even crying with these incredible Australians, I felt proud to be part of their life's journey.

I'm proudest of all in acknowledging the contribution of Maureen Raunic (nee Pomeroy). Maureen, your father, 'Jack', and his entire family, provided the inspirational backdrop to this tragic story. His unforgettable dedication to your family and the Australian Army unfortunately placed him at the epicentre of the tragedy. Your father became a Royal Australian Engineer mentor and icon because his actions spoke louder than words. As a pioneer of military demolitions training, his death was not in vain. The Royal Australian Engineer Corps changed protocols and procedures from Jack's experiences that day; an action that no doubt has saved hundreds of lives since that fateful day. I am confident that he not only left an indelible mark on the country's military engineering fraternity, but also is fondly remembered in Terang as one of their favourite sons. His life may have only been 31 years-short, but he did enough to forge a remarkable engineering legacy. He was a soldier whom I consider to be one of Australia's silent and truly forgotten military heroes.

I'm also proud to acknowledge other direct descendants of the victims and your contribution to this story; Neryl Hogan (nee Nixon); Bobby Hurst; Neil and Susan Dilley; Gordon and Ronda Poschalk and Shirley Booth (nee Partridge) who lost fathers and brothers in the tragedy. To Bruce and Di Bartlett, whose father was the sole miraculous survivor; I can't acknowledge your contribution and thank you enough. Your father Allan was one of Australia's true unsung heroes; your uncle Alf, an inspirational soldier. In every sense of the word, if ever there was just one word, your father was a 'gentleman'. I have no doubt, despite his lifelong struggle with internal grief and painful memories of that fateful day, that everyone he met appreciated his loving, compassionate and gracious spirit of life. I'm also positive that he's passed that spirit and sense of family to his children and grandchildren. May he finally rest in peace.

Acknowledgement should also include the multiple eyewitnesses, treating doctors and all other trainees and staff who assisted in the aftermath of this most horrific event. This includes the men appointed to the Court of Inquiry. Your strength and courage in dealing with the aftermath of such awfulness cannot be repaid in mere words. The Australian public and the entire military community will be forever indebted to you. Your brave compassion, professionalism and dignity afforded the victims under incredible pressure and personal pain hasn't gone unnoticed by surviving families.

For Malcolm Tapscott, the nephew of a brave and courageous AWAS, Sheila Oehm, you've described an aunty who not only witnessed the horror up close, but defiantly battled authorities for more than fifty years to have horror etched in her memory recognised as a significant contributor to her failing health. Sheila epitomized the fighting spirit of WWII veterans and was a true representative of the AWAS spirit of commitment. Her persistence, despite her declining health, resulted in the first public recognition of the tragedy. She knew all too well that many families throughout Australia were still feeling the pain of anonymity; then by doing her best to try and to ease that pain, her work with the former Commandant of Kapooka, Colonel Gordon Hurford, ensured this story became public knowledge after a dormant 50-year period. As a nation, we can simply express to her family a very deep and sincere 'thank you'.

To Helen and Fiona Bichel (nee Merry), you presented me with details of your father and *poppy*, which elevated him amongst some of the country's most respected former soldiers. Along with other influential officers such as Captain Archie Smith, who formally identified the men in a makeshift morgue, your *Poppy's* courage, strength, compassion and sense of obligatory duty to *his men* following the horrific explosion is in itself an inspirational story of courageous leadership under extreme pressure. His career as a soldier, and later a Commissioned Officer, is a remarkable blueprint of leadership for all current and future Australian military leaders to emulate. He was a true leader of men and an inspiration to the tens of thousands of soldiers he trained.

I'd like to particularly acknowledge the immediate family members of the victims, who have since passed. Your resolve, strength and dignity in the aftermath of such horrific circumstances which claimed your sons, husbands and dads, epitomized the might and courage of other military families similarly losing loved ones in the service of defending the nation. Whilst the remainder of the country was preparing to celebrate victory and the imminent end of World War II, signaling the long-awaited return of loved ones, your celebrations were diminished in the knowledge that your sons, husbands, brothers, uncles, nephews and mates, were no longer going to walk through your front door. Sadly you've waited a lifetime—but finally you're eternally reunited.

Those who know the scant details of the 1945 Kapooka tragedy will acknowledge how much I have depended on evidence and work provided by Dr Peter Rushbrook, formerly of Charles Sturt University. Peter, your persistence in having this inglorious incident recognised, and your insightfulness delivered via your 2008 academic paper *Lest We Forget: The Kapooka Tragedy 1945* has provided many individuals—including yours truly— with the information to stir individual and collective emotions and generate national interest in this important event in the country's military past. Along with historian David Mitchelhill-Green who extensively wrote about the tragedy in the 2009 Edition of the UK Military magazine *After The Battle*, you've both supplied the groundwork for others to conduct further research in order to write about this most significant event in Australia's forgotten military past. Your determination in raising the awareness of the tragic events via *The Australian* newspaper is admirable. Without the contribution of you both, who knows how long this tragedy may have remained buried. The families of these men were eternally grateful for your passion and interest.

Concerning the recollections of distant relatives of victims—living and passed—I've relied on family information found in genealogy sources in order to narrate individual family stories. Information hidden in digitalised newspapers of a bygone era, subsequently extracted from online media site *Trove*, provided a wealth of knowledge from news reports covering this tragedy. The online media capability has allowed me contemporary access into personal milestones of the victims and their families. Milestones such as births, deaths and marriages, weren't only notices from the past, but a snapshot of Australian family social culture during the war-years. For the level of personal information I managed to extract, I acknowledge the contribution of every family who bothered to mark personal milestones in their life via local and State newspaper media of their time. The information supplied has been rich in content, and I acknowledge that it was often at a time of great personal grief.

To the *Australian War Memorial*, the *Australian's at War Film Archive* and the *National Archives of Australia*, from the millions of military veterans, we acknowledge and remain humbled at the selfless way you diligently preserve our national fabric. Making records of our forefathers accessible, including details of horrific combat fatalities in personal medical records, serves to remind us all that returning to war, as a solution to the world's problems, is never a solution.

Dark days in Australia's wartime history, especially unwelcome fatal training incidents, don't necessarily demand or receive public attention. However, the rightful recognition of this tragedy by the *Australian War Memorial*, the *1st Recruit Training Battalion* and the *Royal Australian Engineer Foundation*, should be especially mentioned. Your actions in keeping the memory and honor of these 26 fallen Sappers alive ensures this tragedy will no longer 'disappear', once again, into mere folklore. Your unwavering respect for their memory, your recognition of their personal sacrifices, including their contribution to Australia's WWII war effort, will forever remind us all that they were more than just heroic soldiers giving their life for the safety of the nation—they were brave and courageous sons of our nation.

My final but most important thanks and acknowledgement, as always, goes to my family. My wife Dee, and of course our children, Brylee, Dylan and Jakson. They were my guiding light throughout twenty years of Army service and continue to be my inspiration and rock of support during my many hair-brained ideas—like joining the Army. Without each and every one of them, telling this tragic story of love and admiration for family would somehow be meaningless. For your patience, love, and understanding, thank you.

I'd finally like to thank and acknowledge personally the following contributors:

Maroochydore Military Museum and Library
Maroochydore RSL Sub Branch
Colonel Gordon Hurford (Rtd)
NX202751 Sergeant Des Martin
Peter Wyatt
Geoff Hughes
Mark Shanks
Bruce Munchenberg
Tom Locke
Mal Maloney
Tracey and NX205840 Sapper Stan Emery
SX30560 Corporal Keith Kuhn
Jack and VX65073 Gunner Bill Rhodes
Chris Jones and the Vetaffairs Administration Team
VX95842 Sapper Clarence 'Hank' Keenan
VX146042 Sapper Des Surkitt
RAE Foundation
Frank Lawton
Charles Gibson
Australian Army Museum of Military Engineering
Wayne Vost
John Sheahan
Terry and Margaret Dineen
John Jesser
Peter Nolan
Adam Parsons
Malcolm Tapscott
Helen and Fiona Bichel
SX33869 Sapper Patrick 'Paddy' Cranswick
NX180216 Sapper Allan Robert McPaul
Gary Croker

A special mention of thanks:

Maureen Raunic (nee Pomeroy)
Neryl Hogan (nee Nixon)
Di and Bruce Bartlett
Bobby Hurst
Neil and Susan Dilley
Gordon and Ronda Poschalk
Shirley Booth (nee Partridge)
Patricia Anne Schlitz (nee Robson)

Index

1st, RAE Bn 19, 141, 212, 217, 280, 290, 295, 307, 311, 325, 337, 341, 354, 359, 370, 380, 405
2nd, AIF xx-xxviii, 7, 9, 15, 21, 45, 51, 63, 75, 141, 152, 262, 266-267, 277, 279-281, 285-286, 288-294, 299, 302, 306-307, 318, 322, 324, 336, 340, 342, 347-348, 352-353, 357-358, 362, 365, 369, 376, 378-379, 388, 390, 404
54th, Camp Hospital 33, 53, 124, 125, 135, 136, 138, 139, 142, 165, 178, 186, 211, 213, 222, 403
Bailey, Ed 268
Ballingall, Harry 267, 277
Bartlett, Allan 64, 77, 93, 112, 113, 135, 137, 143, 148, 177, 180, 185, 191-193, 195-197, 201, 203, 204, 206, 224-226, 235, 243, 290, 293, 294, 303, 306, 327, 377, 387-392, 397, 399, 400, 404, 405, 422
Bartlett, Alf 'Brick' 178-180, 388, 390, 392
Bennett, Victor 132. 133, 135, 147, 200
Benporath, Gordon 268

Berg, Maurice 20, 125, 187, 188
Blamey, Dolf 411
Blamey, Lady 53, 69, 104, 377, 420
Blamey, Sir Thomas xxii, 214, 310, 411
Boyd, Colin 66, 76, 80, 99, 100, 107-109, 113, 114, 116, 157, 177, 200, 234, 333-337, 340, 343, 354
Carroll, Grace 186, 193
Collins, Joseph 65, 76, 93, 113, 150, 151, 177, 200, 287, 348, 358-363
Conwell, James 22, 74, 75, 79, 80, 90, 91, 95, 98, 99, 106, 123, 124, 162, 198
Cousins, William 'Bill' 23, 58, 70, 77, 79, 80, 90-92, 94, 95, 100, 103, 106, 107, 110-113, 116, 128, 157, 160-162, 166, 177, 189, 191, 193-196, 199, 214, 225, 226, 233-235, 313-320, 398
Cowra, xxi-xxvii, 21, 34, 39, 52, 63-64, 66, 74-76, 157, 264, 267, 280, 285-287, 290, 291, 294-295, 302-303, 307, 310, 324-325, 331, 333, 335-336, 340-341, 343, 348, 353-354, 359, 362, 370, 376, 378-380, 384, 389-390, 393

Cranswick, Patrick 'Paddy' 62, 65, 77, 79, 80, 105, 107, 125, 151, 408, 409, 420
Dedman, John MP 172, 183
Dilley, Norm 75, 77, 93, 110, 111, 113, 151, 176, 200, 287, 326-333, 338, 348, 354, 377, 422
Dodds, Ed 21, 31, 35, 51, 57, 73, 74, 76-79, 93, 100, 106, 107, 110, 117, 122, 130, 131, 133, 139, 140, 143, 156, 192, 195, 198, 199, 202, 210, 223
Dolphin, Arnold xx, xxxii
Doubleday, Maida 144, 147, 150
Doyle, Mayoress 211
Dunphy, Wallis 184, 187, 195, 205
Ellis, Fernleigh 136, 138, 152, 154, 163-165, 167-169, 171, 196, 200, 201, 210, 223
Emery, Stan 30, 38, 39, 42, 43, 68, 106, 114, 369, 371, 408, 416-418, 420
Faull, Joseph 76, 78, 93, 112, 113, 149, 150, 177, 200, 338-343, 354
Flood, Allan 54, 65, 77, 93, 113, 150, 177, 200, 287, 353-360, 362
Forbes, Brig Alexander 184, 185, 193, 205, 206, 208, 209, 223, 224, 406
Fuller, MP 172
Grasby, Denby 64, 76, 93, 103, 112, 113, 148, 176, 200, 287-294, 303, 305, 311, 376, 379, 384, 388, 391, 398
Haig, Sir Douglas 184
Hanchard, Arthur 106
Hardey, Gladys 210
Hickson, Harry (& family) 14-15, 17, 23, 31, 34, 48, 54-56, 126, 164
Hogan, Neryl 401, 416, 418, 422
Holdsworth, George 23, 39, 41, 42, 114

Hurford, Gordon 395-397, 407, 423
Hurley, Colin 65, 76, 80, 99, 100, 107-109, 113, 114, 116, 150, 177, 200, 234, 376-380
Hurst, Kevin 'Brickie' 65, 69, 76, 93, 113, 156, 157, 177, 200, 217, 287, 348-356, 359, 362, 410, 422
Johnson, Harold xx, xxxii
Judd, Lemual 'Henry' 124, 130-132, 134-137, 200
Kapooka, Camp ix-xxix, 13-14, 33, 54, 63, 73, 124, 173-174, 183-184, 213, 234, 280, 340, 343, 370, 380, 393, 395, 397, 405, 411, 420,
Keenan, Clarence 'Hank' 45, 116, 125, 407, 420
Kendall, Colin 23, 44, 46, 75, 80, 98-100, 107, 108, 110, 111, 113, 114, 121, 123, 135, 137, 143, 185, 188, 191, 194-196, 198, 202, 204-206, 224, 225, 232-234, 378, 403, 404
Kuhn, Keith 46, 119, 407, 420
Lamb, Cecil 5, 10
Linthorne, Ron 70, 106, 110-113, 148, 153-155, 158, 176, 200, 222, 265-272, 277, 280
Lloyd, Maj-Gen 208, 211, 214, 216, 219, 226
Loftus, Lloyd 210
Macdonnell, George 20, 24, 125, 136, 152, 186, 187, 202, 216
Mahon, Lawrie 6, 9,
Mather, Les 65, 77, 93, 113, 149, 200, 372-377, 384, 385
McDonald, Brig Warren 19, 20, 24, 73, 183, 186, 187, 216, 219

McFarlane, Doug 21, 73
McNab, Phill 100, 117, 122, 123, 137, 201, 202, 228
McPaul, Allan 65, 69, 126, 217, 354, 410, 420
McRae, Tom 6, 48, 57
Menzies, Robert xix, xx, 253, 254, 299, 329, 411
Merritt, Ivan 64, 76, 77, 93, 103, 113, 137, 143, 163, 166, 171, 173, 176, 195, 200, 290-295, 311, 376, 379, 384, 389, 391, 398
Merry, Edward 20, 21, 24, 35, 39, 47, 48, 51, 59, 117, 123, 124, 125, 130-133, 137, 138, 140, 159, 187, 189, 191, 192, 197, 202-207, 210, 223, 404, 405
Militia, 11, 75, 141, 266, 277, 285, 302, 305-306, 317-318, 322, 335, 339, 347, 361-362, 368, 378-379, 385
Miller, George 133, 135, 148, 200
Mitchelhill-Green, David 396, 400, 424
Moore, Terrence 65, 76, 93, 112, 113, 149, 150, 177, 200, 339-343, 354
Morphy, Stan 64, 76, 93, 113, 150, 176, 200, 308-311, 370
Murdoch, Brig xxiii, xxiv
Musto, Thomas 86, 94, 98, 102, 110, 115, 162, 199, 406
Nixon, Jack 39, 68, 75, 77, 93, 103, 106, 111, 113, 151, 177, 200, 367-372, 401, 408, 416, 418
Oehm, Sheila 85, 94, 98, 102, 110, 124, 131, 162, 202, 215, 228, 395, 406, 407, 410, 422
Parslow, Norm 5

Partridge, Geoff 65, 68, 77, 93, 100, 105, 107, 113, 125, 151, 177, 200, 376, 380-386, 398, 409, 410, 422
Pierce, Kevin 53, 64, 76, 93, 104, 113, 144, 149, 200, 257-264, 287, 288, 405
Platt, Frank 53, 64, 76, 93, 113, 149, 176, 200, 304-308, 311, 376
Platten, Walter 388
Pomeroy, Dorothy (Jack's wife) 5-6, 9-10, 12, 14-15, 17, 33, 54-58, 126, 164, 412
Pomeroy, Herbert 'Jack' xxv, xxvi, xxx, xxi, xxxii, 3-15, 17, 23, 31, 47, 51, 54, 56, 79, 106-108, 120, 126, 128, 139, 153, 157, 159, 164, 166, 169, 187, 189-194, 196, 198-200, 204, 205, 225, 229, 230, 232, 234, 240, 243, 245-253, 257, 271, 319, 354, 368, 394, 395, 398, 401, 408, 412-415, 418, 421
Pomeroy, Maureen (Jack's daughter) 14, 17, 48, 56, 398, 401, 414-415, 418, 421
Pomeroy, Susan (mother) 4, 6-8, 15, 245-251, 259-260, 413
Pomeroy, William (father) 15, 245, 247
Pomeroy, William, Ted, George, Charles & Len (brothers) 3-4, 7-11, 245-252
Poschalk, Ernie 67, 76, 93, 113, 149, 176, 200, 297-304, 308-311, 353, 354, 376, 384, 400, 422
Prior, Brig Claude 136, 152, 183-185, 216,
RAETC, ix, xxiii, 13, 17-18, 21, 23-25, 27, 35, 52, 152, 208-209, 212, 253, 337

Reid, William 'Bill' 75, 77, 106, 110, 111, 113, 137, 143, 163, 166, 171, 173, 176, 195, 200, 283-287, 289, 294, 332, 354
Rhodes, Bill 90, 420
Robson, Edward 'Teddy' 64, 77, 93, 113, 157, 158, 160-162, 166, 177, 200, 287, 344-349, 353, 362
Roach, Jack 80, 411, 412
Ross, Stan 68, 75, 77, 80, 99, 100, 107-109, 113, 114, 116, 156, 166, 177, 200, 234, 363-367
Rushbrook, Dr Peter 400, 424
Sherwood, Police SGT 154, 201, 220
Sinclair, F.R MP 172
Sim, Frank 91, 202, 228
Slade, Clive 184, 187, 195, 204-205
Smith, Archie 80, 136-139, 141, 144, 148, 150-153, 155, 158, 159, 162, 163, 165, 167, 195, 199, 200, 210, 212, 217, 218, 223, 224, 405, 406, 423
Steele, Maj-Gen Clive xxiii, 216
Surkitt, Des 47, 117, 169, 408, 410, 420
Tafe, Roy 22, 31, 37, 44, 46, 51, 62, 68, 74, 75, 79, 80, 85, 90, 91, 95, 98-100, 102, 107, 110, 117, 122-124, 127, 130, 131, 133, 134, 139, 140, 143, 162, 192, 194, 195, 198-203, 205-207, 234, 404
Thomas, Stanley 388
Titus, (family of) 23, 58, 59, 70, 320, 398
Tunley, Les 124, 133, 137, 139, 142, 148, 153, 161, 165-168, 200, 201, 225
Veale, Brig 19
Watson, William 210
Wit, Alf 64, 76, 93, 113, 152, 176, 200, 278-281, 303, 340
Woodmason, Bill 5, 10
Woodmason, Ed 5
Woods, Alf 71, 75, 106, 110, 111, 113, 148, 176, 200, 270, 271-278, 280
Woods, Thomas 'Toddy' 52, 77, 93, 104, 113, 144, 147, 150, 169, 200, 320-326, 398, 399, 408